W9-ABS-002

NONLINEAR DIFFERENTIAL EQUATIONS

INVARIANCE, STABILITY, AND BIFURCATION

ACADEMIC PRESS RAPID MANUSCRIPT REPRODUCTION

PROCEEDINGS OF A SYMPOSIUM BASED ON THE INTERNATIONAL CONFERENCE ON NONLINEAR DIFFERENTIAL EQUATIONS: INVARIANCE, STABILITY, AND BIFURCATION, HELD IN VILLA MADRUZZO, TRENTO, ITALY, FROM MAY 25- 30, 1980.

NONLINEAR DIFFERENTIAL EQUATIONS

Invariance, Stability, and Bifurcation

Edited by

PIERO de MOTTONI

Istituto per le Applicazioni
del Calcolo "Mauro Picone"
CNR
Roma, Italy

LUIGI SALVADORI

Dipartimento di Matematica
Libera Università di Trento
Trento, Italy

ACADEMIC PRESS 1981

A Subsidiary of Harcourt Brace Jovanovich, Publishers

New York London Toronto Sydney San Francisco

ACADEMIC PRESS, INC.
111 Fifth Avenue, New York, New York 10003

United Kingdom Edition published by
ACADEMIC PRESS, INC. (LONDON) LTD.
24/28 Oval Road, London NW1 7DX

Library of Congress Cataloging in Publication Data
Main entry under title:

Nonlinear differential equations.

Proceedings of a conference held in Trento, Italy, May 25-30, 1980.
1. Differential equations, Nonlinear--Congresses.
I. De Mottoni, Piero. II. Salvadori, Luigi.
QA370.N65 515.3'55 81-543
ISBN 0-12-508780-2

PRINTED IN THE UNITED STATES OF AMERICA

81 82 83 84 9 8 7 6 5 4 3 2 1

Contents

Contributors

Numbers in parentheses indicate the pages on which the authors' contributions begin.

Norman W. Bazley (1), *Mathematisches Institut, Universität Köln, Weyertal 86-90, D-5000 Köln, West Germany*

Edoardo Beretta (11), *Istituto di Biomatematica, Via A. Saffi, 1, I-61029, Urbino, Italy*

S. R. Bernfeld (29), *Department of Mathematics, The University of Texas at Arlington, Box 19408, Arlington, Texas 76019*

Marco Biroli (41), *Istituto Matematico, Politecnico di Milano, Via Bonardi, 9, I-20133 Milano, Italy*

Victor I. Blagodatskikh (55), *Steklov Mathematical Institute, 117393, Vsvilova 42, Moscow, U.S.S.R.*

Moses Boudourides (59), *Department of Mathematics, Democritus University of Thrace, School of Engineering, Xanthi, Greece*

Vincenzo Capasso (65), *Istituto di Analisi Matematica, Palazzo Ateneo, I-70121 Bari, Italy*

Silvia Caprino (77), *Istituto di Matematica, Università di Camerino, Via V. Venanzi, I-62032, Camerino, Italy*

M. Cecchi (85), *Istituto di Matematica Applicata, Università di Firenze, Via S. Marta, 3, I-50139 Firenze, Italy*

Nathaniel Chafee (97), *School of Mathematics, Georgia Institute of Technology, Atlanta, Georgia 30332*

Klaus Deimling (129), *Gesamthochschule Paderborn, Fachbereich 17, Mathematik-Informatik, Warburgerstrasse 100, D-4790 Paderborn, West Germany*

Odo Diekmann (133), *Mathematisch Centrum, Kruislaan 413, NL-1098 SJ Amsterdam, The Netherlands*

Giorgio Fusco (145), *Istituto di Matematica Applicata, Università di Roma, Via A. Scarpa, 10, I-00161 Roma, Italy*

Stephan van Gils (133), *Mathematisch Centrum, Kruislaan 413, NL-1098 SJ Amsterdam, The Netherlands*

Jesús Hernández (161), *Departamento de Matemática, Universidad Autónoma, Ciudad Universitaria de Cantoblanco, Madrid 34, Spain*

Nicoletta Ianiro (175), *Istituto di Meccanica e Macchine, Universita dell'Aquila, Monteluco-Roio, I-67100 L'Aquila, Italy*

R. Kannan (183), *Department of Mathematics, The University of Texas at Arlington, Box N 19408, Arlington, Texas 76019*

Nicholas D. Kazarinoff (195), *Department of Mathematics, SUNY at Buffalo, 106 Diefendorf Hall, Buffalo, New York 14214*

Hansjörg Kielhöfer (207), *Institut für Angewandte Mathematik, Universität Würzburg, Am Hubland, D-8700 Würzburg, West Germany*

Klaus Kirchgässner (221), *Mathematisches Institut, Universität Stuttgart, Pfaffenwaldring 57, D-7000 Stuttgart, West Germany*

V. Lakshmikantham (243), *Department of Mathematics, The University of Texas at Arlington, Box 19408, Arlington, Texas 76019*

S. Leela (259), *Department of Mathematics, SUNY at Geneseo, Geneseo, NY 14454*

Carlotta Maffei (175),*Istituto di Matematica, Università di Camerino, Via V. Venanzi, I-62032 Camerino, Italy*

M. Marini (85), *Istituto di Matematica Applicata, Università di Firenze, Via S. Marta, 3, I-50139 Firenze, Italy*

Jean Mawhin (269), *Institut Mathématique, Université de Louvain, Chemin du Cyclotron, 2, B-1348 Louvain-la-neuve, Belgium*

Ju. A. Mitropolsky (283), *Institute of Mathematics, Academy of Sciences of the Ukrainian SSR, Repina, 3, 252601 Kiev, U.S.S.R.*

Piero de Mottoni (327), *I.A.C. "M. Picone", C.N.R., Viale del Policlinico, 137, I-00161 Roma, Italy*

P. Negrini (29), *Istituto di Matematica, Università di Camerino, Via V. Venanzi, I-62032 Camerino, Italy*

L. Salvadori (29), *Dipartimento di Matematica, Libera Università di Trento, I-38050 Povo (Trento), Italy*

Andrea Schiaffino (327, 339), *Istituto Matematico "G. Castelnuovo", Universita di Roma, Piazzale Aldo Moro, 5, I-00185 Roma, Italy*

Alberto Tesei* (339), *I.A.C. "M. Picone", C.N.R., Viale del Policlinico, 137, I-00161 Roma, Italy*

Rosanna Villella-Bressan (347), *Istituto di Analisi e Meccanica Università, Via Belzoni, 7, I-35100 Padova, Italy*

P. L. Zezza (85), *Istituto di Matematica, Università di Siena, Via del Capitano, 15, I-53100 Siena, Italy*

**Present address: Istituto di Matematica Applicata, Università di Roma, Via A. Scarpa, 10, I-00161 Roma, Italy*

Preface

An international conference on Nonlinear Differential Equations: Invariance, Stability, and Bifurcation was held at the Villa Madruzzo, Trento, Italy, May 25–30, 1980. The conference is part of a series of meetings sponsored by the Centro Interuniversitario per la Ricerca Matematica (CIRM) and by the Italian Council for Scientific Research (CNR). It is a pleasure to acknowledge the support received from the sponsoring agencies, which made the conference possible.

The purposes of the conference were to highlight developments in the qualitative theory of nonlinear differential equations, and to promote the exchange of mathematical ideas in stability and bifurcation theory. The mutual interaction and cooperation between qualified researchers, active both in theoretical and applied investigations, proved extremely fruitful and stimulating.

The present volume consists of the proceedings of the conference. It includes papers that were delivered as survey talks by Professors N. Chafee, K. Kirchgässner, V. Lakshmikantham, and Ju. A. Mitropolsky, as well as a number of research reports. A number of contributions focus on the interplay between stability exchange for a stationary solution and the appearance of bifurcating periodic orbits. Another group of papers deals with the development of methods for ascertaining boundedness and stability. Nonlinear hyperbolic equations are considered in further contributions, featuring, among others, stability properties of periodic and almost periodic solutions. Papers devoted to the development of bifurcation and stability analysis in nonlinear models of applied sciences are also included.

We wish to express our appreciation to Mr. A. Micheletti, secretary of CIRM, for assisting us in organizing the conference. A special grant of the CNR, which we warmly acknowledge, made the typing of the proceedings possible. This was carefully carried out at the Centro Stampa KLIM, Rome.

ABSTRACT NONLINEAR WAVE EQUATIONS: EXISTENCE, LINEAR AND MULTI-LINEAR CASES, APPROXIMATION, STABILITY

Norman W. Bazley [1]

Mathematics Institute,
University of Cologne
West Germany

1. INTRODUCTION

We consider the initial value problem for a nonlinear wave equation given by

$$u''(t) + Au(t) + M(u(t)) = 0$$
$$u(0) = \varphi, \quad u'(0) = \psi. \tag{1}$$

Here A is a strictly positive, self-adjoint operator in a separable Hilbert space $\mathcal{H}$, with domain D_A; the initial values satisfy $\varphi \in D_A$, $\psi \in D_{A^{\frac{1}{2}}}$.

The purpose of this article is to survey some recent existence and approximation results for (1). Such problems were first studied by K. Jörgens [9] and I. Segal [14]. In 1970 F. Browder [4] carried through an operator theoretic study of the above equation, which was recently simplified and generalized by E. Heinz and W. von Wahl [8]. These results were in-

---

[1] Supported in part by the European Research Office of the U.S. Army.

ISBN 0-12-508780-2

vestigated in the Diplomarbeit of D. Kremer [11], which motivated other Diplomarbeiten [5,15] and studies [1-3] at the University of Cologne. Related works appear in [6,7,12].

2. EXISTENCE THEORY OF BROWDER - HEINZ - von WAHL

In [8] the following two assumptions, introduced in [4], are used to prove the existence of a strong solution of (1):

<u>Assumption</u>: $M: D_{A^{\frac{1}{2}}} \to \mathcal{H}$ is a mapping defined on all of $D_{A^{\frac{1}{2}}}$; (2)

<u>Assumption A</u>: For every $c > 0$ there exists a $\kappa(c)$ such that

$$\|M(A^{-\frac{1}{2}}u) - M(A^{-\frac{1}{2}}v)\| \leq \kappa(c)\|u - v\| \text{ for } \|u\|, \|v\| \leq c. \tag{3}$$

The first assumption is often easy to verify in applications, even though $A^{\frac{1}{2}}$ itself is usually not known in closed form. However, the assumption (3) that $M(A^{-\frac{1}{2}}u)$ satisfies a Lipschitz condition on spheres is usually more difficult to check. We refer to it as "Assumption A of Browder-Heinz-von Wahl".

When both assumptions (2) and (3) are satisfied we have by [8]:

<u>Theorem</u>. For arbitrary vectors $\varphi \in D_A$ and $\psi \in D_{A^{\frac{1}{2}}}$ there exists a unique solution $u = u(t;\varphi,\psi)$ of (1) in an interval $0 \leq t < \tilde{T}$, and $u(t)$ belongs to the class $C^2([0,\tilde{T}),\mathcal{H}) \cap C^1([0,\tilde{T}), D_{A^{\frac{1}{2}}}) \cap C^0([0,\tilde{T}),D_A)$. If $\tilde{T} < \infty$, $\lim_{t\uparrow\tilde{T}} E(t) = +\infty$, where $E(t) = \|A^{\frac{1}{2}}u(t)\|^2 + \|u'(t)\|^2$.

Thus there exists a unique C^2 solution $u(t)$, which has the representation $u(t) = \sum_{i=1}^{\infty} a_i(t)u_i$, for $\{u_i\}_1^\infty$ a complete orthonormal system in $\mathcal{H}$.

We outline some of the steps used in proving the above theorem. For details, the reader is referred to [8]. There Heinz and von Wahl consider the Banach space $X = C^0([0,T],D_{A^{\frac{1}{2}}})$

with norm $\|u\|_X = \max_{0 \le t \le T} \|A^{\frac{1}{2}}u(t)\|$, and for $u \in X$, $n = 1,2,\ldots,$ they define mappings S_n as

$$u \to S_n(u) = w(t) - \int_0^t \sin(t-s)A^{\frac{1}{2}}(A^{-\frac{1}{2}}M_n(u(s)))ds, \tag{4}$$

where

$$w(t) = \cos tA^{\frac{1}{2}}\varphi + \sin tA^{\frac{1}{2}}(A^{-\frac{1}{2}}\psi).$$

Here, for $u \in D_{A^{\frac{1}{2}}}$, they define $M_n(u) = M(E_n u)$, where E_λ is the resolution of the unity for A. From Assumption A, each S_n has a fixed point $v_n(t)$ with $\|v_n\|_X \le c$, where $c = \|A^{\frac{1}{2}}\varphi\| + \|\psi\| + 1$, and $0 \le t \le T = 1/(2\kappa(c))$. Further, detailed arguments show that the $v_n(t)$ converge to a unique strong solution of (1) in the norm

$$\lim_{n\to\infty} \sup_{0 \le t \le T} E(u(t) - v_n(t)) = 0 \tag{5}$$

where

$$E(u(t)) = \|A^{\frac{1}{2}}u(t)\| + \|u'(t)\|. \tag{6}$$

The same arguments are re-applied at the endpoint T, to obtain a strong solution in a larger time interval. Extension to the maximal interval $0 \le t < \tilde{T}$ is obtained by repeated applications.

3. ASSUMPTION A IN SIMPLE CASES

In this section we present some recent results of K.-G. Strack and the author in some previously overlooked simple cases. Proofs and applications are given in [3].

We first consider the special case of (1) where $M(u) = Bu$, for B an unbounded, linear operator with $D_B \supseteq D_{A^{\frac{1}{2}}}$. We remark that if A is an elliptic differential operator of order 2p, then $D_{A^{\frac{1}{2}}} = \overset{\circ}{H}{}^p$ and the domain condition (2) only allows differential operators of order p or less, in agreement with Goldstein [7]. For this case Assumption A becomes $\|BA^{-\frac{1}{2}}u\| \leq \kappa\|u\|$, which is equivalent to $\|A^{\frac{1}{2}}x\|^2 \geq \varepsilon\|Bx\|^2$, for all $x \in D_{A^{\frac{1}{2}}}$, where $x = A^{-\frac{1}{2}}u$ and $\varepsilon = 1/\kappa^2$. Thus we have

Lemma. Assumption A is satisfied if and only if there is $\varepsilon > 0$ such that $(A^{\frac{1}{2}}x, A^{\frac{1}{2}}x) - \varepsilon(Bx, Bx) \geq 0$ for all x in $D_{A^{\frac{1}{2}}}$. Well known results of Kato [10] lead to the alternative formulation:

Lemma. Let $D_{B^*B} \supseteq D_A$ and suppose there exists a $\delta > 0$ such that $\|B^*Bu\| \leq \delta\|Au\|$ for all u in D_A. Then Assumption A is satisfied if and only if there exists $\varepsilon > 0$ such that $A - \varepsilon B^*B \geq 0$ on D_A.

Similar results hold if M(u) has a Gâteaux derivative M'_u for every u in $D_{A^{\frac{1}{2}}}$ and $\sup\{\|M'_v A^{-\frac{1}{2}}\|;\ v \in D_{A^{\frac{1}{2}}}\} < \infty$. Then we have [16] the global Lipschitz condition

$$\|M(A^{-\frac{1}{2}}u) - M(A^{-\frac{1}{2}}v)\| \leq \left(\sup_{v \in D_{A^{\frac{1}{2}}}} \|M'_v A^{-\frac{1}{2}}\|\right) \cdot \|u - v\|. \tag{7}$$

One such result reads:

Lemma. Let $\varepsilon > 0$ be such that $A - \varepsilon M'^*_v M'_v$ is essentially self-adjoint on each $D_A \cap D_{M'^*_v M'_v}$. Further, let $0 < \varepsilon_o \leq \varepsilon$ be such that $A - \varepsilon_o M'^*_v M'_v \geq 0$ for all $v \in D_{A^{\frac{1}{2}}}$. Then Assumption A holds.

Many nonlinearities arising in mathematical physics are generated by α-linear forms, where α is a positive integer. That is, there exists an α-linear form $\tilde{M}(u_1,\ldots,u_\alpha)$ such that $M(u) = \tilde{M}(u,\ldots,u)$. For this case one can easily show:

Lemma. Let $M(A^{-\frac{1}{2}}u_1,\ldots,A^{-\frac{1}{2}}u_\alpha)$ be a bounded α-linear form. Then Assumption A holds.

4. FAEDO-GALERKIN APPROXIMATIONS

In [1,2] we considered the approximation of a solution $u(t) = \sum_{i=1}^{\infty} \alpha_i(t)u_i$ of (1). Our idea was to consider the time dependent infinite vectors $\vec{\alpha}(t)$ with components $\{\alpha_1(t), \alpha_2(t), \ldots\}$ in the Hilbert space ℓ^2. We further made the

Assumption B: The operator A has a pure point spectrum, $Au_i = \lambda_i u_i$, $i = 1,2,\ldots$, with a complete orthonormal system of eigenvectors $\{u_i\}_1^\infty$.

Then we approximated the first n components of $\vec{\alpha}(t)$ by n-dimensional components $\vec{\alpha}^n(t) = \{\alpha_1^n(t), \ldots \alpha_n^n(t)\}$ of the Faedo-Galerkin approximations

$$P^n v = \sum_{i=1}^{n} \alpha_i^n(t) u_i. \tag{8}$$

Here $P^n v$ satisfies the equations

$$\begin{aligned} &\frac{d^2}{dt^2} P^n v + P^n A P^n v + P^n M(P^n v) = 0, \\ &P^n v(0) = P^n \varphi, \; P^n v'(0) = P^n \psi. \end{aligned} \tag{9}$$

The nonlinear term $M(P^n v)$ has the form

$$M(P^n v) = M\Big(\sum_{j=1}^{n} \alpha_j^n u_j\Big) = \sum_{i=1}^{m(n)} \beta_i(\alpha_1^n, \ldots \alpha_n^n) u_i, \tag{10}$$

where

$$\beta_i = \Big(M\Big(\sum_{j=1}^{n} \alpha_j^n u_j\Big), u_i\Big), \; i = 1, \ldots m(n), \tag{11}$$

and $m(n)$ is either finite or infinite. Then the equations (9) reduce to the system of second order ordinary differential equations

$$\begin{aligned} &\ddot{\alpha}_i^n + \lambda_i \alpha_i^n + \beta_i(\alpha_1^n, \ldots \alpha_n^n) = 0 \\ &\alpha_i^n(0) = (\varphi, u_i), \; \dot{\alpha}_i^n(0) = (\psi, u_i), \; i = 1, \ldots n. \end{aligned} \tag{12}$$

Our principal result is that the Faedo-Galerkin approximations $P^n v$ are identical with $P^n v_n$, the projections of the fixed points v_n of the operators S_n in (4). This follows since $E_{\lambda_n} v = P^n v$ implies that v_n satisfies

$$\frac{d^2 v_n}{dt^2} + Av_n + M(P^n v_n) = 0$$
$$v_n(0) = \varphi, \quad v_n'(0) = \psi. \tag{13}$$

Operating on (13) by P^n and noting that $P^n A = P^n A P^n$ leads to (9). We can thus carry over the convergence results (5) of Browder-Heinz-von Wahl to the estimation of $\{\alpha_1(t), \alpha_2(t) \ldots\}$ by $\{\alpha_1^n(t), \ldots, \alpha_n^n(t)\}$, the solutions of (12). For this purpose we introduce the norms E_n defined by

$$E_n(\vec{\alpha}(t) - \vec{\alpha}^n(t)) = \{\sum_{i=1}^{n} \lambda_i (\alpha_i(t) - \alpha_i^n(t))^2\}^{\frac{1}{2}}$$
$$+ \{\sum_{i=1}^{n} (\alpha_i'(t) - \alpha_i^{n\prime}(t))^2\}^{\frac{1}{2}}, \tag{14}$$

and set $E_\infty(\vec{\alpha}) = E(\vec{\alpha}) = E(u(t))$. Detailed arguments [2] show that convergence in each compact interval holds according to the following

Theorem. For each $\hat{T} < \tilde{T}$ we have

$$\lim_{n \to \infty} \sup_{0 \leq t \leq \hat{T}} E_n(\vec{\alpha}(t) - \vec{\alpha}^n(t)) = 0. \tag{15}$$

It follows that

$$E(\vec{\alpha}(t)) = \lim_{n \to \infty} E_n(\vec{\alpha}^n(t)), \tag{16}$$

for $0 \leq t \leq \hat{T}$. Thus $E_n(\vec{\alpha}^n(t))$, the energy of the approximating systems (12), and $E(\vec{\alpha}(t))$, the energy of the solution (1), blow up and remain small together. In particular, for $\varepsilon > 0$ we have $E(\vec{\alpha}(t)) < \varepsilon$, $0 \leq t < \infty$, if and only if $\lim_{n \to \infty} E_n(\alpha^n(t)) < \varepsilon$, $0 \leq t < \infty$, a fact which will be used in the next section on stability analysis.

Finally, we make two remarks on the above procedure. Firstly, in a qualitative or quantitative analysis of (1) by the system (12) it is extremely useful to have an explicit expression for the functions $\beta_i(\alpha_1^n,\ldots,\alpha_n^n)$ in (11); if so, we say [1] that the nonlinearity $M(u)$ is "reproducing" relative to the sequence $\{u_i\}_i^\infty$. The idea is that many nonlinearities have computable expansion coefficients, when applied to a suitable complete orthonormal system. An explicit example is given in the next section.

The second remark concerns the case when the nonlinearity $M(u)$ is the gradient of a functional $\Phi(u)$; that is, $\Phi(u+h) - \Phi(u) = (M(u),h) + R_u(h)$, where $R_u(h)$ denotes terms of higher order in h. This holds, for example, when $M(u)$ is a cyclically monotone operator. Then the approcimations (12) have the form of a conservative Hamiltonian system of classical mechanics, and

$$\frac{\partial\beta_i}{\partial\alpha_j^n} = \frac{\partial\beta_j}{\partial\alpha_i^n} \qquad 1 \leq i,\ j \leq n.$$

Thus there exists a potential $V_n(\alpha_1^n,\ldots,\alpha_n^n)$ and we can write (12) in the form

$$\begin{aligned} &\ddot{\vec{\alpha}}_n + \operatorname{grad} V_n(\alpha_1^n,\ldots,\alpha_n^n) = 0 \\ &\alpha_i^n(0) = (\varphi,u_i),\ \dot{\alpha}_i^n(0) = (\psi,u_i),\ i = 1,\ldots,n. \end{aligned} \tag{17}$$

5. STABILITY

We announce some new results of K.-G. Strack, who in [15] obtained qualitative results for (1) in special cases. In particular, he considered $u(x,t)$ in $\mathcal{H} = L^2(0,1)$ as a solution of the equation

$$\begin{aligned} &u_{tt} - u_{xx} + u^3 = 0 \\ &u(0,t) = u(1,t) = 0 \\ &u(x,0) = \varphi(x),\ u_t(x,0) = \psi(x). \end{aligned} \tag{18}$$

Here $Au = -u''$, $u(0) = u(1) = 0$, which has a point spectrum diverging to infinity. The eigenfunctions are given by $u_i = \sqrt{2} \sin i\pi x$, $i = 1,2,..$, with corresponding eigenvalues $\lambda_i = i^2\pi^2$, $i = 1,2,...$ The nonlinearity $M(u) = u^3$ is defined on $D_{A^{\frac{1}{2}}}$ and satisfies Assumption A, so that the theorem of Browder-Heinz-von Wahl guarantees the existence of a strong solution $u(t)$.

Strack investigates the stability of the zero solution $u(x,t) = 0$ of (18), that is, $\varphi(x) = \psi(x) = 0$. His analysis makes essential use of the results of Rutkowski [13], who use the reproducing property of the nonlinearity to obtain explicit expansion coefficients $\beta_i(\alpha_1^n,...,\alpha_n^n)$ for (10). This allows the construction of Lyapunov functions for the nth approximation (12). Detailed arguments [15] lead to the following stability theorem for (18).

<u>Theorem</u>. Let $\varepsilon > 0$. There exists a δ, independent of n, such that $\|A\varphi\|_2 + \|\psi\|_2 < \delta$ implies $E_n(\vec{\alpha}^n) < \varepsilon$ for all $n \in \mathbb{N}$, $t \geq 0$.

By the convergence result in (16) we have that the zero solution is stable in the energy norm.

REFERENCES

1. N.W. Bazley: <u>Approximation of Wave Equations with Reproducing Nonlinearities</u>, Nonlinear Analysis TMA <u>3</u>, 539-546, (1979).
2. N.W. Bazley: <u>Global Convergence of Faedo-Galerkin Approximations to Nonlinear Wave Equations</u>, Nonlinear Analysis TMA <u>4</u>, 503-507, (1980).
3. N.W. Bazley, K.-G. Strack: <u>Self-adjoint Sums of Operators and the Theory of Nonlinear Wave Equations</u>, Univ. Cologne Technical Report, 1980.
4. F. Browder: <u>Existence Theorems for Nonlinear Partial Differential Equations</u>. In <u>Global Analysis</u>, Proc. Symp. Pure Math. vol. 16, Providence, AMS, 1970, 1-60.
5. L. Brüll: <u>Diplomarbeit</u>, Universität zu Köln, 1980.

6. R.W. Dickey: Free Vibrations of Dynamic Buckling of the Extensible Beam, J. Math. Anal. Appl. 29, 443-454, (1970).
7. J.A. Goldstein: Semigroups of Second Order Differential Equations, J. Functional Anal. 4, 50-70, (1969).
8. E. Heinz, W.v. Wahl: Zu einem Satz von F.W. Browder über nichtlineare Wellengleichungen, Math. Z. 141, 33-48, (1974).
9. K. Jörgens: Das Anfangswertproblem im Grossen für eine Klasse nichtlinearer Wellengleichungen, Math. Z. 77, 295-308, (1961).
10. T. Kato: Perturbation Theory for Linear Operators, Berlin-Heidelberg-New York, Springer, 1966.
11. D. Kremer: Diplomarbeit, Universität zu Köln, 1980.
12. M. Reed: Abstract Nonlinear Wave Equations, Lecture Notes in Mathem. n° 507, Berlin-Heidelberg-New York, Springer, 1976.
13. P. Rutkowski: Diplomarbeit, Universität zu Köln, 1979.
14. I. Segal: Nonlinear Semigroups, Annals of Math. 78, 339-364, (1963).
15. K.-G. Strack: Diplomarbeit, Universität zu Köln, 1980.
16. J. Weyer: Regularität in der nichtlinearen Spektraltheorie, Proc. Roy. Soc. Edinburgh 83A, 81-91, (1979).

STABILITY PROBLEMS OF CHEMICAL NETWORKS

E. Beretta

Istituto di Biomatematica
Università di Urbino

In this talk I shall deal with the stability problem of the equilibrium, or steady state, of a chemical network. The approach to this problem I will present is the following: once a global or local Liapunov function, for example with analogy to thermodynamics, is chosen, one tries to obtain criteria of simple applicability to set up the chemical network classes for which the Liapunov function assures the global or local asymptotic stability of the equilibrium. According with this approach, I shall first present a synthesis of the theory developed by Horn, Feinberg and Jackson and the main result which is known as the zero deficiency theorem. Then, I shall present the theory which introduces the D-symmetrizability of chemical networks, in which a D-symmetrizable network is singled out by the associated knot graph.

1. DEFINITIONS AND NOMENCLATURE

The H.F.J. theory (Horn and Jackson [7], Horn [8], Feinberg [5]) is by far the most important tool in studying the stability of chemical networks, both because it clearly elucidates the mathematical structure of a chemical network's kinetic equation and because it offers a criterion of simple appli-

ISBN 0-12-508780-2

cability to ascertain the existence and uniqueness of the equilibrium state and its asymptotic stability in the large. In this section, I shall deal with the mathematical structure of the kinetic equation, following the nomenclature proposed by Clarke [3].

The chemical species will be divided into two classes: the internal species which on the network's time scale are the dynamical variables and the external species, which on the same time scale are constant so that they can be omitted as dynamical variables. This means that the external species must be crossed from any reaction in which they take part and the reactions so obtained will be called pseudoreactions. When this crossing out leaves empty one side of a pseudoreaction, the symbol □ will be introduced.

According to Clarke [3], a chemical network is a set of m internal species $A_1,\ldots,A_m$, a set of M pseudoreactions, a matrix of net stoichiometries $\underset{\sim}{\nu}$ and a vector function $\underset{\sim}{r}(\underset{\sim}{c})$ whose jth component gives the velocity of the jth pseudoreaction when the concentration vector $\underset{\sim}{c} \in V^+$, $V = \bar{R}^+_m$ being the kinetic domain. The network's kinetic equation is:

$$\frac{d\underset{\sim}{c}}{dt} = \underset{\sim}{f}(\underset{\sim}{c}) \tag{1}$$

where $\underset{\sim}{f}$, the species formation vector function, is C^1 over V:

$$\underset{\sim}{f}(\underset{\sim}{c}) = \underset{\sim}{\nu}\underset{\sim}{r}(\underset{\sim}{c}). \tag{2}$$

Any pseudoreaction can be represented by an arrow between two complexes:

$$\underset{\sim}{y}_i \xrightarrow{k_{ji}} \underset{\sim}{y}_j, \quad k_{ji} > 0 \quad (i \neq j) \tag{3}$$

where $\underset{\sim}{y}_i \in V$ is the reactant's complex and $\underset{\sim}{y}_j \in V$ is the product's complex and a complex is a linear combination by non-negative integers of vectors of the natural basis of V. When in a pseudoreaction □ appears, the related complex will be the zero vector of V: $\underset{\sim}{0}$.

The column of $\underset{\sim}{\nu}$ corresponding to the pseudoreaction (3) is

$\underset{\sim}{y}_j - \underset{\sim}{y}_i$ and the associated component of $\underset{\sim}{r}(\underset{\sim}{c})$ is: $r_{ji} = k_{ji}\underset{\sim}{c}^{\underset{\sim}{y}_i}$, where $\underset{\sim}{c}^{\underset{\sim}{y}_i} = \prod_{\rho=1}^{m} c_\rho^{y_i^{(\rho)}}$, $y_i^{(\rho)}$ being the ρth component of y_i.
Let us suppose that the M pseudoreactions take place among n distinct complexes. To the chemical network the reaction diagram may be associated by the following rules: every complex is represented by a knot (●); every pseudoreaction $\underset{\sim}{y}_i \to \underset{\sim}{y}_j$ is represented by the directed arc $\overset{i}{\bullet} \longrightarrow \overset{j}{\bullet}$.

The connected components of the reaction diagram are called linkage classes.

Let ℓ be the number of the linkage classes. If every linkage class is strongly connected, then the network is weakly reversible (W.R.).

A subdomain of the kinetic domain V is said to be uniform if, for all the pairs of distinct complexes $\underset{\sim}{y}_i$, $\underset{\sim}{y}_j$, the velocity r_{ij} either is identically vanishing or it is always positive in the subdomain. V^+ is an uniform subdomain of V. In an uniform subdomain the W.R. is preserved.
By the nomenclature introduced for the pseudoreaction, the species formation vectorial function may be written:

$$\underset{\sim}{f}(\underset{\sim}{c}) = \sum_{i,j=1}^{n} r_{ij}(\underset{\sim}{c})(\underset{\sim}{y}_j - \underset{\sim}{y}_i) \tag{4}$$

where $\underset{\sim}{f}(\underset{\sim}{c})$ belongs to the subspace S of V called stoichiometric space and whose dimension is

$$s = \dim S = \operatorname{rank} \underset{\sim}{\nu}. \tag{5}$$

Let W be the real vector space of the complexes whose natural basis $\{\underset{\sim}{w}_1,\dots,\underset{\sim}{w}_n\}$ are the n distinct complexes themselves. If one defines the linear transformation Y: W → V so that $\underset{\sim}{Y}\underset{\sim}{w}_j = \underset{\sim}{y}_j$, j = 1,2,...,n, then

$$\underset{\sim}{f}(\underset{\sim}{c}) = \underset{\sim}{Y}\, \underset{\sim}{g}(\underset{\sim}{c}), \tag{6}$$

where $\underset{\sim}{g}(\underset{\sim}{c})$, the complexes formation vectorial function, lies in the subspace $L(\Delta)$ of W, whose dimension [5] is:

$$\dim L(\Delta) = n - \ell. \tag{7}$$

With reference to (1) and (6), the equilibrium set in V^+ is:

$$E = \{\underset{\sim}{c} \in V^+ : \underset{\sim}{f}(\underset{\sim}{c}) = \underset{\sim}{0}\} \tag{8}$$

and the set of complex balancing is:

$$C = \{\underset{\sim}{c} \in V^+ : \underset{\sim}{g}(\underset{\sim}{c}) = \underset{\sim}{0}\}. \tag{9}$$

In general, $C \subseteq E$. If $C = E \neq \emptyset$ the chemical network is said to be complex balanced (C.B.).
Let us take now the dimension of the subspace $\ker Y \cap L(\Delta)$ of W:

$$\delta = \dim (\ker Y \cap L(\Delta)), \quad \delta \geq 0 \tag{10}$$

which is called "deficiency" of the network.
Since $\underset{\sim}{g}(\underset{\sim}{c}) \in L(\Delta)$, if $c^0 \in E$ then $\underset{\sim}{g}(\underset{\sim}{c}^0) \in \ker Y \cap L(\Delta)$ and a sufficient condition so that the chemical network be C.B. is $\delta = 0$. Because $\delta = \dim L(\Delta) - \dim S$, from (5) and (7) it follows that the sufficient condition for the C.B. of the network is:

$$\delta = (n - \ell) - s = 0. \tag{11}$$

Let us introduce the reaction simplex R. If the dimension of the stoichiometric space is s, (m - s) linear conservation laws, among the components of the vector $\underset{\sim}{c}(t)$ in V, exist [3]. These conservation laws may be represented by an $(m - s) \times m$ matrix $\underset{\approx}{\gamma}$, so that in V

$$\underset{\approx}{\gamma}\,\underset{\approx}{\nu} = \underset{\approx}{0}. \tag{12}$$

From (1), (2) it follows that in S $\underset{\approx}{\gamma}(\underset{\sim}{c}(t) - \underset{\sim}{c}(0)) = \underset{\sim}{0}$, where $\underset{\sim}{c}(0)$ is the initial concentration in $V - \{\underset{\sim}{0}\}$. Since the reaction simplex is $R = (\underset{\sim}{c}(0) + S) \cap V$, its explicit definition is:

$$R = \{\underset{\sim}{c} \in V : \underset{\approx}{\gamma}(\underset{\sim}{c}(t) - \underset{\sim}{c}(0)) = \underset{\sim}{0}\}. \tag{13}$$

Suppose now that there exists a function L: $V \to R$ so that:

(L-1) L is continuous everywhere in V and C^1 in V^+;

(L-2) L is strictly convex in V and L has no minima on ∂V;

(L-3) the vector $\underset{\sim}{\mu}(\underset{\sim}{c}) = \underset{\sim}{\nabla} L$ is such that:

(i) $\underset{\sim}{c}^o \in E$ in V^+ if and only if $\underset{\sim}{\mu}(\underset{\sim}{c}^o) \in S^{\perp}$;

(ii) if $\underset{\sim}{c} \in V^+$, but $\underset{\sim}{c} \notin E$, then the dissipative inequality $\underset{\sim}{\mu}(\underset{\sim}{c}) \cdot \underset{\sim}{f}(\underset{\sim}{c}) < 0$ holds.

Wallwork and Perelson [14] proved that if in (1) $\underset{\sim}{f}$ is a vector field C^1 over R and there exists a vector $\underset{\sim}{\mu}(\underset{\sim}{c})$ which over R satisfies (L-3), (i), (ii), then R is positively invariant under $\underset{\sim}{f}$.

2. THEORY OF HORN, FEINBERG, JACKSON

In the theory developed by Horn, Feinberg, Jackson (H.F.J. theory) is chosen the Liapunov function H: $V \to R$:

$$H(\underset{\sim}{c}) = \sum{}' (\ln c_j - \ln a_j - 1) c_j, \tag{14}$$

where $\underset{\sim}{a}$ is a suitable reference vector in V^+ and $\sum{}'$ runs over all "j" for which $c_j > 0$ [7]. The Liapunov function (14) is quite similar to that introduced by Shear [12] by analogy with the H-Boltzmann function of statistical mechanics. In [13], Shear proved that chemical networks which are detailed balanced (D.B.) at equilibrium, in R have a unique equilibrium state which is globally asymptotically stable.

Horn and Jackson [7] called the Liapunov function (14) "pseudo Helmholtz function" (P.H.F.) and it is easily verified that the P.H.F. satisfies (L-1), (L-2) and furthermore, that $\lim_{\|\underset{\sim}{c}\| \to \infty} H(c) = +\infty$.

The property (L-3) (i), (ii) concerning the vector

$$\underset{\sim}{\nabla} H = \ln \underset{\sim}{c} - \ln \underset{\sim}{a} \tag{15}$$

defines the chemical networks which are Quasi-Termodynamic (Q.T.). For Q.T. chemical networks, besides the positive invariance of R, the existence and uniqueness in R of the equilibrium state is proved from the extremum and convexity properties of the P.H.F., and by applying the Liapunov direct method the global asympotic stability in R of the equilibrium is also proved [7]. The whole set of results established in the H.F.J. theory is represented in the following scheme(see Fig. 1 below).

The main result of the H.F.J. theory is that if a chemical network is C.B. then the network is Q.T. A sufficient condition for C.B. requires that the network's deficiency be equal zero (see (11)). From the implications (4), (2) and (5) of Scheme 2-1, it follows:

Theorem (zero deficiency). If $\delta = 0$, the weak reversibility implies the Q.T. behaviour; violation of weak reversibility implies the violation of Q.T. behaviour.

Let us observe that if a network is D.B. at some concentration $\underset{\sim}{a} \in V^+$, then it is C.B. at the same concentration. Therefore, the implications (1), (2) of Scheme 2-1 imply the Q.T. behaviour. As a consequence, the results obtained by Shear

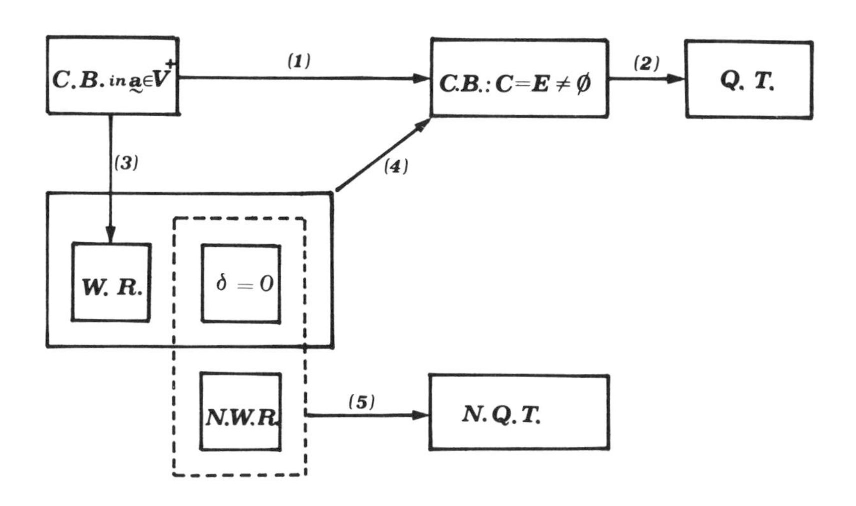

Fig. 1. The Scheme 2-1.

[12,13] are a particular case of the H.F.J. theory.
The zero deficiency theorem offers a criterion of simple applicability to ascertain if a chemical network is Q.T.

To elucidate this point let us take, for instance, the reversible enzyme reaction like Michaelis-Menten:

$$S + E \rightleftharpoons ES \rightleftharpoons EP \rightleftharpoons E + P. \qquad (16)$$

The kinetic domain is $V = \overline{R}^+_5$. By ordering the vector components according to S, P, E, ES, EP, the distinct complexes are:

$$y_1 = \begin{pmatrix}1\\0\\1\\0\\0\end{pmatrix}, \quad y_2 = \begin{pmatrix}0\\0\\0\\1\\0\end{pmatrix}, \quad y_3 = \begin{pmatrix}0\\0\\0\\0\\1\end{pmatrix}, \quad y_4 = \begin{pmatrix}0\\1\\1\\0\\0\end{pmatrix}.$$

The reaction diagram is:

(17)

Therefore $n = 4$, $\ell = 1$. Furthermore, the reaction diagram is reversible and then it is also W.R. It is easily verified that the stoichiometric matrix has rank $s = 3$. Then the network has deficiency $\delta = 0$ and because of W.R. the network is Q.T.

Some criticism. The H.F.J. theory often proves to be ineffective when applied to networks containing pseudoreactions which end to or start from $\square$.
These pseudoreactions, which usually codify the network's exchange with outside, generally have two negative effects:

- they may introduce a N.W.R. linkage class containing the zero complex $\underset{\sim}{0}$. For example, supposing to supply at constant rate the substrate and to remove at a rate proportional to its concentration the product, in (16) the pseudoreactions $P \longrightarrow \square \longrightarrow S$ must be added. In this way we add in (17) the N.W.R. linkage class $\bullet \longrightarrow \underset{\underset{\sim}{0}}{\bullet} \longrightarrow \bullet$.
- they may increase the network's deficiency leading to $\delta > 0$. In fact, also supposing to take reversible the pseudoreactions $P \rightleftharpoons \square \rightleftharpoons S$, we introduce three distinct com-

plexes, and another reversible linkage class. Thus $n = 7$, $\ell = 2$. The new stoichiometric matrix has rank $s > 5$, and therefore $\delta > 0$.

Consider now chemical networks which have $\delta > 0$, and for which the set $E = \emptyset$. Let the subset $\underset{\sim}{g}(V^+)$ of $L(\Delta)$ be the image of V^+ under $\underset{\sim}{g}$. In the vector space W, with reference to the subspaces ker Y, $L(\Delta)$ and to the subset $\underset{\sim}{g}(V^+)$, the situations described in Fig. 2 below are met.

In the case (a) there may be networks, both W.R. and N.W.R., which have a Q.T. behaviour. However, because the condition of C.B. cannot be met for these networks, they cannot be framed within the H.F.J. theory. The case (b) is excluded by the implication (1) of Scheme 2-1. In the case (c) the networks are C.B., and because of the implication (2) of Scheme 2-1, they have a Q.T. behaviour. Because of the implication (3) of Scheme 2-1, this situation occurs only if the network is W.R.

Recently, G. Fichera et al. [6] studied a model of a biochemical system, with deficiency $\delta = 1$, which falls in the case (a). The chemical reactions of the model are:

$$A_1 + A_6 \rightleftharpoons A_2; \quad A_1 \rightleftharpoons A_4 + A_5; \quad A_2 \rightleftharpoons A_3 + A_5;$$

$$A_4 + A_6 \rightleftharpoons A_3.$$

By inspection of the associated reaction diagram it follows that the condition of C.B. at some point $\underset{\sim}{a} \in V^+$ is coincident

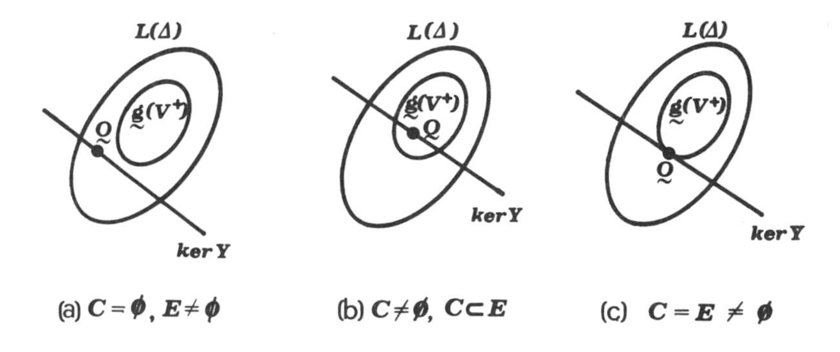

Fig. 2.

with that of D.B. at the same point. Fichera ed al., though leaving the hypothesis of D.B. and therefore that of C.B., were able to prove that there is always a unique and asymptotically stable positive equilibrium point which, under a suitable hypothesis, is asymptotically stable in the large (Q.T. behaviour). Finally, for completeness, I wish to mention that Horn traslated the H.F.J. theory results in terms of graph theory [8-10].

3. D-SYMMETRIZABILITY AND KNOT GRAPHS

Among the possible approaches to the local asymptotic stability of a chemical network's equilibrium, I will expose a theory which introduces the D-symmetrizability of a chemical network [1,2], the D-symmetrizable networks being codified by the associated knot graph.

The hypotheses under which the theory is developed are the following: the internal chemical species are supplied at constant rate and/or removed at a rate proportional to their concentrations. The pseudoreactions among the internal species alone are all reversible. Furthermore, let us suppose that there exists an isolated equilibrium point c^o in some subset Ω of the uniform subdomain V^+. Let $\underset{\sim}{x} = \underset{\sim}{c} - \underset{\sim}{c}^o$, $\underset{\sim}{c}$, $\underset{\sim}{c}^o \in \Omega$ and in Ω let the kinetic equation (1) be approximated by its linear part:

$$\dot{\underset{\sim}{x}} = \underset{\approx}{Q}\,\underset{\sim}{x} \tag{18}$$

where Q is the Jacobian matrix evaluated at equilibrium. If the pseudoreactions among internal species alone are "r", then

$$\underset{\approx}{Q} = \sum_{j=1}^{r} \underset{\approx}{A}^{(j)} + \underset{\approx}{F} \tag{19}$$

where $\underset{\approx}{A}^{(j)}$ is the contribution to $\underset{\approx}{Q}$ of the reversible pseudoreaction "j" and $\underset{\approx}{F}$ is the negative semidefinite diagonal matrix of the monomolecular flow-out. For the definition and properties of the matrices $\underset{\approx}{A}^{(j)}$ see [1].

Among the positive definite quadratic forms I take as Liapunov function:

$$L = \underset{\sim}{x}^T \underset{\approx}{\mathcal{D}}\, x \tag{20}$$

where the real constant diagonal positive matrix $\underset{\approx}{\mathcal{D}}$ is chosen so that

(d-i) $\underset{\approx}{\mathcal{D}}\underset{\approx}{Q} = (\underset{\approx}{\mathcal{D}}\underset{\approx}{Q})^T$;

(d-ii) for all $j = 1,2,\ldots,r$ $\underset{\approx}{\mathcal{D}}\underset{\approx}{A}^{(j)}$ be symmetric and negative semidefinite.

If such a matrix $\underset{\approx}{\mathcal{D}}$ exists, the Jacobian matrix (or the network) is called D-symmetrizable. If $\underset{\approx}{Q}$ is D-symmetrizable the following results hold true:

(A) $$\dot{L} = (\underset{\sim}{\nabla} L)^T \dot{x} = \underset{\sim}{x}^T (\underset{\approx}{\mathcal{D}}\underset{\approx}{Q} + \underset{\approx}{Q}^T \underset{\approx}{\mathcal{D}})\underset{\sim}{x} \tag{21}$$

from which we can define $\underset{\sim}{\nabla} L$:

$$\underset{\sim}{\nabla} L = (\underset{\approx}{\mathcal{D}} + (\underset{\approx}{Q}^{-1})^T \underset{\approx}{\mathcal{D}}\underset{\approx}{Q})\underset{\sim}{x}. \tag{22}$$

The potentiality condition for the Liapunov function (20) (irrotationality of $\underset{\sim}{\nabla} L$) requires the symmetry of

$$\underset{\approx}{\mathcal{D}} + (\underset{\approx}{Q}^{-1})^T \underset{\approx}{\mathcal{D}}\underset{\approx}{Q}. \tag{23}$$

A sufficient condition is the symmetry of $\underset{\approx}{\mathcal{D}}\underset{\approx}{Q}$ (see (d-i)).

(B) A consequence of (d-i), (d-ii) is that:

$$\dot{L} = 2\,\underset{\sim}{x}^T (\underset{\approx}{\mathcal{D}}\underset{\approx}{Q})\underset{\sim}{x} = 2\{\sum_{j=1}^{r} \underset{\sim}{x}^T (\underset{\approx}{\mathcal{D}}\underset{\approx}{A}^{(j)})\,\underset{\sim}{x} + \underset{\sim}{x}^T (\underset{\approx}{\mathcal{D}}\underset{\approx}{F})\underset{\sim}{x}\} \le 0. \tag{24}$$

Then, I can prove the following:

<u>Theorem</u>. If the matrix $\underset{\approx}{Q}$ is non-critical and D-symmetrizable, then the equilibrium state $\underset{\sim}{c}^o$ is asymptotically stable in Ω.

Proof: Since Q is D-symmetrizable, the positive diagonal matrices $\underset{\approx}{\mathcal{D}}^{1/2}$ and $\underset{\approx}{\mathcal{D}}^{-1/2}$ symmetrize the Jacobian matrix by the similarity transformation:

$$\mathcal{D}^{1/2}\mathbf{Q}\,\mathcal{D}^{-1/2} = \mathcal{D}^{-1/2}(\mathcal{D}\,\mathbf{Q})\,\mathcal{D}^{-1/2}.$$

As a consequence, all the characteristic roots of $\mathbf{Q}$ are real. Let us introduce the new variables $\mathbf{y} = \mathcal{D}^{1/2}\mathbf{x}$. Then

$$\dot{L} = 2\,\mathbf{x}^{T}(\mathcal{D}\,\mathbf{Q})\,\mathbf{x} = 2\,\mathbf{y}^{T}[\mathcal{D}^{-1/2}(\mathcal{D}\,\mathbf{Q})\mathcal{D}^{-1/2}]\,\mathbf{y} =$$

$$= 2\,\mathbf{y}^{T}(\mathcal{D}^{1/2}\mathbf{Q}\,\mathcal{D}^{-1/2})\,\mathbf{y}$$

Since from (B) it follows that $\dot{L} \leq 0$, if $\mathbf{Q}$ is non-critical $\dot{L} < 0$ for all $\mathbf{y} \neq \mathbf{0}$ and $\dot{L} = 0$ if and only if $\mathbf{y} = \mathbf{0}$, that is, if and only if $\mathbf{c} = \mathbf{c}^{o}$. Therefore, in Ω, the asymptotic stability of the equilibrium follows. Q.E.D.

The D-symmetrizability of a chemical network may be ascertained through the associated knot graph [1]. To construct the knot graph one must first cross out the external species, and from the pseudoreactions so obtained those which end to or start from □ must be omitted. Then, for every of the remaining r reversible pseudoreactions one must construct the interactant graph:

Definition. We shall call "interactant" the set of all the internal species which take part on the same side of the same pseudoreaction.

Let $\nu_{\rho j} = p_{\rho j} - r_{\rho j}$ the stoichiometric coefficient of A_{ρ} in the jth pseudoreaction. For every pseudoreaction "j" (j = 1,2, ...,r) we have two intearactants:

$$I_{j-1} = \{A_{\rho},\ \rho \in \{1,\ldots,m\}:\ r_{\rho j} \neq 0\} \quad \text{"reactants"}$$

$$I_{j} = \{A_{\rho},\ \rho \in \{1,\ldots,m\}:\ p_{\rho j} \neq 0\} \quad \text{"products"}$$

The jth pseudoreaction is represented by a pair of arrows between I_{j-1} and I_j. One may obtain two interactant graphs:

if $I_{j-1} \cap I_j = \emptyset$, then $I_{j-1} \rightleftarrows I_j$;

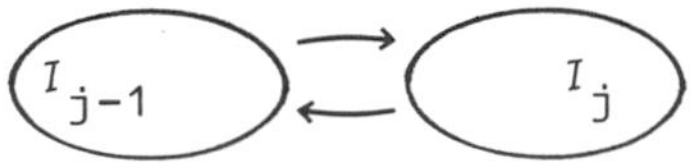

if $I_{j-1} \cap I_j \neq \emptyset$, then I_{j-1} I_j .

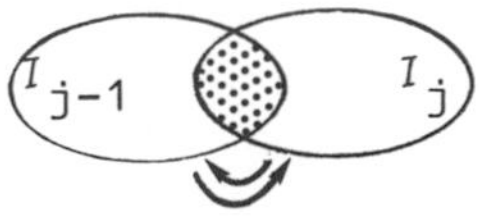

To the second case belong the reactions like autocatalytic. Consider now the connected interactants:

<u>Definition</u>. Two interactants I_i, I_j are connected if $I_i \cap I_j \neq \emptyset$, or if there exists I_k ($k \neq i$, $k \neq j$) such that $I_i \cap I_k \neq \emptyset$, $I_k \cap I_j \neq \emptyset$ and so on.

The knot graph associated with a chemical network is obtained by the following rules [1,2].

(k-1) every maximal set of connected interactans is represented by the same knot (●);

(k-2) every pair of reaction arrows is substituted by one undirected arc.

With reference to the topological classes of knot graphs the following results hold true [1]: all the chemical networks whose associated knot graph is a tree are D-symmetrizable. All the chemical networks whose associated knot graph is a cycle of length ℓ are D-symmetrizable if, at equilibrium, the cyclical relation among forward and reverse reaction rates

$$\prod_{j=1}^{\ell} r_{+jo} = \prod_{j=1}^{\ell} r_{-jo} \qquad (25)$$

holds, where the zero subscripts refer to equilibrium values. For a loop (cycle of length $\ell = 1$) the detailed balance of the pseudoreaction (say the jth) is then implied, that is, $r_{+jo} = r_{-jo}$.

Finally, it is to be mentioned that, from the study of the unbalance of (25) when applied to enzyme reactions, interesting negative graphical criteria for enzyme oscillators may be devised [2].

4. ON THERMODYNAMIC MEANING OF D-SYMMETRIZABILITY

Edelen [4] developped a theory which enables one to interpret in terms of nonlinear phenomenological thermodynamics the properties of a "termodynamically admissible" system, that is, of a system which has an asymptotically stable equilibrium point. Without entering in the details of the Edelen's theory, I will employ some of his ideas to give a possible thermodynamic interpretation of D-symmetrizability. For the chemical network, described by (1), (2), I will define the thermodynamic forces $\underset{\sim}{X} \in R_m$ by the invertible, nonsingular transformation X: V → R_m such that:

$$\underset{\sim}{X} = \underset{\approx}{D}\,\underset{\sim}{c} \tag{26}$$

where $\underset{\approx}{D}$ is a real, constant, symmetric and positive definite matrix.
Let the function $\underset{\sim}{f}^*(\underset{\sim}{X})$ be defined by:

$$\underset{\sim}{f}(\underset{\sim}{c}) = \underset{\sim}{f}(\underset{\approx}{D}^{-1}\underset{\sim}{X}) = \underset{\sim}{f}^*(\underset{\sim}{X}) \tag{27}$$

where $\underset{\sim}{f}(\underset{\sim}{c})$ is the species formation vector function. Owing to (1) and (26), the evolution equation in the space of thermodynamic forces will be:

$$\dot{\underset{\sim}{X}} = \underset{\approx}{D}\,\underset{\sim}{f}^*(\underset{\sim}{X}). \tag{28}$$

Since $\underset{\approx}{D}$ is non-singular, $\underset{\sim}{X}^o$ is an equilibrium point of (28) iff $\underset{\sim}{c} = \underset{\sim}{c}^o$ in Ω. Varying $\underset{\sim}{c}$ in Ω - {$\underset{\sim}{c}^o$}, that is, $\underset{\sim}{X}$ in X(Ω) - {$\underset{\sim}{X}^o$}, then $\underset{\sim}{f}^*(\underset{\sim}{X}) = \underset{\sim}{f}(\underset{\sim}{c}) \neq \underset{\sim}{0}$. As Liapunov function in X(Ω) let us take a function with the structure of an entropy production:

$$\sigma(\underset{\sim}{X}) = \underset{\sim}{J}(\underset{\sim}{X})^T(\underset{\sim}{X} - \underset{\sim}{X}^o) \tag{29}$$

where $\underset{\sim}{J}(\underset{\sim}{X})$ are the conjugate thermodynamic fluxes. Taking the time derivative of (29) along the trajectories of (28), one obtains:

$$\dot{\sigma}(X) = (\underset{\sim}{\nabla}\sigma)^T\dot{\underset{\sim}{X}} = (\underset{\sim}{\nabla}\sigma)^T\underset{\approx}{D}\,\underset{\sim}{f}^*(\underset{\sim}{X}). \tag{30}$$

The negative definiteness of (30) in X(Ω) is then assured choosing

$$\underset{\sim}{\nabla}\sigma = -2\underset{\sim}{f}^*(\underset{\sim}{X}), \tag{31}$$

and the irrotationality condition of $\underset{\sim}{\nabla}\sigma$ requires the symmetry of the matrix: $\partial\underset{\approx}{f}^*(\underset{\sim}{X})/\partial\underset{\sim}{X}$.

Now, suppose that in Ω the species formation vector be approximated by its linear part:

$$f(c) = Q(c - c^{o}), \tag{32}$$

where Q is the network's Jacobian matrix at equilibrium. Then, from (26), (27)

$$f^{*}(X) = Q\, D^{-1}(X - X^{o}). \tag{33}$$

Thus, the irrotationality condition for $\nabla\sigma$, taking into account of (31), (33), leads to:

$$Q\, D^{-1} = (Q\, D^{-1})^{T}. \tag{34}$$

In $X(\Omega)$, the Liapunov function $\sigma(X)$ is approximated by the quadratic form:

$$\sigma(X) = (X - X^{o})^{T}(-Q\, D^{-1})(X - X^{o}), \tag{35}$$

whose positive definiteness follows from the negative definition of the symmetric matrix $Q\, D^{-1}$. By a comparison between (29) and (35) it may be concluded that the thermodynamic fluxes are:

$$J(X) = -f^{*}(X) = -Q\, D^{-1}(X - X^{o}). \tag{36}$$

Then, (33) or (36) establish linear phenomenological equations connecting fluxes and forces and the irrotationality condition (34) ensures the symmetry of the phenomenological coefficients. In accordance with [4], the thermodynamic fluxes (36) together with the symmetry condition (34), are called "Onsager fluxes". Accordingly, the function $\sigma(X)$ in (35) may be interpretated as an entropy production which takes its minimum at equilibrium. This formal analogy with the linear phenomenological thermodynamics resides on the symmetry and negative definiteness of the matrix $Q\, D^{-1}$. It is easy to verify that these requirements are met if Q is non-critical and D-symmetrizable by the symmetric matrix $\mathcal{D} \equiv D$. Finally, from the point of view of stability theory the arguments exposed in this section may be framed within the variable gradient method [2].

CONCLUSIONS

The H.F.J. theory provides criteria, for example the zero deficiency theorem, both for the existence and uniqueness of the equilibrium and for its global asymptotic stability in the reaction simplex R. Furthermore, some implications proved in H.F.J. theory could be used to obtain criteria which hold also for networks with deficiency $\delta > 0$. The class of Q.T. networks which cannot be C.B. in V^+ is of particular interest. This network class needs a different approach than the H.F.J. theory, approach which could be the object of future study.

Concerning the approach which leads to D-symmetrizabile networks, codified by the associated knot graphs, the "a priori" existence of an isolated equilibrium point is required and then the local asymptotic stability is ensured. Though D-symmetrizability gives less information than H.F.J. theory, it applies, however, also to networks with deficiency $\delta > 0$. In the following, I propose an example of D-symmetrizable networks with deficiency $\delta > 0$, but many others may be found. Since m is the dimension of V, the dimension of the stoichiometric space is $s \leq m$. So, it is enough to take networks for which $n - \ell > m$ to obtain $\delta > 0$. Let $\nu_{\rho j}$ be the stoichiometric coefficient of A_ρ in the jth reaction:

$$\nu_{11}A_1 \overset{(1)}{\rightleftharpoons} \nu_{21}A_2;\ \nu_{22}A_2 \overset{(2)}{\rightleftharpoons} \nu_{32}A_3;\ \ldots;$$

$$\nu_{m-1,m-1}A_{m-1} \overset{(m-1)}{\rightleftharpoons} \nu_{m,m-1}A_m;\quad A_m \rightleftharpoons \square \rightleftharpoons A_k$$

$$(1 \leq k < m),$$

where $\nu_{\rho,\rho-1} \neq \nu_{\rho,\rho}$ for all $\rho = 2,\ldots, m-1$.
If $\nu_{k,k-1}, \nu_{k,k}, \nu_{m,m-1}$ are $\neq 1$, then $n = 2m+1$ and $\ell = m$.
If $\nu_{k,k-1} = 1$ (or $\nu_{k,k} = 1$) and $\nu_{m,m-1} \neq 1$, then $n = 2m$ and $\ell = m-1$. In both cases $n - \ell = m+1$ and $\delta > 0$.

Since the pseudoreactions which end to or start from □ must be omitted in the knot graph, this network has an associated knot graph which is a chain:

and therefore is D-symmetrizable.

REFERENCES

1. E. Beretta, F. Vetrano, F. Solimano, C. Lazzari: Some Results about Nonlinear Chemical Systems Represented by Trees and Cycles, Bull. Math. Biol., 41 641-664, (1979).
2. E. Beretta: D-symmetrizability and Stability of Chemical Networks, in Kinetics of Physicochemical Oscillations (Eds. U.F. Franck and E. Wicke), Vol. II, 572-584, Aachen, 1979.
3. B.L. Clarke: Stability of Complex Reaction Networks, Adv. Chem. Phys., 42, 1-213, (1980).
4. D.G.B. Edelen: Asymptotic Stability, Onsager Fluxes and Reaction Kinetics, Int. J. Engng Sci., 11, 819-839, (1973).
5. M. Feinberg: Complex Balancing in General Kinetic Systems, Arch. Rat. Mech. Analysis, 49, 187-194, (1973).
6. G. Fichera, M.A. Sneider, J. Wyman: On the Existence of a Steady State in a Biological System, Atti Acc. Naz. Lincei, Memorie Sci. fisiche matematiche e naturali. S. VIII, Vol. XIV, Sez. III, pp. 1-26, (1977).
7. F. Horn, R. Jackson: General Mass Action Kinetics, Arch. Rat. Mech. Analysis, 47, 81-116, (1972).
8. F. Horn: Necessary and Sufficient Conditions for Complex Balancing in Chemical Kinetics: Arch. Rat. Mech. Analysis, 49, 172-186, (1973).
9. F. Horn: On a Connexion Between Stability and Graphs in Chemical Kinetics. I. Stability and Reaction Diagram, Proc. Roy. Soc. London, A334, 299-312, (1973).

10. F. Horn: On a Connexion between Stability and Graphs in Chemical Kinetics. II. Stability and Complex Graph., Proc. Roy. Soc. London, A334, 313-330, (1973).
11. F. Horn: Stability and Complex Balancing in Mass Action Systems with Three Short Complexes, Proc. Roy. Soc. London, A334, 331-342, (1973).
12. D.B. Shear: An Analog of the Boltzmann H-Theorem (a Liapunov Function) for Systems of Coupled Chemical Reactions, J. Theoret. Biol., 16, 212-228, (1967).
13. D.B. Shear: Stability and Uniqueness of the Equilibrium Point in Chemical Reaction Systems, J. Chem. Phys., 48, 4144-4147, (1968).
14. D. Wallwork, A.S. Perelson: Restrictions on Chemical Kinetic Models, J. Chem. Phys., 65, 284-292, (1976).

STABILITY AND GENERALIZED HOPF BIFURCATION THROUGH A REDUCTION PRINCIPLE

S.R. Bernfeld[1]

The University of Texas at Arlington

P. Negrini[2]

Università di Camerino, Italy

L. Salvadori[1]

Università di Trento, Italy

1. INTRODUCTION

We are interested in obtaining an analysis of the bifurcating periodic orbits arising in the generalized Hopf bifurcation problem in R^n. The existence of these periodic orbits has often been obtained by using such techniques as the Liapunov-Schmidt method or topological degree arguments (see [5] and its references). Our approach, on the other hand, is

---

[1]This research was partially supported by U.S. Army Research Grant DAAG29-80-C-0060.

[2]Work performed under the auspices of Italian Council of Research (CNR)

ISBN 0-12-508780-2

based upon stability properties of the equilibrium point of the unperturbed system. Andronov et. al. [1] showed the fruitfulness of this approach in studying bifurcation problems in R^2 (for more recent papers see Negrini and Salvadori [6] and Bernfeld and Salvadori [2]). In the case of R^2, in contrast to that of R^n, $n > 2$, the stability arguments can be effectively applied because of the Poincaré-Bendixson theory. Bifurcation problems in R^n can be reduced to that of R^2 when two dimensional invariant manifolds are known to exist. The existence of such manifolds occurs, for example, when the unperturbed system contains only two purely imaginary eigenvalues.

In this paper we shall be concerned with the general situation in R^n in which the unperturbed system may have several pairs of purely imaginary eigenvalues. To be more precise, let us consider the differential system

$$\dot{p} = f_o(p), \tag{1.1}$$

where $f_o \in C^\infty[B^n(r_o), R^n]$, $f_o(0) = 0$, and $B^n(r_o) = \{p \in R^n: \|p\| < r_o\}$. Assume the Jacobian matrix $f_o'(0)$ has two purely imaginary eigenvalues $\pm i$ and that the remaining eigenvalues $\{\lambda_j\}_{j=1}^{n-2}$ satisfy $\lambda_j \neq mi$, $m = 0, \pm 1, \ldots$.

For those $f \in C^\infty[B^n(r_o), R^n]$, $f(0) = 0$, which are close to f_o (in an appropriate topology) consider the perturbed system

$$\dot{p} = f(p). \tag{1.2}$$

We are interested in determining the number of nontrivial periodic orbits of (1.2) lying near the origin and having period close to 2π for those f close to f_o.

In approaching this problem, we will consider for any positive integer k the following property:

a) (i) there exist a neighborhood $\mathcal{N}$ of f_o, a neighborhood $\mathcal{U}$ of the origin 0 in R^n and a number $\delta_1 > 0$ such that for every $f \in \mathcal{N}$ there are at most k nontrivial periodic orbits of (1.2) lying in $\mathcal{U}$ whose period is in $[2\pi-\delta_1, 2\pi+\delta_1]$;
(ii) for each integer j, $0 \leq j \leq k$, for each neighborhood $\mathcal{U}_1$ of 0, for each $\delta \in (0, \delta_1)$ and for each neighborhood N of

f_o, $N \subseteq \mathcal{N}$, there exists $f \in N$ such that (1.2) has exactly j nontrivial periodic orbits lying in $\mathcal{U}_1$ whose period is in $[2\pi-\delta, 2\pi+\delta]$.

In contrast to (a) another property which we consider in this paper is:

(A) For any neighborhood N of f_o, for any integer $j \geq 0$, for any neighborhood $\mathcal{U}_1$ of 0, and for any number $\delta > 0$ there exists $f \in N$ such that (1.2) has j nontrivial periodic orbits lying in $\mathcal{U}_1$ whose period is in $[2\pi-\delta, 2\pi+\delta]$.

In R^2, Andronov et. al. [1] proved that property (a) is a consequence of the origin of (1.1) being h-asymptotically stable or h-completely unstable where h is an odd integer and $k = \frac{h-1}{2}$. The origin of (1.1) in R^n is said to be h-asymptotically stable or h-completely unstable if h is the smallest positive integer such that the origin of (1.2) is asymptotically stable (completely unstable) for all f for which $f(p) - f_o(p) = o(\|p\|^h)$ (that is h is the smallest positive integer such that asymptotic stability or complete instability of the origin for (1.1) are recognizable by inspecting the terms up to order h in the Taylor expansion of f_o) (see Negrini and Salvadori [6] for further information on the h-asymptotic stability). In a recent paper Bernfeld and Salvadori [2] in R^2 extended the results of Andronov et. al. [1] by proving property (a) is equivalent to the h-asymptotic stability (h-complete instability) of the origin of (1.1) (where again $k = \frac{h-1}{2}$). It was also shown that property (A) is equivalent to the case in which the origin of (1.1) is neither h-asymptotically stable nor h-completely unstable for any positive integer h.

The problem in R^n was first considered by Chafee [3]. Using the Liapunov-Schmidt method he obtained a determining equation $\psi(\xi, f) = 0$ where ξ is a measure of the amplitude of the bifurcating periodic orbits of (1.2) and f represents the right hand side of (1.2). By assuming that the multiplicity of the zero root of $\psi(\cdot, f_o)$ is a finite number k, he proved that property (a) holds for this k.

Our goal in this paper is to relate the number k in property (a) with the conditional asymptotic stability properties of the origin for a differential system which is close in some sense to the unperturbed system (1.1). These stability properties are precisely the h-asymptotic stability (or h-complete instability) of the origin for a particular differential equation (S_h) in R^2. The construction of (S_h) as well as the recognition of the h-asymptotic stability (or h-complete instability) of the origin of (S_h) can be accomplished by solving linear algebraic systems. Thus, the number k, $k = \frac{h-1}{2}$, can be determined using elementary algebraic techniques. The analysis of our problem is completed by observing that when the origin for (S_h) is neither h-asymptotically stable nor h-completely unstable for every $h > 0$ then the property (A) holds.

The main ingredients of our analysis are: (i) the construction of a quasi-invariant manifold Σ_h for the unperturbed system (1.1); (ii) the use of the Poincaré map along a particular set of solutions of (1.1) which are initially close to Σ_h.

In conclusion, the quantitative problem of determining the number of bifurcating periodic solutions of the perturbed system (1.2) can be reduced to an analysis of the qualitative behavior of the flow near the origin of a two dimensional system appropriately related to the unperturbed system (1.1). In addition, an algebraic procedure allows for a concrete solution to the problem.

In a forthcoming paper, the authors will apply an extension of the Poincaré procedure [8], given by Salvadori [7] in order to compute in certain cases the number k directly from system (1.1).

2. RESULTS

By an appropriate change of coordinates depending on f we may write systems (1.1) and (1.2) respectively in the form

$$\begin{aligned} \dot{x} &= -y + X_o(x,y,z) \\ \dot{y} &= x + Y_o(x,y,z) \\ \dot{z} &= A_o z + Z_o(x,y,z), \end{aligned} \tag{2.1}$$

and

$$\dot{x} = \alpha x - \beta y + X(x,y,z)$$
$$\dot{y} = \alpha y + \beta x + Y(x,y,z) \qquad (2.2)$$
$$\dot{z} = Az + Z(x,y,z)$$

Here α, β are constants, A and A_o are $(n-2) \times (n-2)$ constant matrices, and X, Y, X_o, Y_o belong to $C^\infty[B^n(r_o),R]$ and Z, Z_o belong to $C^\infty[B^n(r_o),R^{n-2}]$. Moreover, X, Y, Z, X_o, Y_o, Z_o are of order greater than one and the eigenvalues of A_o, $\{\lambda_j\}_{j=1}^{n-2}$ satisfy the condition that $\lambda_j \neq mi$, $m = 0, \pm 1, \ldots$. We shall refer to the right hand sides of (2.1) and (2.2) as f_o and f respectively.

We now consider an (n-2) dimensional polynomial of some degree h, $h \geq 1$, given by

$$\phi^{(h)}(x,y) = \phi_1(x,y) + \ldots + \phi_h(x,y) \qquad (2.3)$$

where $\phi_j(x,y)$ is homogeneous of degree j. We attempt to determine $\phi_1,\ldots,\phi_h$ in order to obtain along the solutions of (2.1)

$$\left[\frac{d}{dt}(z - \phi^{(h)}(x,y))\right]_{z=\phi^{(h)}(x,y)} = o(x^2 + y^2)^{h/2}; \qquad (2.4)$$

that is, we have to satisfy

$$\frac{\partial\phi^{(h)}(x,y)}{\partial x}[-y + X_o(x,y,\phi^{(h)}(x,y))] + \frac{\partial\phi^{(h)}(x,y)}{\partial y}[x + Y_o(x,y,\phi^{(h)}(x,y))] = A_o\phi^{(h)}(x,y) + Z_o(x,y,\phi^{(h)}(x,y)) + o(x^2 + y^2)^{h/2}. \qquad (2.5)$$

This implies for every $j \in \{1,\ldots,h\}$, ϕ_j has to satisfy the partial differential equation

$$\frac{\partial\phi_j}{\partial y}x - \frac{\partial\phi_j}{\partial x}y = A_o\phi_j + U_j, \qquad (2.6)$$

where U_j is an (n-2) dimensional homogeneous polynomial of degree j depending on the functions $\phi_1 \ldots \phi_{j-1}$. Under the assumptions on A_0 (2.6) has a unique solution and can be solved recursively by observing that $\phi_1(x,y) \equiv 0$.

The two dimensional surface $z = \phi^{(h)}(x,y)$ is tangent at the origin to the eigenspace corresponding to the eigenvalues $\pm i$. This surface will be called a quasi-invariant manifold of order h.

Given any $h > 0$ define the following two dimensional system

$$\begin{aligned} \dot{x} &= -y + X_0(x,y,\phi^{(h)}(x,y)) \\ \dot{y} &= x + Y_0(x,y,\phi^{(h)}(x,y)). \end{aligned} \tag{S_h}$$

(This is the system referred to in the introduction).

We distinguish the two possible cases:

I. There exists $h > 1$ (and then h must be odd) such that $x \equiv y \equiv 0$ is either h-asymptotically stable or h-completely unstable for (S_h).

II. Case I does not hold.

We are now able to state our main result.

Theorem 1. In case I property (a) holds with $k = \frac{h-1}{2}$. In case II, property (A) holds.

If all the eigenvalues of A_0 have real part not equal to zero, then for every $h > 1$ there exists a C^{h+1} two dimensional center manifold which will be denoted by H_h. We notice that if $z = \phi(x,y)$ is the equation of this center manifold, we can write

$$\phi(x,y) = \phi^{(h)}(x,y) + o(x^2 + y^2)^{h/2}.$$

As corollary of Theorem 1 the following result holds.

Theorem 2. Suppose that all the eigenvalues of A_0 have real part different than zero. Then: (i) if there exists an h (and h must be odd) such that the origin of the unperturbed system (2.1) is either h-asymptotically stable or h-completely unstable on H_h (that is, with respect to initial points on H_h) then

(a) holds with $k = \frac{h-1}{2}$; (ii) if for every $h > 1$ the origin for the unperturbed system (2.1) is neither h-asymptotically stable nor h-completely unstable on H_h then (A) holds.

Under some more particular hypotheses on the eigenvalues of $f_o'(0)$ the stability properties in Theorem 2 can be expressed in terms of the unperturbed system (2.1) without any explicit involvement of H_h. This can be proved by the extension of the Poincaré procedure [8] given by Liapounov [4]. Precisely the following result holds.

Theorem 3. Suppose all the eigenvalues of A_o have negative real part. Then (i) if the origin of the unperturbed system (2.1) is either h-asymptotically stable or h-unstable (in the whole) then (a) holds with $k = \frac{h-1}{2}$; (ii) if for every $h > 1$ the origin for the unperturbed system (2.1) is neither h-asymptotically stable nor h-unstable, then (A) holds.

Notice that we are using the concept of h-unstable whose definition is analogous to that of h-complete instability. A similar theorem can be stated when $f_o'(0)$ has two purely imaginary eigenvalues $\pm i$ and the remaining eigenvalues have positive real part.

3. OUTLINE OF PROOF OF THEOREM 1

We shall only present a sketch of the main ideas used in the proof of Theorem 1. A more comprehensive analysis of our results is in preparation and will be given elsewhere.

Using the transformation

$$\zeta = z - \phi^{(h)}(x,y),$$

we can rewrite the unperturbed system (2.1) as

$$\begin{aligned} \dot{x} &= -y + X_o^{(h)}(x,y,\zeta) \\ \dot{y} &= x + Y_o^{(h)}(x,y,\zeta) \\ \dot{\zeta} &= A_o\zeta + W_o^{(h)}(x,y,\zeta), \end{aligned} \tag{3.1}$$

where $X_0^{(h)}(x,y,0) = X_0(x,y,\phi^{(h)}(x,y))$, $Y_0^{(h)}(x,y,0) = Y_0(x,y,\phi^{(h)}(x,y))$. From (2.4) we observe that $W_0^{(h)}(x,y,0)$ is of order greater than h. Analogously, we can rewrite the perturbed system (2.2) as

$$\begin{aligned} \dot{x} &= \alpha x - \beta y + X^{(h)}(x,y,\zeta) \\ \dot{y} &= \alpha y + \beta x + Y^{(h)}(x,y,\zeta) \\ \dot{\zeta} &= A\zeta + W^{(h)}(x,y,\zeta), \end{aligned} \tag{3.2}$$

where $X^{(h)}(x,y,0) = X(x,y,\phi^{(h)}(x,y))$, $Y^{(h)}(x,y,0) = Y(x,y,\phi^{(h)}(x,y))$ and $X^{(h)}$, $Y^{(h)}$, $W^{(h)}$ are of order ≥ 2. For simplicity, we shall again refer to the right hand sides of (3.1) and (3.2) as f_0 and f respectively.

We now state the following lemma whose proof is based on the implicit function theorem.

Lemma 1. There exist L, ε, $\delta > 0$ and a neighborhood $\bar{N}$ of f_0 such that for every $f \in \bar{N}$ and for every periodic solution $(x(t,x_0,y_0,\zeta_0), y(t,x_0,y_0,\zeta_0), \zeta(t,x_0,y_0,\zeta_0))$ of (3.2) lying in $B^n(\varepsilon)$ whose period is in $[2\pi-\delta, 2\pi+\delta]$ we have $\|\zeta_0\| \leq L(x_0^2 + y_0^2)$.

The substitution

$$x = r\cos\theta, \quad y = r\sin\theta, \quad \zeta = rv, \tag{3.3}$$

into (3.2) gives a system which we write as

$$\begin{aligned} \frac{dr}{d\theta} &= R^{(h)}(\theta,r,v,f) \\ \frac{dv}{d\theta} &= \frac{A}{\beta}v + \eta^{(h)}(\theta,r,v,f), \end{aligned} \tag{3.4}$$

where $R^{(h)}$, $\eta^{(h)} \in C^\infty$. The solutions of (3.4) for which $r(\theta) \neq 0$ for all θ are the orbits of corresponding solutions of (3.2). Moreover, the origin is a solution of both (3.2) and (3.4). We denote by $(r(\theta,c,v_0,f), v(\theta,c,v_0,f))$ the solution of (3.4) passing through $(0,c,v_0)$. When the solutions $(r(\theta),v(\theta))$ of (3.4) are known, the corresponding solutions of (3.2) can be completely determined by solving the equation

$$\frac{d\theta}{dt} = \Theta(\theta, r(\theta), v(\theta), f), \tag{3.5}$$

where Θ is greater than some positive number in a neighborhood of the origin in R^n and for f close to f_o. Every 2π-periodic solution of (3.4), $r(\theta), v(\theta)$ represents a periodic orbit of (3.2) whose period T is given by

$$T = \int_0^{2\pi} \frac{d\theta}{\Theta(\theta, r(\theta), v(\theta), f)} \tag{3.6}$$

We now introduce for system (3.4) properties (a') and (A') which correspond to properties (a) and (A) for system (3.2).

(a') (i) There exist a neighborhood $\mathcal{N}$ of f_o and a neighborhood $\mathcal{U}'$ of $r = 0, v = 0$ such that for every $f \in \mathcal{N}$ there are at most k nontrivial 2π periodic solutions of (3.4) lying in $\mathcal{U}'$.

(ii) For each integer j, $0 \leq j \leq k$, for each neighborhood N of f_o, $N \subseteq \mathcal{N}$, and for each neighborhood $\mathcal{U}_1'$ of $r = 0$, $v = 0$ there exists $f \in N$ such that (3.4) has exactly j nontrivial 2π periodic solution lying in $\mathcal{U}_1'$.

(A') For any neighborhood N of f_o, for any integer $j \geq 0$, and for any neighborhood $\mathcal{U}_1'$ of $r = 0$, $v = 0$ there exists $f \in N$ such that (3.4) has j nontrivial 2π periodic orbits lying in $\mathcal{U}_1'$.

We then have:

<u>Lemma 2</u>. Property (a') implies (a).

In order to prove Lemma 2, it is sufficient in view of Lemma 1, to ascertain the following property: (b') the 2π periodic solutions of (3.4) lying in a fixed neighborhood of $r = 0$, $v = 0$ tend to the origin as $f \to f_o$.

A solution $(r(\theta), v(\theta))$ of (3.4) will be called a $(2\pi, v)$ solution if $v(2\pi) = v(0)$. Every 2π periodic solution is a $(2\pi, v)$ solution but the converse is not, in general, true. In order to find the 2π periodic solutions, we only need to analyze the set of $(2\pi, v)$ solutions. Under our assumptions on A_o we can use the implicit function theorem to derive from the second equation in (3.4) a C^∞ function $\tau(c,f), \tau(0,f) = 0$, such that a solution of (3.4) passing through $(0, c, v_o)$, with $f - f_o$, c, and v_o sufficiently small, is a $(2\pi,v)$ solution if and only if $v_o = \tau(c,f)$. Denote by $(r(\theta,c,f),v(\theta,c,f))$ the $(2\pi,v)$ solution passing through $(0,c,\tau(c,f))$.

Let us write

$$\begin{aligned} r(\theta,c,f) &= u_1(\theta,f)c + \dots + u_h(\theta,f)c^h + o(c^h) \\ v(\theta,c,f) &= \nu_1(\theta,f)c + \dots + \nu_{h-1}(\theta,f)c^{h-1} + o(c^{h-1}) \end{aligned} \tag{3.7}$$

where $u_1(0,f) = 1$, $u_i(0,f) = 0$ for $i > 1$ and

$$\nu_i(0,f) = \nu_i(2\pi,f) \quad \text{for } i \geq 1. \tag{3.8}$$

Consider now the displacement function relative to the $(2\pi,v)$ solutions which is defined as

$$V(c,f) = r(2\pi,c,f) - c. \tag{3.9}$$

Assume Case I. We prove that

$$\frac{\partial^i V}{\partial c^i}(0,f_o) = 0 \text{ for } i = 1 \dots h-1 \text{ and } \frac{\partial^h V}{\partial c^h}(0,f_o) \neq 0. \tag{3.10}$$

Because $\eta^{(h)}(\theta,r,0,f_o)$ is of order $> h-1$, the hypothesis on A_o and the requirement that (3.8) holds implies that $\nu_i(\theta,f_o) \equiv 0$ for $i = 1 \dots h-1$. Thus, in order to compute the functions $u_i(\theta,f_o)$, we may put $v = 0$ into equation (3.4), for $f = f_o$. We then obtain the equation

$$\frac{dr}{d\theta} = R^{(h)}(\theta,r,0,f_o),$$

which is precisely the equation in polar coordinates of the orbits of (S_h). Since $x = y = 0$ is either h-asymptotically stable or h-completely unstable for (S_h) we have

$$u_1(\theta,f_o) \equiv 1,\ u_i(2\pi,f_o) = 0,\ i = 2 \dots h-1,$$

$$u_h(2\pi,f_o) \neq 0,$$

thus implying (3.10) holds (see [6] for more details).

Let us extend the domain of $V(c,f)$ to include negative values of c. Since the origin is a solution of (3.4) for any f, an application of Rolle's Theorem, in view of (3.10), implies that there exist a $\delta > 0$, and a neighborhood N of f_0 such that for any $f \in N$, $V(c,f)$ has at most $h - 1$ nonzero roots lying in $[-\delta,\delta]$. On the other hand, it is easy to recognize that for each positive root of $V(c,f)$ there is a negative root of $V(c,f)$. Thus, there are at most $\frac{h-1}{2}$ 2π periodic solutions of (3.4) lying in a neighborhood U' of $r = 0$, $v = 0$. This proves (a')(i) is satisfied.

Property (a')(ii) can be proved by assuming a particular perturbed system of the form

$$\begin{aligned} \dot{x} &= -y + X_0^{(h)}(x,y,\zeta) + \sum_{i=0}^{(h-3)/2} a_i x(x^2 + y^2)^i \\ \dot{y} &= x + Y_0^{(h)}(x,y,\zeta) + \sum_{i=0}^{(h-3)/2} a_i y(x^2 + y^2)^i \qquad (3.11) \\ \dot{\zeta} &= A_0\zeta + W_0^{(h)}(x,y,\zeta), \end{aligned}$$

where a_i are constants depending on j, $0 \le j \le k$, N and U_1'.

Since the roots of $V(\cdot,f) = 0$ approach zero as $f \to f_0$, property (b') holds. Lemma 2 then implies (a) holds, proving Theorem 1 for case I.

Finally, Case II follows from Case I in the following manner. For any positive integer j we assume in (3.1) $h = 2j + 1$ and use perturbations similar to that used in (3.11). Precisely we consider perturbed systems of the form

$$\begin{aligned} \dot{x} &= -y + X_0^{(h)}(x,y,\zeta) + bx(x^2 + y^2)^{\frac{h-1}{2}} \\ \dot{y} &= x + Y_0^{(h)}(x,y,\zeta) + by(x^2 + y^2)^{\frac{h-1}{2}} \qquad (3.12) \\ \dot{\zeta} &= A_0\zeta + W_0^{(h)}(x,y,\zeta), \end{aligned}$$

where b is a constant. We then have for the corresponding reduced system (S_h) the origin is either h-asymptotically stable (or h-completely unstable) if $b < 0$ (or $b > 0$). Thus we have reduced the problem to Case I. Since j and b are arbitrary, property (A') holds. We immediately have (A), thus concluding the proof of Theorem 1.

REFERENCES

1. A. Andronov, E. Leontovich, I. Gordon, A. Maier: Theory of Bifurcations of Dynamical Systems in the Plane, Israel Program of Scientific Translations, Jerusalem, 1971.
2. S. Bernfeld, L. Salvadori: Generalized Hopf Bifurcation and h-Asymptotic Stability, J. Nonlinear Analysis, T.M.A. (to appear).
3. N. Chafee: Generalized Hopf Bifurcation and Perturbation in a Full Neighborhood of a Given Vector Field, Indiana Univ. Math. J. 27 173-194, (1978).
4. A.M. Liapounov: Problème Général de la Stabilité du Mouvement, Ann. of Math. Studies, 17, Princeton N. J., Princeton University Press, 1947.
5. G. Marsden, M. McCracken: The Hopf Bifurcation and its Applications, Notes in Applied Math. Sci. 19, New York, Springer Verlag, 1976.
6. P. Negrini, L. Salvadori: Attractivity and Hopf Bifurcation, J. Nonlinear Analysis, T.M.A. 3, 87-100, (1979).
7. L. Salvadori: Sulla Stabilità dell'Equilibrio nei Casi Critici, Ann. Mat. Pura Appl. (4) 49, 1-33, (1965).
8. G. Sansone, R. Conti: Nonlinear Differential Equations, New York, Macmillan, 1964.

ALMOST PERIODICITY AND ASYMPTOTIC BEHAVIOR FOR THE SOLUTIONS OF A NONLINEAR WAWE EQUATION

Marco Biroli

Istituto di Matematica
Politecnico di Milano, Italy

1. INTRODUCTION

We first give some preliminaries on almost periodic functions: let be X a Banach space:

Definition 1. We shall say that $f(t): \mathbb{R} \to X$ is almost periodic if to every $\varepsilon > 0$ there corresponds a relatively dense set $\{\tau\}_\varepsilon$ such that

$$\sup_{t \in \mathbb{R}} \|f(t+\tau) - f(\tau)\|_X \leq \varepsilon \ , \text{ for all } \tau \in \{\tau\}_\varepsilon$$

(A set E is said relatively dense if there is $\ell > 0$ such that every interval $(a, a+\ell)$, $a \in \mathbb{R}$ contains some point of E). The interest of almost periodic functions is given be the following result: the space of almost periodic functions coincides with the closure with respect to the uniform convergence on $\mathbb{R}$ of the trigonometric polynomials.
We observe also that the space of almost periodic functions is a Banach space for the norm Sup $\|f(t)\|_X$.
We recall finally two important properties of almost periodic functions:

(A) If f(t) is almost periodic, f(t) has a relatively compact trajectory;

ISBN 0-12-508780-2

(B) Let be $f(t)$ continuous; $f(t)$ is almost periodic iff, for any sequence of real numbers $\{c_n\}$, there exists a subsequence $\{s_n\}$ such that the sequence $\{f(t + s_n)\}$ converges uniformly on $\mathbb{R}$.

In this lecture we consider the nonlinear wawe equation

$$\begin{aligned} & u_{tt}(t,x) - \Delta u(t,x) + \beta(u_t(t,x)) \ni f(t,x) \\ & \quad \text{a.e. in } \Omega \times (0,+\infty) \\ & u(t,x)\big|_{\partial\Omega} = 0 \qquad \text{a.e. in } (0,+\infty) \\ & u(0,x) = u_o(x) \qquad u_t(0,x) = u_1(x) \quad \text{a.e. in } \Omega \end{aligned} \tag{1.1}$$

where β is a maximal monotone graph, $\Omega \subset R^N$ an open bounded set with $\partial\Omega$ smooth.
We are interested in the following problem: is there a solution of (1.1) to which all the solutions of (1.1) are asymptotic for $t \to +\infty$?
If f satisfies suitable almost periodicity hypoteses and β has a suitable polynomial growth to infinity and is uniformly monotone, the answer is affirmative; precisely there is a unique energy-almost periodic solution u to the problem

$$\begin{aligned} & u_{tt}(t,x) - \Delta u(t,x) + \beta(u_t(t,x)) \ni f(t,x) \text{ a.e. in } \Omega \times R \\ & u(t,x)\big|_{\partial\Omega} = 0 \text{ a.e. in } R \end{aligned} \tag{1.2}$$

and all the solutions of (1.1) are asymptotic to u for t going to $+\infty$.
A result of this type was first shown by G. Prouse [5], and after improved by the author [2] in the case $\beta(0) = 0$ and β continuous. We describe here a recent of M. Biroli, A. Haraux, [3], concerning the case β not continuous.

2. ASYMTOTIC BEHAVIOR

(H_1) We assume either

$$\dim \Omega = N \geq 2 \text{ and } |\beta^{o}(w)| \leq C_1 |w|^k + C_2$$

where $\beta^{o}(w) = g$ if $g \in \beta(w)$ and $|g| \leq |f|$ $\forall f \in \beta(w)$ and

$k < \dfrac{N+2}{N-2}$ if $N > 2$, $k < +\infty$ if $N = 2$, or

$$\dim \Omega = 1 \text{ and } 0 \in \operatorname{int} D(\beta).$$

(H_2) We assume $0 \in D(\beta)$ $\beta(0) = 0$

(H_3) If $f \in \beta(u)$, $g \in \beta(v)$ then $(f - g)(u - v) \geq \alpha |u - v|^2$, $\alpha > 0$.

Let f be S^2-almost periodic, f_t S^2-bounded in $L^2(\Omega)$ $(f(\cdot+\eta,x))$, $(f_t(\cdot+\eta,x))$ is almost periodic (bounded) in $L^2(0,1;\, L^2(\Omega))$ and $u(t,\cdot)$ be a (weak) solution to (1.2) which is almost periodic in energy (i.e., $u(t,\cdot)$ is almost periodic in $H_o^1(\Omega)$ and $u_t(t,\cdot)$ is almost periodic in $L^2(\Omega)$).

Lemma 1. If v is a (weak) solution to (1.1) with v, v_t bounded in energy on $(0,+\infty)$ we have

$$\lim_{t\to+\infty} (v(t) - u(t)) = 0 \text{ in energy.}$$

Proof: Let w be v - u; we have

$$\|w(t)\|_E^2 \leq \|w(0)\|_E^2 - 2\int_0^t \|w_t(t)\|_{0,2}^2 \, dt \tag{2.1}$$

From (2.1), w being bounded in energy on $(0,+\infty)$, we have

$$\lim_{i\to+\infty} w_t(t + i) = 0 \quad \text{in} \quad L^2(0,1;\, L^2(\Omega)) \tag{2.2}$$

From the hypothesis above we have (possibly after extracting a subsequence)

$$\lim_{i\to+\infty} u(t + i) = \tilde{u}(t), \qquad \lim_{i\to+\infty} v(t + i) = \tilde{v}(t) \tag{2.3$_1$}$$

in $L^2(0,1; H^1_0(\Omega))$;

$$\lim_{i\to+\infty} u_t(t+i) = \tilde{u}_t(t), \quad \lim_{i\to+\infty} v_t(t+i) = \tilde{v}_t(t) \qquad (2.3_2)$$

in $L^2(0,1; L^2(\Omega))$.
We suppose now, the sake of simplicity, dim $\Omega = N > 2$ (the proof in the case dim $\Omega = 1$ is substantially the same modulo some tecnical refinements of the method).
From Sobolev embedding theorems we have that $g_1(t+i) = [-(u_{tt} - \Delta u) + f](t+i) \in \beta(u_t(t+i))$, $g_2(t+i) = [-(v_{tt} - \Delta v) + f](t+i) \in \beta(v_t(t+i))$ are bounded in $L^2(0,1; L^2(\Omega))$ then we have, possibly after extracting a subsequence,

$$\lim_{i\to+\infty} g_k(t+i) = h_k(t) \quad \text{in } L^2(0,1; L^2(\Omega)) \qquad (2.4)$$

where $h_1 \in \beta(\tilde{u}_t)$, $h_2 \in \beta(\tilde{v}_t)$, $k = 1,2$.
From (2.3_1) (2.3_2) (2.4), if we denote $\tilde{w} = \tilde{v} - \tilde{u}$, we have

$$\Delta\tilde{w} = h_1 - h_2 = h. \qquad (2.5)$$

where $h = h(x)$ depends only on x.
We show now

$$h(x) = 0 \qquad (2.6)$$

Since h is in $L^2(\Omega)$, we have $\tilde{w} \in H^2(\Omega) \cap H^1_0(\Omega)$; then there is $E \subset R$ with meas$(R - E) = 0$ such that for all $t \in E$ there is $\Omega_1(t) \subset \Omega$ with meas $(\Omega_1(t))$ = meas (Ω), so that $x \in \Omega_1(t)$ implies $\Delta\tilde{w}(x) = h_1(t,x) - h_2(t,x) = h(x)$.
We observe also that there is $\Omega_2 \subset \Omega$ such that meas(Ω_2) = meas(Ω) and for all $x \in \Omega_2$ $\tilde{u}_t(t,x)$, $\tilde{v}_t(t,x)$ are defined almost everywhere in t and can be extended to R as locally uniformly continuous functions.
If $h(x) \neq 0$ there is $\eta > 0$ such that

$$\text{meas}(|\Delta\tilde{w}| \geq \eta) > 0$$

then if $\{t_n\} \subset E$ is dense in E

$$\text{meas}(x \in \Omega_2 \bigcap_{n>0} \Omega_1(t_n),\ |\Delta\tilde{w}(x)| > \eta) = \text{meas}(\Omega_3) > 0$$

The set of the reals λ such that $\beta(\lambda)$ contains a segment with measure $\geq \eta$ is a set T for which 0 is not a accumulation point; let be $\varepsilon = \text{dist}(0,T) > 0$.
If $x \in \Omega_3$ we have $|\tilde{u}_t(t_n,x)| \geq \varepsilon\ \forall\, n \in N \Rightarrow |\tilde{u}_t(t,x)| \geq \varepsilon$ $\forall\, t \in R$, then there is $\Omega_4 \subset \Omega_3$ such that $\text{meas}(\Omega_4) \geq \frac{1}{2}$ $\text{meas}(\Omega_3)$ and

$$\text{either } x \in \Omega_4 \Rightarrow \tilde{u}_t(t,x) \geq \varepsilon$$

$$\text{or } \quad x \in \Omega_4 \Rightarrow \tilde{u}_t(t,x) \leq -\varepsilon$$

In the first case we have

$$\int_{\Omega_4} \tilde{u}_t(t,x)\,dx \geq \varepsilon\ \text{meas}(\Omega_4) = \gamma > 0 \Rightarrow \int_{\Omega_4} \tilde{u}(t,x)\,dx \geq \gamma t - C$$

then, being $\tilde{u}$ bounded in $L^2(\Omega)$ on $(0,+\infty)$, we get a contradiction; analogously we get a contradiction in the second case; then we have $\text{meas}(\Omega_3) = 0$ which is equivalent to $h(x) = 0$.

<u>Remark 1</u>. We observe that the proof of the lemma 1 uses only the hypotheses (H_1) (H_2).

<u>Lemma 2</u>. Let be $(u_o,u_1) \in H_o^1(\Omega) \times L^2(\Omega)$ and u the solution to (1.1) relative to (u_o,u_1); then u is bounded in energy on $(0,+\infty)$.

We give, for the sake of simplicity, the proof in the case $\dim(\Omega) > 2$.
We observe that from (H_2) we have $|\beta^o(w)| \geq \alpha|w|$; let us define

$$E(t) = \|\nabla u(t,\cdot)\|_{0,2}^2 + \|u_t(t,\cdot)\|_{0,2}^2, \quad g = f - u_{tt} + \Delta u$$

In the proof we can suppose u to be a strong solution to (1.1) a regularisation procedure gives the result in the general case.
We have

$$E'(t) \leq 2(f(t) - g(t),\ u_t(t))$$

Let be $\bar{t}$ fixed and $T > 0$; we distinguish two cases

$$\text{(I) } E(\bar{t}+T) \le E(\bar{t}) \qquad \text{(II) } E(\bar{t}+T) \ge E(\bar{t})$$

In case (II) we have

$$\int_{\bar{t}}^{\bar{t}+T} (f(t) - g(t),\ u_t(t))dt \le 0,$$

then

$$\alpha\int_{\bar{t}}^{\bar{t}+T} \|u_t(t)\|_{0,2}^2\ dt \le \int_{\bar{t}}^{\bar{t}+T} (f(t),\ u_t(t))dt. \tag{2.7}$$

From (2.7) we have

$$\int_{\bar{t}}^{\bar{t}+T} \|u_t(t)\|_{0,2}^2\ dt \le M, \tag{2.8}$$

then

$$\int_{\bar{t}}^{\bar{t}+T} (g(t),\ u_t(t))dt \le M_1. \tag{2.9}$$

From (2.8), (2.9) $|E(t) - E(s)|$ is a priori bounded in $[\bar{t},\bar{t}+T]$, then

$$\operatorname{Sup}_{t\in[\bar{t},\bar{t}+T]} E(t) \le E(t^*) + K_1, \quad t^* \in [\bar{t},\bar{t}+T]. \tag{2.10}$$

We can choose

$$E(t^*) = \frac{1}{T}\int_{\bar{t}}^{\bar{t}+T} E(t)dt \le K_2 + \frac{1}{T}\int_{\bar{t}}^{\bar{t}+T} \|\nabla u(t)\|_{0,2}^2\ dt \tag{2.11}$$

From (2.10), (2.11)

$$\operatorname{Sup}_{t\in[\bar{t},\bar{t}+T]} E(t) \le K_3 + \frac{1}{T}\int_{\bar{t}}^{\bar{t}+T} \|\nabla u(t)\|_{0,2}^2\ dt. \tag{2.12}$$

Now, multiplying the equation (1.1) by u we get easily

$$\int_{\bar{t}}^{\bar{t}+T} \|\nabla u(t)\|_{0,2}^2\ dt \le -\ [u(t),u_t(t))]_{\bar{t}}^{\bar{t}+T} + \int_{\bar{t}}^{\bar{t}+T} \|u_t(t)\|_{0,2}^2 dt$$
$$+ \int_{\bar{t}}^{\bar{t}+T} (f(t) - g(t),\ u(t))dt$$

then if λ_o = imbedding constant of $H_o^1(\Omega)$ into $L^2(\Omega)$

$$\int_{\bar t}^{\bar t+T} \|\nabla u(t)\|_{0,2}^2 dt \leq 2\left(\int_{\bar t}^{\bar t+T} (\|u_t(t)\|_{0,2}^2 + \|f(t)\|_{0,2}^2)dt + \right.$$

$$\left. \operatorname{Sup}_{t\in[\bar t,\bar t+T]} E(t)\right) + 2\left(\int_{\bar t}^{\bar t+T} (g(t),u(t))dt\right.$$

Using Sobolev embedding theorem

$$\|g(t)\|_{-1,2} \leq C_1(1 + \|u_t(t)\|_{0,2}^2 + (g(t),u(t)))$$

then choosing $T = 4\lambda_o$

$$\int_{\bar t}^{\bar t+T} \|\nabla u(t)\|_{0,2}^2 dt \leq C_2 + C_3(\operatorname{Sup}_{t\in[\bar t,\bar t+T]} E(t))^{1/2} +$$

$$\frac{T}{2}\operatorname{Sup}_{t\in[\bar t,\bar t+T]} E(t)$$

Dividing by T and using (2.12) we obtain in case II

$$|E(t)| \leq M_2; \tag{2.14}$$

therefore, in both cases I and II we have

$$|E(t)| \leq \operatorname{Sup}(M_2, \operatorname{Sup}_{t\in[0,T]} E(t)).$$

By standards methods, [1], we can conclude:

Corollary 1. The problem (1.2) has a solution u bounded in energy.

Remark 2. We observe that the proofs of lemma 2, coroll. 1 use the hypothesis (H_1) and the relation $|\beta^o(w)| \geq \alpha|w|$, $\alpha > 0$.

Lemma 3. Let (u_o,u_1) belong to $H^2(\Omega) \cap H_o^1(\Omega) \times H_o^1(\Omega)$ and u be the solution of (1.1) relative to (u_o,u_1); then u, u_t are bounded in energy on $(0,+\infty)$.

Proof: For h > 0 we set $v^h(t) = h^{-1}(u(t + h) - u(t))$; we have

$$v_{tt}^h - \Delta v^h + \varphi^h = f^h (f^h(t) = h^{-1}(f(t+h) - f(t)))$$

with

$$\varphi^h = f^h + \Delta v^h - v_{tt}^h \in h^{-1}(\beta(u_t(t+h)) - \beta(u_t(t)))$$

We define $E_1^h(t) = \|\nabla v^h(t)\|_{0,2}^2 + \|v_t^h(t)\|_{0,2}^2$; we have

$$(E_1^h(t))' = 2(f^h(t) - \varphi^h(t), v_t^h(t)).$$

We fix temporarily two numbers τ, $\ell > 0$ and we observe that

$$E_1^h(\tau + 1) - E_1^h(\tau) = 2\int_\tau^{\tau+\ell} (f^h(t) - \varphi^h(t), v_t^h(t))dt.$$

We distinguish two possibles cases:

$$\text{(I)}\ E_1^h(\tau+1) \leq E_1^h(\tau) \qquad \text{(II)}\ E_1^h(\tau+1) \geq E_1^h(\tau)$$

In the case (II) we have easily

$$\int_\tau^{\tau+1} \|v_t^h(t)\|_{0,2}^2\, dt \leq C_1(\ell); \quad \int_\tau^{\tau+\ell} (\varphi^h(t), v_t^h(t))dt \leq C_2(1) \tag{2.15}$$

then $\forall\ t_1, t_2 \in [\tau, \tau+\ell]$

$$|E_1^h(t_1) - E_1^h(t_2)| \leq C_3(\ell) \tag{2.16}$$

We observe now that for $h \to 0$, $v^h \to u_t$ in $L^2_{loc}(0,+\infty;H^1(\Omega))$ and $v_t^h \to u_{tt}$ in $L^2_{loc}(0,+\infty;L^2(\Omega))$.

Let be $E_1(t) = \|\nabla u_t(t)\|_{0,2}^2 + \|u_{tt}(t)\|_{0,2}^2$; we have $E_1^h \to E_1$ in $L^2_{loc}(0,+\infty)$.

Let M be the set

$$M = \{t\varepsilon(0,+\infty)\ \lim_{h\to 0} E_1^h(t) = E_1(t) \text{ and } \lim_{h\to 0} E_1^h(t+4) = E_1(t+4)\}:$$

we have $\text{meas}((0,+\infty) - M) = 0$.

Let $\bar{t} \in M$ be fixed: we distinguish two cases

$$\text{(I) } E_1(\bar{t} + 4) \le E_1(\bar{t}) \qquad \text{(II) } E_1(\bar{t} + 4) \ge E_1(\bar{t}).$$

We consider the case (II), for $0 < h < \delta$, $\tau = \bar{t}$ $\ell = 4$ we have (2.15), (2.16).
Going to limit $h \to 0$ we have obtain

$$\int_t^{t+4} \|u_{tt}(t)\|_{0,2}^2 \, dt \le C_1; \; \left|E_1(\theta) - \frac{1}{2}\int_{t+1}^{\bar{t}+3} E_1(t)dt\right| \le C_2, \tag{2.17}$$

for almost all $\theta \in [\bar{t}, \bar{t}+4]$ and for $\theta = \bar{t}$, $\bar{t}+4$.
From the inequalities (2.17) we have

$$\mathrm{Sup}(E_1(\bar{t} + 4), \text{ ess. } \mathrm{Sup}_{\theta\in[\bar{t},\bar{t}+4]} E_1(\theta)) \le C_3 + \frac{1}{2}\int_{t+1}^{t+3} \|u_t(t)\|_{0,2}^2 \, dt$$

Since u is a strong solution,

$$\varphi(w_1) - \varphi(u_t(t)) \ge (f(t) + \Delta u(t) - u_{tt}(t), w - u_t(t))$$

for all $w \in H_o^1(\Omega)$, where

$$\varphi(w) = \begin{cases} \int_\Omega j(w(x))dx & \text{if } j(w(x)) \in L^1(\Omega) \\ +\infty & \text{otherwise} \end{cases}$$

$\partial j = \beta$, $j(0) = 0$.
We take $w = 0$: then

$$\int_{\bar{t}}^{\bar{t}+4} \varphi(u_t(t))dt \le E(u(\bar{t})) - E(u(\bar{t} + 4)) + \int_{\bar{t}}^{\bar{t}+4} (f(t), u_t(t))dt + C_4$$

We choose now $w = u(t + h)$ and we divide by h; we have

$$h^{-1}(\varphi(u_t(t + h)) - \varphi(u_t(t))) \ge h^{-1}(f(t) - u_{tt}(t), u_t(t + h) - u_t(t)) + h^{-1}(\nabla u(t), \nabla u_t(t + h) - \nabla u_t(t)).$$

Then for $\bar{t} \le t_1 < t_2 \le \bar{t} + 4$

$$- \int_{t_1}^{t_2} (\nabla u(t),\ h^{-1}(\nabla u_t(t+h) - \nabla u_t(t)))dt \leq \int_{t_1}^{t_2} (u_{tt}(t) - f(t),\ h^{-1}(u_t(t+h) - u_t(t)))dt + h^{-1}\int_{t_2}^{t_2+h} \varphi(u_t(t))dt - h^{-1}\int_{t_1}^{t_1+h} \varphi(u_t(t))dt$$

If we choose t_2 as a Lebesgue point of $\varphi(u_t(t))$ we have

$$\limsup_{h\to 0} - \int_{t_1}^{t_2} (\nabla u(t),\ h^{-1}(\nabla u_t(t+h) - \nabla u_t(t)))dt \leq \int_{t_1}^{t_2} (u_{tt}(t) - f(t),\ u_{tt}(t))dt + \varphi(u_t(t_2))$$

We have also

$$\int_{t_1}^{t_2} (\nabla u(t),\ h^{-1}(\nabla u_t(t+h) - \nabla u_t(t)))dt = [(\nabla u(t),\ h^{-1}(\nabla u(t+h) - \nabla u(t))]_{t_1}^{t_2} + \int_{t_1}^{t_2} (\nabla u_t(t),\ h^{-1}(\nabla u(t+h) - \nabla u(t)))dt.$$

For $h \to 0$ the right hand side converges to

$$[(\nabla u(t),\ \nabla u_t(t))]_{t_1}^{t_2} - \int_{t_1}^{t_2} \|\nabla u_t(t)\|_{0,2}^2\ dt \tag{2.18}$$

for almost every t_1 and t_2, then from the last three relations and (2.17)

$$\int_{t_1}^{t_2} \|\nabla u_t(t)\|_{0,2}^2\ dt \leq [(\nabla u(t),\ \nabla u_t(t))]_{t_1}^{t_2} + \int_{t_1}^{t_2} (u_{tt}(t) - f(t),\ u_{tt}(t))dt + \varphi(u_t(t_2))$$

$$\int_{t_1}^{t_2} \|\nabla u_t(t)\|_{0,2}^2\ dt \leq [(\nabla u(t),\ \nabla u_t(t))]_{t_1}^{t_2} + C + \varphi(u_t(t_2))$$

Since from lemma 2 u is bounded in $H_0^1(\Omega)$, we have

$$\int_{t_1}^{t_2} \|\nabla u_t(t)\|_{0,2}^2 \, dt \leq C(\varepsilon) + \varphi(u_t(t_2)) + \tag{2.19}$$
$$\varepsilon \,\text{ess.Sup}_{\theta\in[\bar{t},\bar{t}+4]} E_1(\theta)$$

and this inequality holds for every t_1 and almost every t_2. Tale now $t_2 = \sigma \in [\bar{t}+2, \bar{t}+4] t_1 = \sigma - 2$, integrating (2.19) on $[\bar{t}+2, \bar{t}+4]$ with respect to σ:

$$\int_{t+2}^{t+4} \left(\int_{\sigma-2}^{\sigma} \|\nabla u_t(t)\|_{0,2}^2 dt\right) d\sigma \leq 2[C(\varepsilon) +$$
$$\varepsilon \text{ ess.Sup}_{\theta\in[\bar{t},\bar{t}+4]} E_1(\theta)] + \int_{\bar{t}+2}^{\bar{t}+4} \varphi(u_t(\sigma)) d\sigma .$$

By elementary considerations, it is easily seen that

$$\int_{t+2}^{t+4} \left(\int_{\sigma-2}^{\sigma} \|\nabla u_t(t)\|_{0,2}^2 dt\right) d\sigma \geq \int_{\bar{t}+1}^{\bar{t}+3} \|\nabla u_t(t)\|_{0,2}^2 dt,$$

then

$$\int_{\bar{t}+1}^{\bar{t}+3} \|\nabla u_t(t)\|_{0,2}^2 dt \leq 2C(\varepsilon) + \varepsilon \text{ ess.Sup}_{\theta\in[\bar{t},\bar{t}+4]} E_1(\theta) +$$
$$\int_{\bar{t}+2}^{\bar{t}+4} \varphi(u_t(\sigma)) d\sigma$$

Takin $\varepsilon = 1/2$ we have

$$\text{Sup}(E_1(\bar{t}+4), \text{ ess.Sup}_{\theta\in[\bar{t},\bar{t}+4]} E_1(\theta)) \_ C_4 +$$
$$\frac{1}{2}\text{ess.Sup}_{\theta\in[\bar{t},\bar{t}+4]} E_1(\theta) + \frac{1}{2}\int_{\bar{t}+2}^{\bar{t}+4} \varphi(u_t(\sigma)) d\sigma \leq$$
$$C_5 + \frac{1}{2}\text{ess.Sup}_{\theta\in[\bar{t},\bar{t}+4]} E_1(\theta),$$

then

$$E_1(\bar{t} + 4) \leq C_6;$$

thus for $\bar{t} \in M$

$$E_1(\bar{t} + 4) \leq \text{Sup}(E_1(\bar{t}), C_6) \tag{2.20}$$

and (2.20) easily implies

$$\text{ess.Sup}_{t\geq 0} E_1(t) \leq \text{Sup}(C_7, \text{Sup}_{t\in[0,4]} E_1(t))$$

By standards methods, [1], we can conclude:

Corollary 2. The problem (1.2) has a solution u bounded in energy with u_t bounded in energy.

Lemma 4. The problem (1.2) has at most one solution bounded in energy with u_t bounded in energy.
Let be u and v two solutions to (1.2) bounded in energy with u_t, v_t bounded in energy, by the same methods of the lemma 1 we have

$$\lim_{t\to\pm\infty} v(t) - u(t) = 0 \text{ in energy} \tag{2.21}$$

Let be $E(t) = \|\nabla(v(t) - u(t))\|^2_{0,2} + \|v_t(t) - u_t(t)\|^2_{0,2}$; we observe that E(t) is not increasing: the, if $E(\bar{t}) \geq \varepsilon > 0$ it follows $E(t) \geq \varepsilon$ for all $t \leq \bar{t}$.
From (2.21) we have $E(t) = 0$, hence $v(t) - u(t) = 0$.
By standards methods, [1], we can now establish:

Lemma 5. There is a unique solution u to (1.2) which is almost periodic in energy with u_t bounded in energy.
From lemmas 1, 3 every solution v to (1.1) with initial data (v_o,v_1) $v_o \in H^2(\Omega) \cap H^1_o(\Omega)$, $v_1 \in H^1_o(\Omega) \times L^2(\Omega)$; for every $n > 0$ there is (v_{on},v_{1n}) smooth such that

$$\|\nabla v_o - \nabla v_{on}\|^2_{0,2} + \|v_1 - v_{1n}\|^2_{0,2} \leq n^{-1}$$

Then, if v_n is the solution to (1.1) relative to (v_{on},v_{1n}), we have

$$\|v(t) - v_n(t)\|^2_{0,2} + \|v_t(t) - v_{nt}(t)\|^2_{0,2} < n^{-1}$$

then

$$\limsup_{t\to+\infty}(\|\nabla v(t) - \nabla u(t)\|^2_{0,2} + \|v_t(t) - u_t(t)\|^2_{0,2}) \leq n^{-1},$$

As $n > 0$ is arbitrary,

$$\lim_{t\to+\infty} (\|\nabla v(t) - \nabla u(t)\|_{0,2}^2 + \|v_t(t) - u_t(t)\|_{0,2}^2) = 0.$$

So we have the following result:

Theorem. The problem (1.2) has a unique solution u which is almost periodic in energy with u_t bounded in energy; if v is a solution to (1.1) relative to the data $(v_o, v_1) \in H_o^1(\Omega) \times L^2(\Omega)$ we have

$$\lim_{t\to+\infty} (v - u) = 0 \quad \text{in energy}$$

Remark. By the same methods of the lemmas 2, 3, we can prove the boundness of the trajectories of the first order system

$$\begin{cases} u_t + v_x + \lambda(u^3 - u) = f \\ v_t + u_x = 0 \qquad \text{on } R \times]-1,1[, \ \lambda > 0 \end{cases}$$

where u, v are 2-periodic in x, u is off an v is even.

OPEN PROBLEMS. (I) In case β is fast increasing, nothing is known on the existence of a bounded or almost periodic solution to (1.2).

(II) In case β has polynomial growth of order k (finite) G. Prodi [4] has shown that if f is periodic of period T, there is a periodic solution to (1.2) of periodic T, but nothing is known on the asymptotic behavior of the solutions to (1.1) for $t\to+\infty$.

REFERENCES

1. L. Amerio, G. Prouse: Abstract Almost Periodic Functions and Functional Equations, Van Nostrand Reinhold (1971).
2. M. Biroli: Bounded or Almost Periodic Solutions of the Nonlinear Vibrating Membrane Equation, Ric. di Mat. 22, 190-202, (1973).

3. M. Biroli, A. Haraux: Asymptotic Behavior for an Almost Periodic Strongly Dissipative Wave Equation, J. Diff. Eq., to appear.

4. G. Prodi: Soluzioni periodiche dell'equazione delle onde con termine dissipativo non lineare, Rend. Sem. Mat. Padova 36, 37-49, (1966).

5. G. Prouse: Soluzioni quasi periodiche dell'equazione non omogenea delle onde con termine dissipativo non lineare, I, II, III, IV, Rend. Acc. Naz. Linc. 38, 804-807; 39 11-18, 155-160, 240-244, (1965).

DIFFERENTIABILITY OF THE SOLUTIONS WITH RESPECT TO THE INITIAL CONDITIONS

V.I. Blagodatskikh

Steklow Mathematical Institute
Moscow, URSS

Consider an ordinary differential equation

$$\dot{x} = f(t,x), \tag{1}$$

where $f\colon E^1 \times E^n \to E^n$ is a given function. Assume that the function $f(t,x)$ is measurable in t, lipschitzean in x with constant $\ell(t)$ and satisfies the restriction $|f(t,x)| \leq k(t)$, where the functions $\ell(t)$ and $k(t)$ are integrable on the given interval $I = [t_0,t_1]$. Let $x(t,x_0)$ be a solution of the equation (1) with the initial condition $x(t_0,x_0) = x_0$. We shall consider the dependence of the solution $x(t,x_0)$ on the initial value x_0.

We define the generalized derivative $K(f(t,x);\, a)$ of the function $f(t,x)$ at the point x in the direction a as the set of all partial limits of the sequence $\left\{\frac{f(t,x+\lambda a) - f(t,x)}{\lambda}\right\}$ when $\lambda \to +0$.
Hence we have

$$K(f(t,x);a) = \lim_{\lambda \to +0} \frac{f(t,x+\lambda a) - f(t,x)}{\lambda}.$$

It is very easy to prove that $K(f(t,x);a)$ is a nonempty compact set in the space E^n for any given $(t,x,a) \in E^1 \times E^n \times E^n$. The multivalued function $K(f(t,x);a)$ is lipschitzean in a; and for

ISBN 0-12-508780-2

any given function $x(t)$, absolutely continuous on I, $K(f(t,x(t);a)$ is measurable in t.

In a similar way we can define the generalized derivative $K(x(t,x_0);h)$ of the solution $x(t,x_0)$ at the point x_0 in the direction h.
So we put

$$K(x(t,x_0);h) = \lim_{\varepsilon\to+0} \frac{x(t,x_0 + \varepsilon h) - x(t,x_0)}{\varepsilon},$$

where the limit is taken in the space $C(I)$.

<u>Theorem</u>. If $\delta x(t) \in K(x(t,x_0);h)$, then $\delta x(t)$ is a solution of the differential inclusion

$$\delta\dot{x} \in K(f(t,x(t,x_0));\delta x) \tag{2}$$

on the interval I with the initial condition $\delta x(t_0) = h$, that is $\delta x(t)$ is absolutely continuous and the inclusion $\delta\dot{x}(t) \in K(f(t,x(t,x_0));\delta x(t))$ holds for almost all $t \in I$.

Remark that in the case the function $f(t,x)$ is differentiable in x, then the differential inclusion (2) coincides with the classsical system in variations

$$\delta\dot{x} = \frac{\partial f(t,x(t))}{\partial x}\delta x. \tag{3}$$

But in our case the differential inclusion (2) is not already linear in δx although it is positive homogeneous in δx.

It is known that under the given assumptions the function $f(t,x)$ is differentiable for almost all $x \in E^n$. Hence for almost all $x \in E^n$ the differential inclusion (2) coincides with the equation (3). But the solution $x(t,x_0)$ can reach the set where the derivative does not exist and we have to consider the inclusion (2).

For example, the function $f: E^1 \to E^1$ given by the relations

$$f(x) = \begin{cases} x - \dfrac{1}{2^k}, & \text{if } x \in \left[\dfrac{1}{2^k}, \dfrac{3}{2^{k+1}}\right], \\ -x + \dfrac{1}{2^{k-1}}, & \text{if } x \in \left[\dfrac{3}{2^{k+1}}, \dfrac{1}{2^{k-1}}\right], \end{cases}$$

$k = 1,2,\ldots,$ and $f(0) = 0$, is differentiable in the direction $a = 1$ everywhere except the point $x = 0$. The generalized derivate $K(f(0);a)$ is the set $[0,\frac{1}{3}]a$. Hence, for the solution $x(t,0) = 0$ we have the differential inclusion in variations

$$\delta x \in [0,\tfrac{1}{3}]\,\delta x.$$

The proof of this theorem is based on results of the paper [1].

REFERENCES

1. V.I. Blogodatskikh: *On the Differentiability of the Solutions with Respect to the Initial Conditions*, Differencialnye Uravnenija *9*, n. 12, 2136-2140, (1973).

SOME REMARKS ON BOUNDEDNESS AND ASYMPTOTIC EQUIVALENCE OF ORDINARY DIFFERENTIAL EQUATIONS

Moses Boudourides

Department of Mathematics
Democritus University of Thrace
Xanthi, Greece

Consider the following linear and nonlinear equations

$$y' = A(t)y \tag{1}$$

$$x' = A(t)x + f(t,x), \tag{2}$$

where y, x, A, f are functions in a Banach space E and the independent variable $t \in J = [0,\infty)$. It is our purpose to find: (i) sufficient and necessary conditions for the existence of bounded solutions of (2) and (ii) sufficient conditions for the existence of a homeomorphism between bounded solutions y, x of (1), (2) respectively, such that

$$\lim_{t\to\infty} \|y(t) - x(t)\| = 0,$$

in which case we say that there exists an asymptotic equivalence between (1) and (2).

Let $L = L(E)$ denote the space of strongly measurable functions $u: J \to E$, Bochner integrable on every finite subinterval $J' \subset J$, with the topology of the convergence in the mean on every such J'.

ISBN 0-12-508780-2

We assume that $A \in L(\tilde{E})$, where $\tilde{E}$ denotes the space of continuous linear mappings $E \to E$. Let $Y(t)$ be the fundamental solution of (1) such that $Y(0) = I$.

Let E_1 be the subspace of E consisting of all the points of E which are values for $t = 0$ of bounded solutions of (1).

We assume that E_1 can be complemented in E by a closed subspace E_2 and we denote by P_1, P_2 the corresponding projections of E onto E_1, E_2.

For any $t, s \in J$ we define

$$H(t,s) = \begin{cases} Y(t)P_1Y^{-1}(s), & \text{for } 0 \leqq s \leqq t \\ -Y(t)P_2Y^{-1}(s), & \text{for } s \geqq t \geqq 0. \end{cases}$$

Let $B(R)$ be a Banach space (with norm $\|\cdot\|_{B(R)}$) of functions $u: J \to R$ such that:

(i) $B(R) \subset L(R)$ and $B(R)$ stronger that $L(R)$ (i.e. the norm topology of $B(R)$ is stronger than the topology of $L(R)$);

(ii) $B(R)$ is not stronger than $L^1(R)$;

(iii) $L_c^\infty(R) \subset B(R)$ ("c" denotes compact support);

(iv) if $u \in B(R)$, $v: J \to R$ measurable and $|v(t)| \leqq |u(t)|$ for $t \in J$, then $v \in B(R)$ and $\|v\|_{B(R)} \leqq \|u\|_{B(R)}$.

By $B = B(E)$ we denote the Banach space of all strongly measurable functions $u: J \to E$ such that $\|u\| \in B(R)$, with the norm $\|u\|_B = \| \|u\| \|_{B(R)}$.

To the function space B we correspond the associate space B^*, defined as the space of measurable functions $v: J \to E$ such that

$$\sup\{\int_0^\infty \|u(s)\| \|v(s)\| \, ds : u \in B, \ \|u\|_B \leqq 1\} < \infty ,$$

with this sup as its norm. It is proved in [4] that the following "Hölder's Inequality" holds: if $u \in B$, $v \in B^*$, then $\|u\| \|v\| \in L^1(R)$ and

$$\int_0^\infty \|u(s)\| \|v(s)\| ds \leqq \|u\|_B \|v\|_{B^*}.$$

By $F(B)$ we denote the set of functions $f: J \times E \to E$ such that

(i) $f(t,x)$ is strongly measurable in t for each fixed $x \in E$ and continuous in x for each fixed $t \in J$;

(ii) there exists $\lambda \in B(R)$ such that for all $u, v \in E$, $t \in J$

$$\|f(t,u) - f(t,v)\| \leqq \lambda(t)\|u - v\|;$$

(iii) $f(\cdot,0) \in B$.

Let $C = C(E)$ denote the space of bounded continuous functions $u: J \to E$ with the norm $\|u\|_C = \sup\{\|u(t)\|: t \in J\}$. The following lemma can be easily proved.

Lemma 1. If $f \in F(B)$ and $x \in C$, then $f(\cdot,x(\cdot)) \in B$.

Now we can give the following necessary and sufficient conditions for the existence of bounded solutions of (2).

Theorem 2. (2) has at least one bounded solution for any $f \in F(B)$ if and only if there exists a constant $K > 0$ such that

$$\|H(t,\cdot)\|_{B*} \leqq K, \text{ for } t \in J,$$

$$K\|\lambda\|_{B(R)} < 1.$$

Proof. The necessity is proved as in [2]. For the sufficiency, we consider the following map on C

$$Tx(t) = \int_0^\infty H(t,s)f(s,x(s))ds,$$

which is easily seen to be a contraction in C. Therefore, there exists a fixed point $x \in C$ of T, which is clearly a solution of (2).

The next theorem gives sufficient conditions for the existence of asymptotic equivalence of (1) and (2).

Theorem 3. Suppose that

(i) $f \in F(B)$;

(ii) there exists a constant $K > 0$ such that

$$\|H(t,\cdot)\|_{B^*} \leqq K, \quad \text{for } t \in J,$$

$$K\|\lambda\|_{B(R)} < 1;$$

(iii) $\lim_{t\to\infty}\|Y(t)P_1\| = 0;$

(iv) for any $c \geqq 0$

$$\lim_{t\to\infty} \int_t^\infty \|H(t,s)\|\{c\lambda(s) + \|f(s,0)\|\}ds = 0.$$

Then there exists a homeomorphism between the bounded solutions y, x of (1), (2) respectively such that

$$\lim_{t\to\infty}\|y(t) - x(t)\| = 0.$$

Proof: Let C_1, $C_2 \subset C$ be the sets of bounded solutions of (1) and (2) respectively. Clearly $C_1(0) = E_1$ and $C_2 \neq \emptyset$ (because of Thm. 2).

First, suppose that $y \in C_1$ is given. Then define a mapping T_y on C by

$$T_y x(t) = y(t) + Tx(t),$$

where T is the contraction in C defined in the proof of Thm. 2. Thus there exists a unique fixed point x of T_y in C, which is easily seen to solve (2), i.e. $x \in C_2$. Moreover, the mapping $y \mapsto x = T_y x$ is continuous.

On the other hand, given any $x \in C_2$, define

$$y(t) = x(t) - Tx(t).$$

Clearly $y \in C_1$ and the mapping $x \mapsto y$ is continuous.

Finally, there exists a $t_1 \in J$ such that

$$\|x(t) - y(t)\| \leqq \int_0^{t_1} \|H(t,s)f(s,x(s))\|ds +$$

$$+ \int_{t_1}^{\infty} \|H(t,s)\| \ \|f(s,x(s))\|ds \leqq$$

$$\leqq \|Y(t)P_1\| \int_0^{t_1} \|Y^{-1}(s)f(s,x(s))\|ds +$$

$$+ \int_{t_1}^{\infty} \|H(t,s)\| \ \|f(s,x(s))\|ds$$

and t_1 can be chosen sufficiently large so that the second integral is arbitrarily small (because of (iv)), while the first integral is negligible as $t \to \infty$ (by (iii)). Thus the relation of asymptotic equivalence is satisfied and the proof is completed.

Theorem 2 for the case $B = L^p(R^n)$, $1 \leqq p \leqq \infty$, was considered by Staikos [5]. Theorem 3 for the cases $B = L^1(R^n)$ and $B = L^p(R^n)$, $1 < p \leqq \infty$, was considered by Brauer and Wong [1] and Talpalaru [7], respectively. Note that in the case $B = L^p(E)$, $1 < p \leqq \infty$, condition (iii) is implied by (ii) and in the case $B = L^p(E)$, $1 \leqq p < \infty$, condition (iv) is implied by (i) and (ii). However these theorems hold for more general Banach spaces than the L^p spaces. For example, the case that B is an Orlicz space was considered by Szufla [6]. Another class of function spaces is the following. By $M^{p,r} = M^{p,r}(E)$, $1 \leqq p$, $r \leqq \infty$, we denote the space of all strongly measurable functions $u: J \to E$ such that the restriction of u on each subinterval $[n,n+1]$ is L^p and the sequence $\{a_n\}$ is L^r, where

$$a_n = \left(\int_n^{n+1} \|u(\tau)\|^p d\tau\right)^{1/p}, \text{ for } 1 \leqq p < \infty,$$

$$a_n = \text{ess sup}\{\|u(\tau)\|: n \leqq \tau \leqq n+1\}, \text{ for } p = \infty.$$

Thus the norm of $u \in M^{p,r}$ is

$$\|u\|_{p,r} = \left(\sum_{n=0}^{\infty} a_n^r\right)^{1/r}, \text{ for } 1 \leqq r < \infty,$$

$$\|u\|_{p,\infty} = \sup\{a_n: n = 0,1,2,\ldots\}.$$

Clearly, the dual of $M^{p,r}$ is $M^{q,s}$, where $p^{-1} + q^{-1} = 1$ and $r^{-1} + s^{-1} = 1$. Moreover, $M^{p,p} = L^p$, $L^p \subset M^{p,r}$ for $p \leqq r$ and $M^{p,r} \subset L^p$ for $r \leqq p$. Note that Theorem 2 for the case $B = M^{p,\infty}$, $1 < p < \infty$, was considered by Lovelady [3].

REFERENCES

1. F. Brauer, J.S.W. Wong: On the Asymptotic Relationship Between Solutions of Two Systems of Ordinary Differential Equations, J. Diff. Equations, 6, 527-543 (1969).
2. R. Conti: On The Boundedness of Solutions of Ordinary Differential Equations, Funkc. Ekvac., 9, 23-26 (1966).
3. D.L. Lovelady: Nonlinear Stepanoff-Bounded Perturbation Problems, J. Math. Anal. Appl., 50, 350-360 (1975).
4. J.L. Massera, J.J. Schäffer: Linear Differential Equations and Function Spaces, New York and London, Academic, (1966).
5. V.A. Staikos: A Note on the Boundedness of Solutions of Ordinary Differential Equations, Boll. Un. Mat. Ital., (4) 1, 256-261 (1968).
6. S. Szufla: On the Boundedness of Solutions of Ordinary Differential Equations in Banach Spaces, Bull. Acad. Polon. Sci. Sér. Sci. Math. Astr. Phys., 17, 745-750 (1969).
7. P. Talpalaru: Quelques problèms concernant l'équivalence asymptotique des systèmes différentiels, Boll. Un. Mat. Ital., (4) 4, 164-186 (1971).

PERIODIC SOLUTIONS FOR A SYSTEM OF NONLINEAR DIFFERENTIAL EQUATIONS MODELLING THE EVOLUTION OF ORO-FAECAL DISEASES[1]

V. Capasso

Istituto di Analisi Matematica
and
Istituto di Matematica Applicata
Università di Bari

1. INTRODUCTION

In paper [1] a mathematical model was proposed for the cholera epidemic which spread in the European Mediterranean regions in the summer of 1973. That model is based on a system of two ordinary differential equations which describe respectively the evolution of the bacteria population in sea waters and the evolution of the human infective population in a urban community.

If we denote by $x_1(t)$ the bacteria population and by $x_2(t)$ the infective population, the two equations may be written in the following form

[1]Work performed under the auspices of the GNFM-CNR, in the context of the Program of Preventive Medicine (Project MPP1), CNR (Italy).

ISBN 0-12-508780-2

$$\begin{cases} \frac{d}{dt} x_1 = - a_{11}x_1 + a_{12}x_2 \\ \\ \frac{d}{dt} x_2 = - a_{22}x_2 + N\beta p f(x_1) \end{cases} \tag{1}$$

subject to suitable initial conditions.

The meaning of the other parameters will be found in [1], while the parameter p in (1) denotes the probability (per unit time) that an individual have a bacteria carrying meal. In [1] such a quantity had been considered, for simplicity, to be time invariant, and this made sense since the cholera epidemic lasted for a very short period of time. On the contrary, if we refer to other oro-faecal diseases such as typhoid fever, viral hepatitis, etc. we need to remark that during a year such a parameter p has in general seasonal fluctuations. Hence it is more appropriate, if we wish to apply model (1) to the whole class of diseases with an oro-faecal transmission, to make the assumption that p is given by the sum of two terms (see also [5]):

$$p(t) = p_o + p_1(t). \tag{2}$$

the first of which, p_o denotes a constant (in time) average level, while the second, $p_1(t)$ may be thought of as a periodic perturbation, with some period $T > 0$, which takes into account seasonal variations around the average level p_o.

The problem we consider here is to establish if such a periodic $p_1(t)$ induces a periodic behaviour of the solutions of system (1), with the same period T of the perturbation.

A behaviour of this kind has been revealed by the statistical analysis of the experimental data relative to the typhoid fever in the town of Bari, Southern Italy [2], where to a mean endemic level (with its own time dependency), we can see a superimposed periodic fluctuation of a seasonal kind. The interpretation of this periodicity (with period one year) is related, according to the authors, to the seasonal variations of the parameter p as in (2) since this describes the eating habits of the human population in that area.

This paper contains some preliminary results of the analysis which is under way of system (1) with (2). In particular here the existence of periodic solutions with the same period T of the perturbation $p_1(t)$ has just been shown. This confirms in part the above said conjecture as far as the biological meaning of system (1) is concerned. Only the case in which the corresponding autonomous system ($p(t) = p_o$) admits a non trivial endemic level is contained here; anyhow this is of primary interest from an epidemiological point of view. Other details and more complete analysis of the problem will be given in a forthcoming paper [4].

2. DEFINITIONS AND BASIC RESULTS

As announced in the introduction the objective of this paper is to study the following system of differential equations

$$\begin{aligned} \frac{d}{dt} x_1(t) &= - a_{11}x_1(t) + a_{12}x_2(t) \\ & \qquad\qquad\qquad\qquad , \quad t > 0 \\ \frac{d}{dt} x_2(t) &= - a_{22}x_2(t) + p(t)h(x_1(t)) \end{aligned} \tag{3}$$

subject to the initial conditions

$$\begin{aligned} x_1(0) &= x_2^o \geq 0 \\ x_2(0) &= x_2^o \geq 0 \end{aligned} \tag{3_o}$$

not both identically zero.

Equation (2) will be supposed to hold with $p_o > 0$ and

$$p_1(t) = p_1(t + T) \tag{4}$$

will be assumed to be a continuous (and periodic of period $T > 0$) function of the time t such that

$$p(t) = p_o + p_1(t), \quad t \geq 0 \tag{5}$$

It will be further assumed (as in [1]; see also [3]) that a_{11}, a_{12}, a_{22} are all positive quantities and that $h(\cdot)$ satisfies the following hypotheses

(i) if $0 < z' < z''$ then $0 < h(z') \leq h(z'')$

(ii) $h(0) = 0$

(iii) h is continuous up to its second derivative

(iv) $h''(z) < 0$ for any $z \geq 0$.

It will be convenient in the following to write system (3), (3_o) with a vector notation

$$\frac{d}{dt} x(t) = f(x(t)) + g(t;x(t)), \quad t > 0 \tag{6}$$

$$x(0) = x^o = \begin{pmatrix} x_1^o \\ x_2^o \end{pmatrix} \tag{6_o}$$

where

$$f(x) =: \begin{pmatrix} f_1(x_1,x_2) \\ f_2(x_1,x_2) \end{pmatrix} = \begin{pmatrix} -a_{11}x_1 + a_{12}x_2 \\ -a_{22}x_2 + p_o h(x_1) \end{pmatrix} \tag{7}$$

and

$$g(t;x) =: \begin{pmatrix} g_1(t;x_1,x_2) \\ g_2(t;x_1,x_2) \end{pmatrix} = \begin{pmatrix} 0 \\ p_1(t)\ h(x_1) \end{pmatrix} \tag{8}$$

It follows from the assumptions that $g(t;x)$ is a continuous function of the time t, periodic with period T:

$$g(t + T;x) = g(t;x) \tag{9}$$

The following theorem can be proved.

<u>Theorem 1</u>. Under the assumptions (i)-(iv) on h and (5) on $p(t)$, if $x_i^o \geq 0$, $i = 1,2$, then a unique solution exists for problem (3), (3_o) such that

$$x_i \in C([0,+\infty), \mathbb{R}_+) \cap C^1((0,+\infty), \mathbb{R}_+).$$

Furthermore if a $j \in \{1,2\}$ exists such that $x_j^o > 0$ then $x_i(t) > 0$ $(i = 1,2)$ for any $t > 0$.

Proof: The non negativity of the solution (and the last statement) follows from the fact that (see e.g. [7, p. 270]).

$$f_i(x) + g_i(t;x) > 0 \quad \text{if} \quad x_i = 0,\ x_j > 0 \quad \text{for } j \neq i \quad (i,j = 1,2) \tag{10}$$

The solution can be extended to all times $t \geq 0$ since, as it will be shown in the proof of Theorem 4, under the assumptions of this theorem it is bounded in a compact set of $\mathbb{K} =: \mathbb{R}_+ \times \mathbb{R}_+$.

If we now limit ourselves to consider the autonomous system

$$\frac{d}{dt} x(t) = f(x(t)), \quad t > 0 \tag{11}$$

in [1] (see also [3]) the following theorem has been proved.

Theorem 2. Let the assumptions (i)-(iv) hold and let

$$\theta =: \frac{p_o h'(0) a_{12}}{a_{11} a_{22}} \in \mathbb{R}_+. \tag{12}$$

a) It $0 < \theta < 1$ then a unique equilibrium point exists for system (11) in the non negative cone $\mathbb{K}$. This point is the origin $0 =: (0,0)$ which is globally asymptotically stable in the whole cone $\mathbb{K}$.

b) If $1 < \theta$ then two equilibrium points exist for system (11) in $\mathbb{K}$. These points are the origin 0 and the point $Q =: (x_1^Q, x_2^Q)$ given by the unique non trivial solution of system

$$\begin{aligned} -a_{11}u_1 + a_{12}u_2 &= 0 \\ -a_{22}u_2 + p_o h(u_1) &= 0 \end{aligned} \tag{13}$$

In this case the origin 0 is an unstable equilibrium point while Q is globally asymptotically stable in $\mathbb{K} - \{0\}$.

The case $\theta = 1$ needs further informations about the behaviour of the function h and anyhow it is a limit case with not too much interest in applications.

3. EXISTENCE OF PERIODIC SOLUTIONS

The case $\theta > 1$ will be considered here. In such a case, as already seen in Section 2, the autonomous system (11), corresponding to $p(t) = p_o$, admits a non trivial endemic level Q which is globally asymptotically stable in the whole cone $\mathbb{K} - \{0\}$. We shall see how this situation is modified when $p(t) = p_o + p_1(t)$ and $p_1(t)$ is a periodic function of the time with period $T > 0$.

It will be convenient to apply a translation of the two axes of the phase space (x_1, x_2) in such a way that the new origin is set in the point Q:

$$\begin{aligned} y_1 &= x_1 - x_1^Q \\ y_2 &= x_2 - x_2^Q \end{aligned} \tag{14}$$

System (3) becomes

$$\begin{aligned} \frac{dy_1}{dt} &= -a_{11}y_1 + a_{12}y_2 \\ \frac{dy_2}{dt} &= -a_{22}y_2 + p_o\bar{h}(y_1) + p_1(t)\bar{\bar{h}}(y_1) \end{aligned} \tag{15}$$

where

$$\bar{h}(y_1) = h(y_1 + x_1^Q) - h(x_1^Q) \tag{16}$$

$$\bar{\bar{h}}(y_2) = h(y_1 + x_1^Q) \tag{17}$$

or, in vectorial form

$$\frac{dy}{dt} = \bar{f}(y) + \bar{g}(t;y) \tag{18}$$

where $y =: (y_1, y_2)$ and

$$\bar{f}(y) =: \begin{pmatrix} \bar{f}_1(y_1,y_2) \\ \bar{f}_2(y_1,y_2) \end{pmatrix} = \begin{pmatrix} -a_{11}y_1 + a_{12}y_2 \\ -a_{22}y_2 + p_o\bar{h}(y_1) \end{pmatrix} \tag{19}$$

$$\bar{g}(t;y) =: \begin{pmatrix} 0 \\ p_1(t)\bar{\bar{h}}(y_1) \end{pmatrix} \tag{20}$$

Observe now that for the unperturbed autonomous system

$$\frac{dy}{dt} = \bar{f}(y) \tag{21}$$

a Liapounov function of a quadratic type can be built up [8, p. 284] (see also [10, Tome 2, p. 31] and [9, p. 31]).
In fact under our assumptions on the function h, we may state that system (21) can be rewritten in the following form:

$$\begin{aligned} \frac{dy_1}{dt} &= -a_{11}y_1 + a_{12}y_2 \\ \frac{dy_2}{dt} &= \bar{a}_{21}y_1 - a_{22}y_2 + Y_2(y_1) \end{aligned} \tag{22}$$

where

$$\bar{a}_{21} =: p_o\bar{h}'(0) = p_o h'(x_1^O) \tag{23}$$

and

$$Y_2(y_1) =: p_o(\bar{h}(y_1) - \bar{h}'(0)y_1) \tag{24}$$

is the nonlinear part associated with the function $\bar{h}(\cdot)$ whose expansion contains terms at least of the second order.
As seen in [1], the matrix

$$A = : \begin{pmatrix} -a_{11} & a_{12} \\ \bar{a}_{21} & -a_{22} \end{pmatrix} \tag{25}$$

has both the eigenvalues with negative real part. Hence [8] (see also [10, Tome 2, p. 31]) one and only positive definite quadratic form $V : \mathbb{R}^2 \to \mathbb{R}$ exists such that

$$\frac{\partial V}{\partial y} \cdot Ay = U(y) \tag{26}$$

where $U : \mathbb{R}^2 \to \mathbb{R}$ is a negative definite quadratic form that one may choose for example to be

$$U(y_1, y_2) = -(y_1^2 + y_2^2). \tag{27}$$

Still following [8,10] it can be shown that the derivative of $V(\cdot)$ along the trajectories of system (22), which is given by

$$\dot{V}(y) = U + \frac{\partial V}{\partial y_2} Y_2(y_1) \tag{28}$$

maintains itself negative definite in a suitable neighborhood of the origin since U and V are quadratic forms and Y_2 is at least of the second order.

To conclude, the following theorem can be stated:

Theorem 3. Under the above assumptions, a positive definite quadratic form $V: \mathbb{R}^2 \to \mathbb{R}$ and a $\gamma > 0$ exist such that for suitable functions $a, b, c \in K$ and for any $y \in B_\gamma(0)$ the following inequalities hold:

(i) $a(\|y\|) \leq V(y) \leq b(\|y\|)$

(ii) $\dot{V}(y) \leq -c(\|y\|)$

where $\dot{V}$ denotes the derivative of V along the trajectories of (21). (Here $a \in K$ means that $a: \mathbb{R}_+ \to \mathbb{R}_+$ is a continuous and decreasing function with $a(0) = 0$ (see e.g. [6]), and $\|\cdot\|$ denotes e.g. the euclidean norm in $\mathbb{R}^2$).

The following Theorem 4 is a consequence of known results [11, Thm. 15.9]. However, it will be proved in a direct and easy way using Theorem 3 above.

Theorem 4. If $\max_{t \in [0,T]} |p_1(t)|$ is sufficiently small, a non trivial periodic solution exists for system (18) and hence for system (6) with the same period T of the perturbation $p_1(t)$.

Proof: Let $V(\cdot)$ and $\gamma > 0$ be defined as in Theorem 3. If now $\gamma_1 \in (0,\gamma)$, for $y \in B_\gamma(0) - B_{\gamma_1}(0)$ it will be $0 < a(\gamma_1) \leq V(y)$. Hence if we set $h =: a(\gamma_1)$, for a fixed $\lambda \in (0,h)$

$$S =: \{y | V(x) < \lambda\} \subset \bar{B}_{\gamma_1}(0). \tag{29}$$

Furthermore, since $V(\cdot)$ is a positive definite quadratic form which is continuous in $\bar{B}_{\gamma_1}(0)$ with its first order partial derivatives, S is a convex neighborhood of 0, and some $M > 0$ exists such that, for any $y \in \bar{B}_{\gamma_1}(0)$:

$$\left\|\frac{\partial V}{\partial y}\right\| \leq M. \tag{30}$$

Going back now to the perturbed system (18) we can state that the time derivative of $V(\cdot)$ along the trajectories of that system is given by

$$\dot{V}_p(t;y) = \frac{\partial V}{\partial y} \cdot \bar{f}(y) + \frac{\partial V}{\partial y} \cdot \bar{g}(t;y) = \dot{V}(y) + \frac{\partial V}{\partial y} \cdot \bar{g}(t;y) <$$

$$< -c(\|y\|) + M\|\bar{g}(t;y)\| \tag{31}$$

at any time $t > 0$ where the solution of system (18) exists.

Since S is a neighborhood of the origin, a $\nu > 0$ will exist such that $B_\nu(0) \subset S$. Now if $y \in G =: S - B_\nu(0)$ it will be $\nu < \|y\|$ and at such points

$$\dot{V}_p(t;y) \leq -c(\nu) + M\|\bar{g}(t;y)\|. \tag{32}$$

It follows from (32) that if $\max_{t \in [0,T]} |p_1(t)|$ is sufficiently small we weill have $\|\bar{g}(t;y)\| = |p_1(t)||\bar{\bar{h}}(y_1)| < \frac{c(\nu)}{2M}$, for $y \in G$, $t > 0$, whence

$$\dot{V}_p(t;y) \leq -c(\nu) + \frac{c(\nu)}{2} < 0 \tag{33}$$

at any time $t > 0$ for which the solution exists in G.

This implies in particular that any trajectory $y_p(t;y_o)$ of system (18) corresponding to an initial condition $y_o \in S$ must be contained in S, which is then a positively invariant subset of $\mathbb{K}$. The trajectory will be bounded and hence it will be defined for all positive times $t \geq 0$.

We can then state that the translation operator (Poincaré map) $\varphi: S \to S$ such that

$$\forall\, y_0 \in S: \varphi(y_0) = y_p(T; y_0)$$

is continuous in y_0 on a compact and convex subset of $\mathbb{K} \subset \mathbb{R}^2$. It follows from Browder's fixed point principle that φ has at least one fixed point y_0 in S such that

$$y_p(T; y_0) = y_0. \tag{34}$$

Now the perturbed system (18) is periodic of period T and (34) implies that it admits at least a periodic solution with the same period T of the perturbation $p_1(t)$ as we wanted to show.

Acknowledgement: It is a pleasure to acknowledge useful discussions with Prof. L. Salvadori of the University of Trento.

REFERENCES

1. V. Capasso, S.L. Paveri-Fontana: A Mathematical Model for the 1973 Cholera Epidemic in the European Mediterranean Region, Rev. Epidém. et Santé Publ. 27, 121-132 (1979). Erratum, ibidem, to appear.
2. V. Capasso, E. Grosso, G. Serio: I modelli matematici nella indagine epidemiologica II. Il tifo addominale. Studio delle serie temporali, Annali Sclavo 22, 189-206, (1980).
3. V. Capasso: Mathematical Models for Infectious Diseases (In Italian), Quaderni dell'Istituto di Analisi Matematica, Bari, 1980.
4. V. Capasso: Qualitative Analysis of a System of Differential Equations Modelling Oro-Faecal Diseases. To appear.

5. K. Dietz: The Incidence of Infectious Diseases under the Influence of Seasonal Fluctuations, in Lecture Notes in Biomathematics Vol. 11, Berlin, Springer-Verlag, 1976.
6. W. Hahn: Stability of Motion, Berlin, Springer-Verlag, 1967.
7. M.A. Krasnosel'skii: Positive Solutions of Operator Equations, Groningen, Noordhoff, 1964.
8. A. Liapounoff: Problème Général de la Stabilité du Mouvement, Princeton, Annals of Mathematics Studies, 1952.
9. N. Rouche, P. Habets, M. Laloy: Stability Theory by Liapounov's Direct Method, Berlin, Springer-Verlag, 1977.
10. N. Rouche, J. Mawhin: Equations Différentielles Ordinaires, Paris, Masson et Cie, 1973.
11. T. Yoshizawa: Stability Theory and the Existence of Periodic Solutions and Almost Periodic Solutions, Berlin, Springer, 1975.

GENERALIZED HOPF BIFURCATION

Silvia Caprino

Istituto Matematico
Università di Camerino, Italy

INTRODUCTION

This paper is based on a joint work with P. Negrini [6].

Consider the one parameter family of ordinary differential equations

$$\dot{z} = f(\mu, z) \qquad (*)$$

$$f \in C^h[\mathbb{R}\times\mathbb{R}^2, \mathbb{R}^2], \; h \geq 3, \; f(\mu,0) = 0$$

and suppose the origin to be an isolated critical point.

It is known [1,2] that, under some conditions on f, a sudden change in the asymptotic properties of the flow near $z = 0$ when μ crosses the critical value $\mu = 0$, gives rise to an annulus, whose boundary consists of two closed orbits around the origin. To be more precise, if for $\mu = 0$ the origin is asymptotically stable and for $\mu > 0$ is completely unstable, then the annulus is asymptotically stable that is it "inherits" the origin's previous property.

If we want attractive periodic orbits (instead of attractive annuli), we need in general a stronger property than only the attractivity of the origin for $\mu = 0$. It has been shown [3] that for $\mu > 0$ the bifurcated periodic orbits are all at-

ISBN 0-12-508780-2

tracting if (i) a suitable transversality condition is satisfied; (ii) the attractivity of the origin for $\mu = 0$ is recognizable upon truncation of $f(0,z)$ up to some order k (k-asymptotic stability). This paper generalizes the results of [3] to the case in which the right-hand side of (*) has a possibly vanishing linear part. To do this, suitable hypotheses are made on f, which reduce to the classical ones for ordinary Hopf bifurcation in the case that a linearization of the problem is possible. To recognize the k-asymptotic stability (or k-complete unstability) of the origin of the unperturbed system , we construct an auxiliary function having special properties along the solutions. This method, due to Malkin [4], is unfortunately less handy than that employed in [3], which is essentially the purely algebraic procedure due to Liapunov and Poincaré [5].

1. PRELIMINARIES

Let $D \subset \mathbb{R}^2$ a neighborhood of the origin and $I \subset \mathbb{R}$ an interval of the real line including 0. Let us consider the differential system:

$$\begin{aligned} \dot{x} &= X_m(\mu,x,y) + \ldots + X_{m+\ell}(\mu,x,y) + X(\mu,x,y) \\ \dot{y} &= Y_m(\mu,x,y) + \ldots + Y_{m+\ell}(\mu,x,y) + Y(\mu,x,y) \end{aligned} \tag{1.1}$$

where $(\mu,x,y) \in I \times D$, $m \geq 1$, $\ell \geq 3$; X_s and Y_s are homogeneous polynomials of degree s in (x,y), whose μ-dependent coefficients belong to $C^{m+\ell}[I,\mathbb{R}]$; X and $Y \in C^{m+\ell}[I \times D,\mathbb{R}]$ start with terms of order m+1 in (x,y). We make the following assumption:

(H1) m is odd and the form of degree m+1: $Y_m x - X_m y$ is sign definite at $\mu = 0$.

With this hypothesis, introducing polar coordinates, we may write:

$$\frac{d\rho}{d\theta} = \frac{R(\mu,\theta)}{G(\mu,\theta)} + F(\mu,\rho;\theta) \tag{1.2}$$

where R, G are homogeneous polynomials of degree m+1 in ($\sin\theta$, $\cos\theta$) and F is of order ≥ 2 in ρ.
For each $\mu \in I$ the solution of (1.2) with initial condition $\rho(\mu,c,0) = c$ is given by:

$$\rho(\mu,c,\theta) = \sum_{i=1}^{m+\ell-1} c^i u_i(\mu,\theta) + f(\mu,c,\theta) \tag{1.3}$$

f being of order $\geq m+\ell$ in c. The u_i's are such that $u_1(\mu,0) = 1$ and $u_i(\mu,0) = 0$ for $i \neq 1$. From (1.2), (1.3) we immediately get

$$u_1(\mu,\theta) = \exp\left[\int_0^\theta \frac{R(\mu,\xi)}{G(\mu,\xi)}\, d\xi\right] \tag{1.4}$$

We make the further assumptions:

(H2) $\displaystyle\int_0^{2\pi} \frac{R(0,\theta)}{G(0,\theta)}\, d\theta = 0$

(H3) (Generalized transversality hypothesis):

$$\int_0^{2\pi} \left\{\frac{\partial}{\partial\mu}\left[\frac{R(\mu,\theta)}{G(\mu,\theta)}\right]\right\}_{\mu=0} d\theta > 0$$

(1.5) <u>Remark</u>. In case m = 1, (1.1) becomes:

$$\dot{x} = \alpha(\mu)x - \beta(\mu)y + \phi(\mu,x,y)$$

$$\dot{y} = \alpha(\mu)y + \beta(\mu)x + \psi(\mu,x,y)$$

where $\alpha(\mu)$ and $\beta(\mu)$ are the real and imaginary part respectively of the couple of eigenvalues associated with the linearized system and ϕ, ψ are of order ≥ 2 in (x,y). In this setup, hypotheses (H1), (H2), (H3) generalize to the case m > 1 the usual assumptions in Hopf bifurcation theory, namely $\beta(0) \neq 0$, $\alpha(0) = 0$ and $\alpha'(0) > 0$ respectively.

2. EXISTENCE OF PERIODIC SOLUTION

For $(\mu,c) \in I \times D$ let us define the following "displacement function":

$$D(\mu,c) = \rho(\mu,c,2\pi) - c = \sum_{i=1}^{m+\ell-1} c^i u_i(\mu,2\pi) + f(\mu,c,2\pi) - c. \tag{2.1}$$

Setting:

$$\tilde{D}(\mu,c) = \sum_{i=1}^{m+\ell-1} c^{i-1} u_i(\mu,2\pi) + \frac{f(\mu,c,2\pi)}{c} - c \quad \text{if } c \neq 0 \tag{2.2}$$

$$\tilde{D}(\mu,0) = 0$$

it follows:

$$D(\mu,c) = c\tilde{D}(\mu,c). \tag{2.3}$$

Definition (2.1) is well posed in a neighborhood of $\mu = 0$, $c = 0$, by (H1). It turns out that looking for periodic solutions of (1.1) is equivalent to looking for nontrivial zeros of D.
One has from (H2), (H3):

$$\tilde{D}(0,0) = 0, \quad \frac{\partial\tilde{D}}{\partial\mu}(0,0) = \frac{du_1}{\partial\mu}(0,2\pi) > 0. \tag{2.4}$$

So we have the following result:

(2.5) <u>Generalized Hopf's theorem</u>:
Under Hypotheses (H1), (H2), (H3) for (1.1) there exist two numbers ε, $\sigma > 0$ and a function $\mu \in C^{m+\ell-1}[[0,\varepsilon[,\mathbb{R}]$, $\mu(0) = 0$, such that for each $c \in]0,\varepsilon[$ and $\mu \in]-\sigma,\sigma[$, the orbit of (1.1) through $x = c$, $y = 0$ is closed if and only if $\mu = \mu(c)$.

3. m+N-ASYMPTOTIC STABILITY. THE METHOD OF MALKIN

Let us consider system (1.1) for $\mu = 0$:

$$\begin{aligned} \dot{x} &= X_m(0,x,y) + \dots + X_{m+\ell}(0,x,y) + X(0,x,y) \\ \dot{y} &= Y_m(0,x,y) + \dots + Y_{m+\ell}(0,x,y) + Y(0,x,y) \end{aligned} \tag{3.1}$$

In polar coordinates it may be written:

$$\begin{aligned} \dot{\rho} &= \rho^m R(0,\theta) + \dots + \rho^{m+\ell} R_{m+\ell+1}(0,\theta) + R(0,\rho,\theta) \\ \dot{\theta} &= \rho^{m-1} G(0,\theta) + \dots + \rho^{m+\ell-1} G_{m+\ell 1}(0,\theta) + G(0,\rho,\theta) \end{aligned} \tag{3.2}$$

We want to recall a method due to Malkin for the analysis of the asymptotic stability of the null solution of (3.1). This method is a generalization of the procedure of Liapunov-Poincaré [5] employed in [3]. The idea of this procedure is to construct, whenever possible, a function defined as

$$V(\rho,\theta) = \rho^2 \varphi_2(\theta) + \dots + \rho^\alpha \varphi_\alpha(\theta), \tag{3.3}$$

where α is even, $\varphi_i \in C^1[\mathbb{R},\mathbb{R}]$ are 2π-periodic for $i = 2,\dots,\alpha$ and such that V has the following property along the solutions of (3.2):

$$\dot{V}(\rho,\theta) = g\rho^{m+\ell-1} + \nu(\rho,\theta) \tag{3.4}$$

$g \neq 0$ constant, ν of order $\geq m + \alpha$ in ρ and 2π-periodic in θ. If such α and g exist, the method is said to be finite. (For a better understanding of the method, see [4,6]).

For each $N \leq \ell$, we give now the following

(3.5) Definition: The null solution of (3.1) is said to be m+N-asymptotically stable (resp. m+N-completely unstable) if: (i) the null solution of the system

$$\begin{aligned} \dot{x} &= X_m(0,x,y) + \dots + X_{m+N}(0,x,y) + \xi(x,y) \\ \dot{y} &= Y_m(0,x,y) + \dots + Y_{m+N}(0,x,y) + \tau(x,y) \end{aligned} \tag{3.6}$$

is asymptotically stable (resp. completely unstable) for any choice of ξ, $\tau \in C[D,\mathbb{R}]$ of order $> m+N$;

(ii) N is the least integer for which (i) is true.

We give now a theorem, whose proof can be found in [6].

(3.7) Theorem. A necessary and sufficient condition for the null solution of (3.1) to be $m+N$-asymptotically stable (resp. $m+N$-completely unstable) is that Malkin's methods if finite with $\alpha = N + 2$ and $g < 0$ (resp. $g > 0$).

(3.10) Remark. Suppose that the origin of (3.1) is $m+N$-asymptotically stable. Let us consider a solution (1.3) for $\mu = 0$ and c sufficiently small. If we evaluate the increment of V along this solution between $\theta = 0$ and $\theta = 2\pi$ by means of both (3.3), (3.4), we see that in (1.3) the u_i's are 2π-periodic for $i = 1,\dots,N$ and u_{N+1} is non periodic; moreover $u_{N+1}(0,2\pi)$ has the same sign as g.

(3.11) Remark. From the definition of D it follows that for $s \geq 1$:

$$\frac{\partial^s D}{\partial c^s}(0,0) = s!u_s(0,2\pi).$$

These two remarks give the following result:

(3.12) Theorem. Under the hypotheses (H1), (H2), (H3), a necessary and sufficient condition for the origin of (3.1) to be $m+N$-asymptotically stable (resp. $m+N$-completely unstable) is that

$$\frac{\partial^s D}{\partial c^s}(0,0) = 0 \text{ for } 1 \leq s \leq N \text{ and } \frac{\partial^{N+1} D}{\partial c^{N+1}}(0,0) < 0 \text{ (resp. } > 0)$$

4. ATTRACTIVITY OF THE PERIODIC ORBITS

Denote by $F^N_{m+1} = F(X_{m+1},\dots,X_{m+N}; Y_{m+1},\dots,Y_{m+N})$ the set of couples of functions $(P,Q) \in C^{m+N}[I \times D,\mathbb{R}]$ such that $[P(\mu,x,y)]_m = [Q(\mu,x,y)]_m = 0$; $[P(0,x,y)]_i = X_i(0,x,y)$; $[Q(0,x,y)]_i = Y_i(0,x,y)$ for $i = m+1,\dots,m+N$.

(The symbol $[.]_i$ denotes the i-th term in the MacLaurin expansion).

Let $\mu_{P,Q}$ be the function of theorem (2.5) relative to the system:

$$\begin{aligned} \dot{x} &= X_m(\mu,x,y) + P(\mu,x,y) \\ \dot{y} &= Y_m(\mu,x,y) + Q(\mu,x,y) \end{aligned} \tag{4.1}$$

(4.2) <u>Definition</u>. The cycles $(c,\mu(c))$ of (1.1) are said to be m+N-asymptotically stable (resp. m+N-completely unstable) if: (i) for each couple $(P,Q) \in F^N_{m+1}$ there exists an $\varepsilon_{P,Q}$ such that the cycles $(c,\mu_{P,Q}(c))$, $c \in [0,\varepsilon_{P,Q}]$ of (4.1) are asymtotically stable (resp. completely unstable);
(ii) N is the least integer for which (i) is true.

We shall denote by $D_{P,Q}$ the displacement function for (4.1). From the identity

$$D_{P,Q}(\mu_{P,Q}(c),c) = 0 \tag{4.3}$$

it follows that if

$$\frac{\partial^i D_{P,Q}}{\partial c^i}(0,0) = 0 \qquad \text{for } 1 \leq i \leq s < \ell, \text{ then} \tag{4.4}$$

$$\frac{\partial^j \mu_{P,Q}}{\partial c^j}(0) = 0 \qquad 1 \leq j \leq s-1 \text{ and}$$

$$\frac{\partial^s \mu_{P,Q}}{\partial c^s}(0) = \frac{-1}{s}\left[\frac{\partial^{s+1} D_{P,Q}}{\partial c^{s+1}}(0,0) \Big/ \frac{\partial \tilde{D}_{P,Q}}{\partial \mu}(0,0)\right] \tag{4.5}$$

The method of Malkin and the characterization of m+N-asymptotic stability given in Th. (3.7), allow us to give the following Theorem, whose proof can be found in [6].

(4.6) <u>Theorem</u>. Under the Hypotheses (H1), (H2), (H3), the two following propositions are equivalent:
(i) the origin for (3.1) is m+N-asymptotically stable (resp. m+N-completely unstable);
(ii) there exists an $\eta > 0$ such that the family of cycles

$(c,\mu(c))$, $c \in [0,\eta[$, appears for $\mu > 0$ (resp. $\mu < 0$). These cycles are m+N-asymptotically stable (resp. m+N-completely unstable).

REFERENCES

1. N. Chafee: The Bifurcation of One or More Closed Orbits from An Equilibrium Point of An Autonomous Differential System, Jour. Diff. Equ. 4, 661-679, (1968).
2. F. Marchetti, P. Negrini, L. Salvadori, M. Scalia: Liapunov Direct Method in Approaching Bifurcation Problems, Ann. Mat. Pura e Appl. (IV) 108, 211-225, (1976).
3. P. Negrini, L. Salvadori: Attractivity and Hopf Bifurcation, Nonlinear Anal: TMW 3, 87-99, (1979).
4. I.G. Malkin: Stability and Dynamical Systems, Providence, R.I. AMS, 1962.
5. M.A. Liapunov: Problème Général de la Stabilité du Mouvement, Ann. Math. Stud. Princeton Univ. Press, Princeton, 1969.
6. S. Caprino, P. Negrini: Attractivity Properties of Closed Orbits in A Case of Generalized Hopf Bifurcation, to appear on Jour. of Math. Anal. and Appl.

BOUNDARY VALUE PROBLEMS FOR NONLINEAR DIFFERENTIAL EQUATIONS ON NON-COMPACT INTERVALS

M. Cecchi
M. Marini

Istituto di Matematica Applicata
Università di Firenze

P.L. Zezza

Istituto di Matematica
Università di Siena

In this note we study a non linear differential system with general boundary conditions on the right open interval $[a,b)$ $(-\infty<a<b\leq+\infty)$. Different BVPs on a compact interval have been deeply studied: we recall only the survey works of L. Cesari [1] and of J. Mawhin [8]. On non compact interval results have been obtained mainly by admissibility theory (Massera-Schaeffer (1966), Corduneanu (1973)). Here we shall use topological methods and we shall reduce the problem to the search of fixed points of an operator M. In this way some results have been obtained by A.G. Kartsatos [6], [7] and by the authors [2], [3]. We note moreover that under rather general hypotheses the operator M is compact, which is not obvious because we want solutions wich are defined and bounded on a non compact interval: the Ascoli-Arzelà theorem therefore fails.

ISBN 0-12-508780-2

1.

In the following we consider the function spaces:

$$C^i = \{x(t)\,|\,x\colon [a,b) \to R^n,\ x^{(i)} \text{ is continuous}\}$$

$$BC = \{x(t)\,|\,x \in C,\ \sup_{a \le t < b} |x(t)| < +\infty\}$$

$$BC_\ell = \{x(t)\,|\ x \in BC, \text{ there exists } \lim_{t \to b^-} x(t) = \ell_x,\ |\ell_x| < +\infty\}$$

$$BC_o = \{x(t)\,|\ x \in BC_\ell,\ \ell_x = 0\}$$

and we shall indicate with $|\cdot|$ and $\|\cdot\|$ the euclidean norm in R^n and the norm in BC respectively.

Let us consider the boundary value problem (BVP)

$$x' = A(t)x + f(t,x) \tag{1}$$

$$Tx = r, \tag{2}$$

where:

a) $A(t)$ is a $n \times n$ matrix, continuous for $t \in [a,b)$ and such that the linear system

$$y' = A(t)y \tag{3}$$

is stable, i.e. the space D of solutions of (3) is contained in BC;

b) $f\colon [a,b) \times R^n \to R^n$ is a continuous function;

c) $T\colon \text{dom } T \subseteq BC \to R^m$ $(m \le n)$ is a linear continuous operator such that $T(D) = R^m$.

Let

$$L\colon \text{dom } L \subseteq BC \to C \times R^m$$

$$L\colon x \to (x' - A(t)x, Tx)$$

where

$$\text{dom } L = BC \cap C^1 \cap \text{dom } T$$

and

$$N: \text{dom } N \subset BC \to C \times R^m$$

$$N: x \to (f(\cdot,x),r).$$

Then (1)-(2) can be equivalently written as

$$Lx = Nx. \tag{4}$$

Remark: In general, L is not a Fredholm operator because $[a,b)$ is not compact.

Theorem 1. (Equivalence Theorem) Eq. (4) is equivalent to

$$\begin{cases} x = Mx \\ x \in A \end{cases}$$

where

$$A = N^{-1}(\text{Im}L)$$

$$M: x \to Px + K_p Nx$$

$P: BC \to \text{Ker } L$ is a projection onto Ker L

$$K_p = (L_{\text{dom } L \cap \text{Im}(I-P)})^{-1}$$

It can be shown that for (1)-(2) the operator M, defined above, becomes

$$Mx = Px + X(t)JT_o^{-1}(r - T\int_a^t X(t)X^{-1}(s)f(s,x(s))ds) + \int_a^t X(t)X^{-1}(s)f(s,x(s))ds$$

where $X(t)$ is a fundamental matrix of (3), J is a suitable immersion of R^m into R^n, T_o is a non singular $m \times m$ submatrix of $TX(t)$ which depends on P, and T_o^{-1} its inverse.

It is clear that the existence of fixed points of M (solutions of (1)-(2)) depends both on the set A and on the compactness of M. In this connection, let $V \subset R^n$ be an open set, and $S \subset V$ a closed set with non void interior. We call:

$$S_1 = \{c \in R_o^+ \text{ such that there is some } d \in S\colon |d| = c\}$$

Theorem 2. (Compactness Theorem). Let $h(t,u) \in C[[a,b) \times V, R^n]$ such that

$$|h(t,u)| \le g(t,|u|) \quad t \in [a,b),\ |u| \in S_1 \tag{6}$$

where $g\colon [a,b) \times S_1 \to R_o^+$ is such that

i) $g(t,v)$ is continuous in v for all t;

ii) $g_\mu(t) = \max_{v \in [0,\mu] \cap S_1} g(t,v)$ is integrable on $[a,b)$ and $\int_a^b g_\mu(t)dt < +\infty$ for all $\mu \in R^+$.

If $K_1\colon BC \to BC$ is a bounded linear operator, then the operator

$$K\colon x \to K_1[\int_a^t h(s,x(s))ds]$$

$$K\colon \text{dom } K \subset BC \to BC$$

where

$$\text{dom } K = \{x \in BC,\ x(t) \in S, \text{ for all } t \in [a,b)\}$$

is continuous and compact.

Proof: Let us consider the operator $\tilde{K}\colon \text{dom } K \to BC$

$$\tilde{K}x = \int_a^t h(s,x(s))ds. \tag{7}$$

We have $\text{Im } \tilde{K} \subseteq BC_\ell$. In fact, let $x \in \text{dom } K$, from (6):

$$\int_a^t |h(s,x(s))|ds \le \int_a^t g(s,|x(s)|)ds \le \int_a^t g_{\|x\|}(s)ds < +\infty;$$

Then (7) is absolutely convergent and the limit for $t \to b^-$

exists. We recall that a set $\Phi \subset BC_\ell$ is relatively compact if and only if it is bounded, equicontinuous and uniformly convergent in the following sense: for each $\varepsilon > 0$ there exists $\delta(\varepsilon)$ such that

$$|g(t) - \lim_{t\to b^-} g(t)| < \varepsilon \quad \text{for all } t > \delta(\varepsilon),\ g \in \Phi.$$

Let us now show that the operator $\tilde{K}$ is continuous and compact.

—Continuity of $\tilde{K}$: Let $\{x_n\} \in \text{dom } K$ such that $x_n \to x \in \text{dom } K$. We have

$$|h(t,x_n(t)) - h(t,x(t))| \leq |h(t,x_n(t))| + |h(t,x(t))| \leq g(t,|x_n(t)|) + g(t,|x(t)|); \tag{8}$$

for n sufficiently large and for $t \in [a,b)$:

$$|x_n(t)| \leq \|x_n\| \leq \|x\| + \varepsilon;$$

from (8) we infer

$$|h(t,x_n(t) - h(t,x(t)| \leq g_{\|x\|+\varepsilon}(t) + g_{\|x\|}(t).$$

From the hypotheses and the Lebesgue dominated convergence theorem the claim follows.

—Compactness of $\tilde{K}$: Consider a bounded set Θ, $\Theta \subset \text{dom } K$; there exists $\gamma \in R^+$ such that $\|x\| \leq \gamma$ for all $x \in \Theta$; let us show that $\tilde{K}(\Theta)$ is relatively compact.

—Equiboundedness: It follows from

$$|\tilde{K}x| \leq \int_a^t |h(s,x(s)|ds \leq \int_a^t g(s,|x(s)|)ds \leq \int_a^t g_\gamma(s)ds < +\infty.$$

—Equicontinuity: Let t_1, $t_2 \in [a,b)$, $t_1 < t_2$, if $x \in \Theta$ we have

$$|\tilde{K}x(t_2) - \tilde{K}x(t_1)| \leq \int_{t_1}^{t_2} g(s,|x(s)|)ds \leq \int_{t_1}^{t_2} g_\gamma(s)ds;$$

From the convergence of the last integral, the statement follows. The proof of equiuniform convergence is similar. It is now easy to see that the operator $K = K_1 \circ \tilde{K}$ is continuous and compact.

Remark. Condition ii) of Theorem 2 is verified if

$$g(t,v) = \alpha(t)v + \beta(t) \qquad v \in S_1$$

with α, β positive, integrable on $[a,b)$, or if

$$g(t,v) = \alpha(t)\frac{1}{v} + \beta(t) \qquad v \in S_1, \quad 0 \notin V$$

with the same hypotheses as above.

2.

In the following a Banach space B is said to be "ADMISSIBLE FOR (1)" if every solution of (1) belongs to B. We have:

Theorem 3. Suppose that for (1)-(2) the following hypotheses hold:

a) $A(t)$ is a continuous $n \times n$ matrix defined on $[a,b)$ such that the linear system (3) is stable, i.e. the space D of its solutions is contained in BC.

b) Let $f \in C[[a,b) \times R^n, R^n]$ be such that

$$|X^{-1}(t)f(t,u)| \leq \alpha(t)|u| + \beta(t)$$

where $\alpha(t)$, $\beta(t)$ are non negative real continuous functions, defined on $[a,b)$, satisfying

$$\int_a^b \alpha(t)dt < +\infty, \qquad \int_a^b \beta(t)dt < +\infty.$$

c) T is a linear, bounded operator such that $TX(t)$ has (maximum) rank m.

If dom T = BC, then BC is admissible for (1), the operator M is defined on BC, its image is contained in BC and it is continuous and compact.

If, eventually, the space D is contained in BC_ℓ and dom $T = BC_\ell$ then BC_ℓ is admissible for (1), the operator M is defined on BC, its image is contained in BC_ℓ and it is continuous and compact.

Proof: Let dom T = BC; we have, for all $t \in [a,b)$:

$$\left|\int_a^t X(t)X^{-1}(s)f(s,x(s))ds\right| \leq |X(t)|\int_a^t |X^{-1}(s)f(s,x(s))|ds <$$

$$< +\infty$$

then BC is admissible for (1). Because dom T = BC, from (5) M is defined on BC.
The continuity and compactness of M follows from Theorem 2 with

$$h(t,u) = X^{-1}(t)f(t,u).$$

If, eventually, $D \subset BC_\ell$ we have $\lim_{t\to b^-} X(t) = W$. Then Im $M \subset BC_\ell$, the last claim follows reasoning as above.

The conditions of the last theorem concern simultaneously the behaviour of the solutions of the homogeneous system (3) and of the perturbing term $f(t,x)$.

We solve now the problem with distinct hypotheses on the linear and the nonlinear parts.

Theorem 4. Suppose that for the system (1)-(2) the following hypotheses hold:

a) $A(t)$ is a continuous $n \times n$ matrix, defined on $[a,+\infty)$ such that the linear system (3) is L^p-stable, i.e. for any fundamental matrix $X(t)$ there is $h > 0$ such that

$$\int_a^t |X(t)X^{-1}(s)|^p ds \leq h, \text{ for all } t \in [a,b).$$

b) $f \in C\,[[a,+\infty)xR^n,R^n]$ is such that

$$|f(t,u| \leq \alpha(t)|u| + \beta(t)$$

where α, β are non negative real continuous functions defined on $[a,+\infty)$ such that

$$\alpha,\ \beta \in L^q[a,+\infty), \qquad \frac{1}{p}+\frac{1}{q}=1.$$

c) T is a linear continuous operator such that dom $T = BC_0$ and $TX(t)$ has rank (maximum) m.

d) If $p = 1$ ($q = \infty$) $\lim_{t\to+\infty} \beta(t) = 0$ and $h \sup_{a\le t<+\infty} \alpha(t) < 1$.

e) If $p = \infty$ ($q = 1$) $\lim_{t\to+\infty} X(t) = 0$.

Then BC_0 is admissible for (1), M is defined on BC_0, its image is contained in BC_0, it is continuous and compact in a set $\Omega \subseteq BC_0$.

Proof: Let $1 < p < +\infty$. From a) and b) we have

$$|x(t)| \le k|X^{-1}(a)x(a)| + \int_a^t |X(t)X^{-1}(s)|\alpha(s)|x(s)|ds + \int_a^t |X(t)X^{-1}(s)|\beta(s)ds$$

and from Hölder's inequality

$$|x(t)| \le c + \int_a^t |X(t)X^{-1}(s)|\alpha(s)|x(s)|ds.$$

The boundedness of $x(t)$ follows from Gronwall's Lemma and Hölder's inequality.

Moreover choosing t_1 sufficiently large:

$$|X(t)\int_a^t X^{-1}(s)f(s,x(s))ds| \le |X(t)|\,|\int_a^{t_1} X^{-1}(s)f(s,x(s)ds| +$$

$$|x(t)|\,[\int_{t_1}^t |X(t)X^{-1}(s)|^p ds]^{\frac{1}{p}} \cdot [\int_{t_1}^t [\alpha(s)]^q ds]^{\frac{1}{q}} +$$

$$[\int_{t_1}^t |X(t)X^{-1}(s)|^p ds]^{\frac{1}{p}} \cdot [\int_{t_1}^t [\beta(s)]^q ds]^{\frac{1}{q}} \le$$

$$|X(t)|\ \ |\int_{t_1}^t X^{-1}(s)f(s,x(s))ds| + 2\varepsilon.$$

Recalling that (3) is asymptotically stable we conclude that BC_0 is admissible for (1) (asymptotic stability follows from

L^p-stability in $[a,+\infty)$).
It is also clear that M maps BC_o into BC_o.

Let us show K_pN is continuous and compact. The continuity follows from the continuity of f applying the Lebesgue dominated convergence theorem to the sequence $X(t)X^{-1}(s)[f(s,x_n) - f(s,x)]$, using once more Hölder's inequality.
As for the compactness of K_pN, we have to show that $K_pN(\Phi)$ is compact in BC_o for each bounded $\Phi \subset BC_o$. Let $x \in \Phi$ and $\|x\| \leq \mu$:

—Equiboundedness: It follows from

$$\|K_pNx\| \leq \mu\int_a^{+\infty} |X(t)X^{-1}(s)|\alpha(s)ds + \int_a^{+\infty}|X(t)X^{-1}(s)|\beta(s)ds \leq \mu k_1 + k_2;$$

—Equicontinuity: Let $t_1 < t_2$:

$$\left|\int_a^{t_2} X(t_2)X^{-1}(s)f(s,x(s))ds - \int_a^{t_1} X(t_1)X^{-1}(s)f(s,x(s))ds\right|$$

$$\leq |X(t_2) - X(t_1)| \int_a^{t_1} |X^{-1}(s)f(s,x(s))|ds +$$

$$\left[\int_a^{t_2}|X(t_2)X^{-1}(s)|^p ds\right]^{\frac{1}{p}} \cdot \mu\left[\int_{t_1}^{t_2}[\alpha(s)]^q ds\right]^{\frac{1}{q}} +$$

$$\left[\int_a^{t_2}|X(t_2)X^{-1}(s)|^p ds\right]^{\frac{1}{p}} \cdot \left[\int_{t_1}^{t_2}[\beta(s)]^q ds\right]^{\frac{1}{q}}$$

and as $t_1 \to t_2$ the statement follows.
The equiuniform convergence can be proved as the admissibility of BC_o. For the cases $p = 1$, $p = +\infty$ see [2].

Remark. The operator M is defined for every x in BC, but we do not need this because every solution of (1) belongs to BC_o.

Admissibility of BC_o for (1) is proved, in special cases, also in [4] and [5].

In Theorem 3 and 4 the perturbing term $f(t,x)$ is defined, as to the second variable, in the whole R^n. We examine now the case when $f(t,x)$ is defined and continuous, as the second variable, in a proper subset of $R^n - \{0\}$.

Defining

$$\Theta_r^R = \{u \in R^n | r \leq |u| \leq R\} \quad r, R \in R^+$$

$$\Delta_r^R = \{x \in BC | x(t) \in \Theta_r^R, \text{ for all } t \in [a,b)\}$$

we have the following:

Theorem 5. Suppose that for the system (1)-(2) the following hypostheses hold:

a) $A(t)$ is a continuous $n \times n$ matrix, defined on $[a,b)$, such that the linear system (3) is strongly stable

b) $f \in C[[a,b) \times \{R^n - \{0\}\}, R^n]$ and

$$|f(t,u)| \leq \alpha(t) \frac{1}{|u|} + \beta(t)$$

where α, β are real continuous non negative functions such that

$$\int_a^b \alpha(t)dt < +\infty, \qquad \int_a^b \beta(t)dt < +\infty$$

c) T is a linear continuous operator such that dom T = BC and $TX(t)$ has rank (maximum) m.

Then BC is admissible for system (1), the operator M is defined on Δ_r^R, its image is contained in BC and it is continuous and compact.

Proof (sketch): The proof of admissibility of BC for (1) is similar to that of the preceeding theorems. As to the continuity and compactness of M it is enough to observe that from the hypotheses of strong stability one gets $|X^{-1}(t)| < H_1$ for all $t \geq a$ and then

$$|X^{-1}(t)f(t,u)| \leq H_1|f(t,u)|.$$

The statement follows from Theorem 2.

Remark. When f is defined in the whole of R^n with respect to x, in order to get existence theorems we have only to add some a priori bound and to use the continuation theorem proved in [9]. When M is defined in Δ_r^R, since Δ_r^R has not the fixed point property whith respect to continuous and compact maps, existence theorems can be obtained using the degree theory, noting that (1)-(2) can be transformed into

$$\begin{cases} z' = g(t,z) \\ T_1 z = r \end{cases}$$

because (3) is strongly stable.

Remark. The results here obtained can be easily extended to the nonlinear Boundary Value Problem

$$\begin{cases} x' = A(t)x + f(t,x) \\ Tx = h(x) \end{cases}$$

where h is a continuous function h: dom $T \subset BC \to R^m$ such that

$$|h(u)| \leq h_1 \|u\| + h_2 \qquad h_1, h_2 \in R^+.$$

REFERENCES

1. L. Cesari: Functional Analysis, Non Linear Differential Equations and the Alternative Method, In: Nonlinear Functional Analisys and Differential Equations, ed.by L. Cesari, R. Kannan, J.D. Schuur, New York, Dekker, 1-197,1976.
2. M. Cecchi, M. Marini, P.L. Zezza: Linear Boundary Value Problems for Systems of Ordinary Differential Equations on Non Compact Intervals, Part I - II, Ann. Mat. Pura e Appl. (IV) Vol. CXXIII (1980) 267-285, Vol. CXXIV (1980) 367-379.
3. M. Cecchi, M. Marini, P.L. Zezza: A Compactness Theorem for Integral Operators and Applications, Int. Symp. on Functional Diff. Eq. and Appl. - Sao Carlos (Brasil) 2-7/VII/79 - Springer, Lectures Notes in Math. n° 799, 1980 .

4. R. Conti: Linear Differential Equations and Control, Institutiones Math., New York, Academic Press, 1976.
5. W.A. Coppel: Stability and Asymptotic Behaviour of Differential Equations, Heath Math. Monographs, Boston, 1965.
6. A.G. Kartsatos: The Leray Schauder Theorem and the Existence of Solutions to Boundary Value Problems on Infinite Intervals, Ind. Un. Math. J. 23, 1021-1029, (1974).
7. A.G. Kartsatos: The Hildebrandt-Graves Theorem and the Existence of Solutions of Boundary Value Problems on Infinite Intervals, Math. Nachr. 67, 91-100, (1976).
8. J. Mawhin: Topological Degree Methods in Nonlinear Boundary Value Problems, Reg. Conf. Series in Math. n. 40, Providence, Amer. Math. Soc., 1979.
9. P.L. Zezza: An Equivalence Theorem for Nonlinear Operator Equations and An Extension of the Leray-Schauder's Continuation Theorem, Boll. U.M.I. (5), 15-A, 545-551, (1978).

THE ELECTRIC BALLAST RESISTOR: HOMOGENEOUS AND NONHOMOGENEOUS EQUILIBRIA

Nathaniel Chafee[1]

School of Mathematics
Georgia Institute of Technology
Atlanta, Georgia

INTRODUCTION

In this paper we shall present a mathematical study of the electric <u>ballast resistor</u>. The ballast resistor is a device consisting of a straight segment of very thin wire surrounded by a gas having a fixed temparature $a > 0$. Trough the wire there passes an electric current I. This current is produced by a voltage V representing a difference in electric potential between the two endpoints of the wire. We suppose that either the current I or the voltage V is held constant. The passage of current through any portion of the wire generates heat. Depending on nearby temperatures, that heat may diffuse into neighboring portions of the wire or it may flow into the surrounding gas. Thus, the distribution of temperature u along the given wire reflects a complicated pattern of heat flow. We are interested in that distribution of temperature.

[1] Research partially supported by the United States National Science Foundation (Grant No. MGS78-05988).

ISBN 0-12-508780-2

Our goal is, under suitable hypotheses, to determine the existence and stability properties of temperature equilibria for the ballast resistor. By such equilibria we mean temperature distributions along the given wire which do not vary with time. Of particular interest to us are nonhomogeneous equilibria, that is, temperature equilibria which are not constant along the wire. Thus, under appropriate hypotheses, we wish to know whether or not there can appear stable, nonhomogeneous temperature equilibria.

Another question which interests us is, as the time t approaches $+\infty$, what is the asymptoric behaviour of any arbitrary distribution of temperature u? Must such a distribution approach an equilibrium, or can a more complicated behaviour occur?

The ballast resistor has a long history and much of that history concerns problems of the type just indicated. The earliest reference we know is a thesis by Gifford and Page [19] written in 1902. In that thesis the authors very briefly allude to the presence of temperature inhomogeneities in a ballast resistor [19, p. 99]. A later but by no means contemporary work is a paper by Busch [5] published in 1921.That paper includes a theoretical discussion of nonhomogeneous temperature equilibria [5, pp. 437-444]. Regarding the current literature we mention papers by Mazur *et al* [4,27,29] and a note by Landauer [25]. The papers [4,27] contain extensive treatments of both the existence and the actual structure of nonhomogeneous equilibria.

In the present work we also will study the existence of nonhomogeneous temperature equilibria. However, we will give greater attention to their stability properties with respect to disturbances of initial data. A major theme in our investigation will be the roles played by two contrasting hypotheses already mentioned in our description of the ballast resistor. The first hypothesis is that the current I is held constant; the second, that the voltage V is held constant. The distinction between these two hypotheses will drastically influence our results concerning the stability of temperature equilibria.

Our study of the ballast resistor requires a mathematical model describing the temperature distribution u along the given wire. We formulate such a model in Section 1 below. It takes the form of a nonlinear parabolic initial-boundary value problem in one space variable, displacement along the wire. Our approach to studying this problem is to interpret it as a nonlinear semigroup acting on an appropriate infinite dimensional phase space. Such an interpretation provides a convenient setting for our investigations of stability and asymptotic behavior. Also, it makes available to us important notions coming from the theory of dynamical systems and from Liapunov's theory of stability, and we shall exploit many of those same notions.

This approach we are adopting is already well known. Indeed, several authors [1-3, 7-9, 14-17, 20, 21, 31] have followed a similar line of reasoning in their investigations of qualitative behavior for partial differential equations, functional differential equations, abstract semigroups, and evolution systems. Our present investigation lies in that same vein. The organization of our work is as follows.

In Section 1, as already stated, we introduce our mathematical model for the ballast resistor. The form of this model varies according to whether we assume the hypothesis of constant current or the hypothesis of constant voltage. In Sections 2, 3, and 4 we treat the case of constant current and in Section 5 and 6 the case of constant voltage.

Section 2 itself deals with the interpretation of our given initial-boundary value problem as a nonlinear semigroup. Here we borrow extensively from papers by Hale [20] and Pazy [30].

In Section 3, under the hypothesis of constant current, we study the asymptotic behavior as the time t goes to $+\infty$ of an arbitrary temperature distribution u in the ballast resistor. Our main result, Theorem 3.2, asserts that, if every temperature equilibrium is in a certain sense isolated, then our arbitrary temperature distribution u approaches one of these equilibria as $t \to +\infty$. Almost all the material in Section 3 comes from an earlier work [9] by ourselves.

In Section 4 we study stability properties of individual temperature equilibria. Our principal result, Theorem 4.2, is perhaps surprising. Essentially, it states that under the hypothesis of constant current any nonhomogeneous temperature equilibrium must be unstable. This theorem has a small history, which we indicate in Section 4.

In Section 5 we turn our attention to the case of constant voltage and then examine again the asymptotic behavior as $t \to +\infty$ of an arbitrary temperature distribution u in the ballast resistor. We construct a number $V_* > 0$ with the property that, for any voltage V satisfying $0 < V < V_*$, every temperature distribution u converges as $t \to +\infty$ to a single homogeneous equilibrium $u = \overline{\psi}$, whose value depends on V. Thus, we can say that this single equilibrium $\overline{\psi}$ is a global attractor. The precise statement appears in Theorem 5.3.

In Section 6 we examine the local stability properties of $\overline{\psi}$ as the parameter V increases beyond V_*. Under suitable hypotheses we show that, as V increases through a threshhold V_0, $V_* \leq V_0 \leq +\infty$, the equilibrium $\overline{\psi}$ passes from a state of being asymptotically stable to a state of being unstable. Under further hypotheses we show that this transition is accompanied by a stationary bifurcation. Indeed, as V increases beyond V_0, two stable nonhomogeneous temperature equilibria $\tilde{\psi}$ bifurcate from $\overline{\psi}$.

Theorem 6.5 at the end of Section 6 contains the exact formulation of these results. Directly before stating that theorem we briefly discuss the significance of its hypotheses relative to the ballast resistor.

Our work in Section 6 leans heavily on the theory of stationary bifurcations developed by Crandall and Rabinowitz in [12, 13].

Thus, in the case of constant voltage, one can obtain stable nonhomogeneous temperature equilibria. As we have already pointed out, in the case of constant current, any nonhomogeneous equilibrium is unstable. The two results stand in sharp contrast.

1. THE MATHEMATICAL MODEL

Our task in this Section is to formulate suitable mathematical relations governing an arbitrary temperature distribution u along the given wire in our ballast resistor. We can imagine that this wire lies along the unit interval [0,1] on an x-axis and, hence, we can regard the temperature u as a function $u(x,t)$ depending on the position x, $0 \leq x \leq 1$, along the wire and on the time t, $0 \leq t < +\infty$. In all that follows we shall restrict u to vary on the interval $(0,+\infty)$.

We shall assume that, along the given wire, the temperature u obeys the following equation:

$$u_t = u_{xx} - g(u) + I^2 r(u) \qquad (1.1a)$$
$$(0 < x < 1,\ 0 < t < +\infty).$$

Here, $g(u)$ represents the amount of heat flowing per unit length of wire and per unit time from the wire into the gas. $r(u)$ represents the resistivity of the wire, that is, its electrical resistance per unit length. As the notation indicates, we are assuming that $g(u)$ and $r(u)$ are functions, suitably smooth, depending on u and u only. Already we have introduced the quantity I signifying the electric current passing through the wire. The term $I^2 r(u)$ represents the amount of heat generated per unit length and per unit time by the current I. The term u_{xx} takes into account the diffusion of heat along the wire. Later we formulate specific hypotheses governing $g(u)$, $r(u)$, and I.

Eq. (1.1a) describes the behavior of u along the given wire, i.e., between its two endpoints. We shall assume that, at those endpoints themselves, u satisfies the relations

$$u_x(x,t) = 0 \quad \text{at} \quad x = 0,1 \qquad (0 < t < +\infty). \qquad (1.1b)$$

These equations assert that there is no net flow of heat through the endpoints of the wire.

By $\phi(x)$, $0 \leq x \leq 1$, we shall mean an initial distribution of temperature along the wire specified at the initial time

$t = 0$. We shall always require that $\phi(x) > 0$ on $0 \leq x \leq 1$. Thus, we have

$$u(x,0) = \phi(x) > 0 \qquad (0 < x < 1). \tag{1.1c}$$

Eqs. (1.1), that is, (1.1a,b,c) constitute a parabolic initial-boundary value problem. However, Eq. (1.1a) is incomplete in the sense that we have not yet specified the dependence of I on x, t, or u. This dependence is defined by hypotheses we are now going to impose on I.

To begin, we shall assume that at each time t, $0 \leq t < +\infty$, the current I is the same at all points x along the given wire. Standard terminology allows us to refer to this condition as <u>electroneutrality</u> along the wire.

Next, we consider the voltage V representing the difference in electric potential between the two endpoints of the given wire. By Ohm's Law and our hypothesis of electroneutrality we can write

$$V = I\int_0^1 r(u(x,t))dx, \tag{1.2}$$

where u(x,t) is the temperature distribution described in Eqs. (1.1).

Our treatment of Eqs. (1.1) will be divided into two separate cases corresponding to two distinct hypotheses pertaining to I. The first of these hypotheses is that I is equal to a positive constant, independent of x, t, or u. The second of these hypotheses is that V is equal to a positive constant, independent of x, t, or u. Thus, we can henceforth speak of the cases of constant current and of constant voltage.

We have now stated all our hypotheses concerning I. In the case of constant current the quantity I^2 in Eq. (1.1a) has, of course, a fixed positive value. In the case of constant voltage the same quantity I^2 is given by

$$I^2 = V(\int_0^1 r(u(x,t))dx)^{-2}, \tag{1.3}$$

where V^2 has a fixed positive value. Under (1.3) Equation (1.1a) becomes an integro-differential equation.

Next, we must state our hypotheses governing the functions g and r appearing in Eq. (1.1a). Those hypotheses are as follows.

(H_1) g and r are real-valued functions defined and C^4-smooth on the interval $(0,+\infty)$.

(H_2) for all $u \in (0, +\infty)$ we have $g'(u) > 0$ and $g''(u) \geq 0$. Moreover, there exists a unique number $a > 0$ such that $g(a) = 0$. Thus, $g(u) < 0$ if $0 < u < a$ and $g(u) > 0$ if $a < u < +\infty$.

(H_3) There exist constants $M, M' > 0$ such that $0 < r(u) < M$ and $0 < r'(u) < M'$ for all $u \in (0, +\infty)$.

Concerning these hypotheses we wish to make several remarks. First, the quantity a in (H_2) represents the temperature of the gas surrounding the wire in our ballast resistor. Second, for our work in Sections 2 through 5 below it suffices that g and r be C^2-smooth on $(0, +\infty)$. Only in Section 6 do we fully exploit the C^4-smoothness stipulated in (H_1).

Next, our formulation of (H_2) is motivated by Newton's Law of cooling and the Stefan-Boltzmann Law of radiation transfer. Our formulation of (H_3) is partially motivated by considerations appearing in [4, p. 361].

2. THE CASE OF CONSTANT CURRENT: A NONLINEAR SEMIGROUP

We are supposing that I in Eq. (1.1a) is a positive constant. With that in mind we define a C^4-smooth function f mapping $(0, +\infty)$ into the real line R by setting $f(u) = -g(u) + I^2 r(u)$ for all $u \in (0, +\infty)$. Here, g and r are as in (1.1a). This allows us to rewrite Eqs. (1.1) as

$$u_t(x,t) = u_{xx}(x,t) + f(u(x,t)) \quad (0 < x < 1,\ 0 < t < +\infty) \tag{2.1a}$$

$$u_x(0,t) = u_x(1,t) = 0 \qquad (0 < t < +\infty) \tag{2.1b}$$

$$u(x,0) = \phi(x) > 0 \qquad (0 < x < 1) \qquad (2.1c)$$

We shall study (2.1) in the present section and in Sections 3 and 4 below.

First, we want to represent (2.1) as an abstract initial value problem. This necessitates an appropriate geometrical setting.

By X we shall mean the space of all C^1-smooth functions ϕ mapping the unit interval [0,1] into R such that $\phi'(0) = \phi'(1) = 0$. On X we shall impose the C^1-supremum norm, which henceforth we denote by $\| \ \|$. X is a Banach space under $\| \ \|$. We shall also introduce an auxiliary norm on X, namely, the C^0-supremum norm, which we denote by $\| \ \|_0$.

For any function $\psi \in X$ and any number $\mu > 0$ we let $B(\psi;\mu)$ denote the open ball in X relative to $\| \ \|$ having center ψ and radius μ.

Next, we define an operator A on X by setting $A\phi = \phi''$. Here, we take the domain D(A) for A to be the set of all $\phi \in X$ such that ϕ is C^3-smooth on [0,1] and $\phi'''(0) = \phi'''(1) = 0$. Thus, A is a densely defined closed linear operator from X into X.

One can show that A generates a holomorphic semigroup $\{T(t)\}$ of bounded linear operators $T(t)$, $0 \leq t < +\infty$, mapping X into X. Also, arguing as in [9, p. 114], one can prove that, if $t > 0$ and if S is any set in X which is bounded relative to $\| \ \|_0$, then the image $T(t)S$ is precompact in X relative to $\| \ \|$.

Now, let Ω be the domain in X consisting of all $\phi \in X$ such that $\phi(x) > 0$ for every $x \in [0,1]$. We define a C^1-smooth function $q\colon \Omega \to X$ by setting $q(\phi) = f(\phi(\cdot))$ for all $\phi \in \Omega$, where f is as in (2.1). In place of (2.1) we consider the Cauchy problem

$$\dot{u} = Au + q(u) \qquad (0 < t < +\infty) \qquad (2.2a)$$

$$u = \phi \quad \text{at} \quad t = 0 \qquad (2.2b)$$

were ϕ is any element in Ω.

By a <u>strict solution</u> of (2.2) one means a function u continuously mapping an interval of the form $[0,s)$, $0 < s \leq +\infty$, into Ω such that (i) $u(0) = \phi$; (ii) $u(t) \in D(A) \cap \Omega$ for all $t \in (0,s)$; (iii) $\dot{u}(t)$ exists and is continuous at every $t \in (0,s)$; (iv) u satisfies (2.2a) on $(0,s)$.

Since $\{T(t)\}$ is holomorphic on X and since q is C^1-smooth on Ω, we can invoke some results given by Pazy [30, p. 27, Theorem 3.1; p. 31, Theorem 5.2] and assert that, for each $\phi \in \Omega$, Eqs. (2.3) have a unique noncontinuable strict solution $u = u(t;\phi)$ defined on an interval $[0,s(\phi))$ with $0 < s(\phi) \leq +\infty$.

We can now define a nonlinear strongly continuous semigroup $\{U(t)\}$ on Ω by setting $U(t)\phi = u(t;\phi)$ for each $\phi \in \Omega$ and $t \in [0,s(\phi))$. In the sequel, when speaking of the solution $u(x,t;\phi)$ for (2.1), we shall mean the strict solution $u(t;\phi)$ of (2.2).

Using a variation of constants formula for (2.2) one can prove that the solution $u(t;\phi)$ of (2.2) is continuous with respect to ϕ in the following sense: For any $\phi \in \Omega$, $t_1 \in (0,s(\phi))$, and $\varepsilon > 0$, there exists a $\delta > 0$ such that, if $\phi_1 \in B(\phi;\delta) \cap \Omega$, then $s(\phi_1) > t_1$ and $\| u(t;\phi_1) - u(t;\phi) \| < \varepsilon$ for every $t \in [0,t_1]$.

Thus, in a suitable fashion, the quantity $U(t)\phi$ varies continuously with resepct to ϕ on Ω. This allows us to exploit certain ideas coming from the theory of dynamical systems. We proceed as follows.

For any $\phi \in \Omega$ we can speak of the orbit $\gamma(\phi)$ for (2.1). By definition, $\gamma(\phi)$ is the set in Ω given by $\gamma(\phi) = \{U(t)\phi: 0 \leq t < s(\phi)\}$. We want to know the circumstances under which such an orbit is precompact in Ω. Using arguments from [30, pp. 28 - 29] and [9, p. 114], one can establish the following proposition.

Proposition 2.1. If $\gamma(\phi)$ is an orbit of (2.1) whose closure $\overline{\gamma(\phi)}$ relative to $\| \ \|$ lies in Ω and if $\gamma(\phi)$ is bounded relative to $\| \ \|_0$, then $s(\phi) = +\infty$ and $\gamma(\phi)$ is precompact in Ω relative to $\| \ \|$.

Next we turn our attention to the notions of positive invariance and invariance. A set $S \subseteq \Omega$ is positively invariant with respect to (2.1) if for each $\phi \in S$ we have $s(\phi) = +\infty$ and $\gamma(\phi) \subseteq S$. The concept of invariance is more subtle than this and requires a premilinary definition.

By a <u>solution of</u> (2.1) <u>on</u> $[0,1] \times (-\infty,+\infty)$ we mean a function $\tilde{u}$ mapping $[0,1) \times (-\infty,+\infty)$ into R such that for each $\tau \in (-\infty,+\infty)$ we have (i) $\tilde{u}(\cdot,\tau) \in \Omega$; (ii) $s(\tilde{u}(\cdot,\tau)) = +\infty$; (iii) $U(t)\tilde{u}(\cdot,\tau) = \tilde{u}(\cdot,\tau+t)$ for every $t \in [0,+\infty)$.

We say that a set $S \subseteq \Omega$ is <u>invariant</u> with respect to (2.1) if for each $\phi \in S$ there exists a solution $\tilde{u}$ of (2.1) on $[0,1] \times (-\infty,+\infty)$ such that $\tilde{u}(\cdot,0) = \phi$ and $\tilde{u}(\cdot,t) \in S$ for all $t \in (-\infty,+\infty)$.

Let $\phi \in \Omega$ and suppose that $s(\phi) = +\infty$. Then, by the <u>ω-limit set</u> of $u(x,t;\phi)$ we mean the set $\omega(\phi)$ given by

$$\omega(\phi) = \bigcap_{\tau>0} \text{closure } \{U(t)\phi: \tau \leq t < +\infty\},$$

where the closure here is with respect to $\| \ \|$.

From Proposition 2.1 and from a result well known in the theory of dynamical systems [20, p. 48, Lemma 3] we obtain the following theorem, which concludes the work of this section.

<u>Theorem 2.2</u>. If $\gamma(\phi)$ is an orbit of (2.1) whose closure $\overline{\gamma(\phi)}$ relative to $\| \ \|$ lies in Ω and if $\gamma(\phi)$ is bounded relative to $\| \ \|_0$, then $s(\phi) = +\infty$ and $\omega(\phi)$ is nonempty, compact and connected relative to $\| \ \|$, and invariant with respect to (2.1). Furthermore, $U(t)\phi \to \omega(\phi)$ as $t \to +\infty$, the convergence here being relative to $\| \ \|$.

3. THE CASE OF CONSTANT CURRENT: ASYMPTOTIC BEHAVIOR

We are going to construct a positively invariant set for (2.1). From Hypotheses (H_2) and (H_3) in Section 1 and from our definition of f at the beginning of Section 2 we obtain the existence of two numbers b_1, b_2 such that

$$\begin{aligned} &0 < a < b_1 \leq b_2 < +\infty \\ &f(b_1) = f(b_2) = 0 \\ &f(u) > 0 \quad \text{for} \quad 0 < u < b_1 \\ &f(u) < 0 \quad \text{for} \quad b_2 < u < +\infty. \end{aligned} \tag{3.1}$$

Not let Ω^* be the set of all $\phi \in \Omega$ such that $b_1 \leq \phi(x) \leq b_2$ for all $x \in [0,1]$. Clearly, Ω^* is bounded relative to $\| \ \|_0$ and Ω^* is closed relative to $\| \ \|$. We have the following theorem.

Theorem 3.1. The set Ω^* is positively invariant with respect to (2.1). Furthermore, if ϕ is any element in Ω, then $s(\phi) = +\infty$ and the solution $u(x,t;\phi)$ of (2.1) has a nonempty ω-limit set $\omega(\phi) \subseteq \Omega^*$. This set $\omega(\phi)$ is compact and connected with respect to $\| \ \|$. Also, $\omega(\phi)$ is invariant with respect to (2.1), and $U(t)\phi \to \omega(\phi)$ as $t \to +\infty$, the convergence here being relative to $\| \ \|$.

Proof. If $\phi \in \Omega^*$, then, using (3.1) and a comparison argument for the parabolic equation (2.1a), one can show that $\gamma(\phi) \subseteq \Omega^*$. From this and Theorem 2.2 it follows that $s(\phi) = +\infty$. Thus, Ω^* is positively invariant.

Now suppose that ϕ is any element in Ω. Then, once again using (3.1) and a comparison argument, one can show that $\gamma(\phi)$ is bounded relative to $\| \ \|_0$, $s(\phi) = +\infty$, and

$$\limsup_{t \to +\infty} (\sup\{u(x,t;\phi) : 0 \leq x \leq 1\}) \leq b_2$$

$$\liminf_{t \to +\infty} (\inf\{u(x,t;\phi) : 0 \leq x \leq 1\}) \geq b_1 .$$

With these assertions and with Theorem 2.2 one can easily complete the required proof.

The next question is, what can we say about the orbits lying in any one of the sets $\omega(\phi)$ described in Theorem 3.1? To answer this question we sketch the following argument.

Let F be any C^1-smooth function mapping $(0,+\infty)$ into R such that $F'(u) = f(u)$ for all $u \in (0,+\infty)$. Define a continuous function $W: X \to R$ by setting

$$W(\phi) = \int_0^1 \{\tfrac{1}{2}\phi'(x)^2 - F(\phi(x))\}dx \qquad (\phi \in X). \tag{3.2}$$

Through a suitable calculation one can show that, along any solution $u = u(x,t;\phi)$ of (2.1), we have

$$\dot{W}(u) = -\int_0^1 u_t^2 dx \qquad (0 < t < +\infty). \tag{3.3}$$

Indeed, this calculation is analogous to one given in [7, p. 26]. Thus, W is monotone non-increasing along each orbit of (2.1) and hence we say that W is a Liapunov function for (2.1).

From Theorem 3.1 we recall that, for any $\phi \in \Omega$, the set $\omega(\phi)$ is invariant with respect to (2.1). Combining this result and Eq. (3.3) with an important principle coming from the theory of dynamical systems [20, p. 50, Theorem 1] we conclude that, for any $\phi \in \Omega$, the set $\omega(\phi)$ consists solely of equilibria ψ for (2.1), that is, elements $\psi \in \Omega$ such that $U(t)\psi = \psi$ for all $t \in [0,+\infty)$. Clearly, the equilibria of (2.1) are precisely those C^2-smooth functions ψ in Ω which satisfy the relations

$$\begin{aligned} &\psi''(x) + f(\psi(x)) = 0 \qquad (0 \leq x \leq 1) \\ &\psi'(0) = \psi'(1) = 0. \end{aligned} \tag{3.4}$$

In Theorem 3.1 we have pointed out that each set $\omega(\phi)$ is connected relative to $\| \ \|$. Thus, we arrive at the following theorem, which is the main result for this section.

Theorem 3.2. Suppose that each equilibrium of (2.1) is isolated relative to $\| \ \|$. Then, for any $\phi \in \Omega$, there exists a unique equilibrium ψ of (2.1) such that $\|U(t)\phi - \psi\| \to 0$ as $t \to +\infty$. Moreover, ψ has the property that $b_1 \leq \psi(x) \leq b_2$ for all $x \in [0,1]$, where b_1, b_2 are as in (3.1).

We close this section with some historical remarks. Theorem 3.2 and the work leading up to it are a recapitulation of more general results presented by ourselves in [9]. Recently Matano has proved a result [26,p. 222, Theorem A] by which in Theorem 3.2 one can discard the hypothesis that each equilibrium of (2.1) is isolated relative to $\| \ \|$.

Several authors [3, 7-9, 18, 32] have obtained results similar to Theorem 3.2 above for various nonlinear parabolic initial-boundary value problems. These same authors in the works just cited have employed Liapunov functions similar or identical to the function W introduced in (3.2).

4. THE CASE OF CONSTANT CURRENT: STABILITY AND INSTABILITY OF EQUILIBRIA

We consider any equilibrium $\psi = \psi(x)$, $0 \le x \le 1$, of (2.1). We say that this equilibrium is homogeneous if $\psi'(x) \equiv 0$ on $[0,1]$. Otherwise, we say that it is nonhomogeneous.

Any equilibrium ψ of (2.1) must satisfy Eqs. (3.4) as well as, by Theorem 3.1, the condition $b_1 \le \psi(x) \le b_2$ for all $x \in [0,1]$, where b_1, b_2 are as in (3.1). Therefore, the homogeneous equilibria of (2.1) are precisely those constant functions $\psi = \psi(x) \equiv b$ on $[0,1]$ whose constant value b satisfies the conditions $b_1 \le b \le b_2$ and $f(b) = 0$, where f is as in (2.1) and (3.4).

The nonhomogeneous equilibria of (2.1) are more difficult to determine. A suitable approach is to investigate (3.4) by means of phase-energy methods. For problems of the general type (3.4) such an analysis appears in [9, pp. 128-130]. For our present purposes we can say that the hypotheses we have formulated in Section 1 are such as to neither preclude nor guarantee the existence of nonhomogeneous equilibria for (2.1).

Given any equilibrium ψ of (2.1), homogeneous or nonhomogeneous, we can speak of its stability properties in the sense of Liapunov. Thus, we say that ψ is stable if for every $\varepsilon > 0$ there exists a $\delta > 0$ such that for each $\phi \in B(\psi;\delta) \cap \Omega$ we have $U(t)\phi \in B(\psi;\varepsilon)$ for all $t \in [0,+\infty)$. In the contrary case we say that ψ is unstable. We call ψ asymptotically stable if ψ is stable and if, in addition, there exists a $\delta_0 > 0$ such that for each $\phi \in B(\psi;\delta_0) \cap \Omega$ we have $\|U(t)\phi - \psi\| \to 0$ as $t \to +\infty$.

To investigate the stability properties of a given equilibrium ψ, we will introduce the eigenvalue problem associated with the linearization of Eq. (2.2a) about ψ. Let $\mu > 0$ be such that $B(\psi;\mu) \subset \Omega$. We define a closed linear operator $A_1 : D(A) \to X$ and a C^1-smooth function $q_1 : B(0;\mu) \to X$ by setting

$$A_1\phi = A\phi + q'(\psi)\phi \qquad (\phi \in D(A))$$

$$q_1(\psi) = q(\psi + \phi) - q(\psi) - q'(\psi)\phi \qquad (\phi \in B(0;\mu)).$$

With u as in (2.2a) we set $v = u - \psi$. In a small neighborhood of ψ we can then rewrite (2.2a) as

$$\dot{v} = A_1 v + q_1(v). \tag{4.1}$$

The required eigenvalue problem is

$$A_1 v = \lambda v \qquad (v \in D(A),\ v \neq 0). \tag{4.2}$$

We can express (4.2) in the more classical form

$$\begin{aligned} &v''(x) + f'(\lambda(x))v(x) = \lambda(x) \qquad (0 \leq x \leq 1) \\ &v'(0) = v'(1) = 0. \end{aligned} \tag{4.3}$$

It is well known [11, p. 212, Theorem 2.1] that the eigenvalues of (4.3) form a sequence $\{\lambda_j\}_{j=1}^{+\infty}$ of real numbers with $\lambda_1 > \lambda_2 > \ldots$ and $\lambda_j \to -\infty$ as $j \to +\infty$ and that the corresponding eingenfunctions $v_j(x)$ each have exactly $j - 1$ zeros in the interval [0,1]. This brings us to the following proposition.

Proposition 4.1. Let ψ be an equilibrium of (2.1) and, as above, let $\lambda_1, \lambda_2, \ldots$ be the eigenvalues of (4.3) with $\lambda_1 > \lambda_2 > \cdots$. Then, ψ is asymptotically stable if $\lambda_1 < 0$ and ψ is unstable if $\lambda_1 > 0$.

We give only a hint of the proof. It can be shown that the operator A_1 in (4.1) generates a compact holomorphic semigroup $\{J(t)\}$ on X. Also, it is easy to verify that $q_1(0) = 0$ and $q_1'(0) = 0$. Hence, exploiting ideas set forth in [10, pp. 314-318], one can arrive at the assertions required by our proposition.

Proposition 4.1 can be described as a statement of the principle of linearized stability for Eqs. (2.1). For other statements of this principle relative to various parabolic problems we mention the papers by Keilhofer [23,24] and the lecture notes by Henry [21, Section 5.1].

Now we want to use Proposition 4.1 to determine the stability properties of homogeneous and nonhomogeneous equilibria of (2.1). The case of homogeneous equilibria is trivial: A homoge neous equilibrium $\psi(x) \equiv b$ is asymptotically stable or is unstable according to whether $f'(b) < 0$ or $f'(b) > 0$. The case of nonhomogeneous equilibria is more interesting.

Theorem 4.2. Any nonhomogeneous equilibrium ψ of (2.1) is unstable.

This theorem is a slight modification of a result appearing in [9, p. 130, Theorem 6.2] for a parabolic problem of the form (2.1) in one space dimension. In [6, p. 268, Theorem 2] Casten and Holland generalized that result to the case of a semilinear parabolic equation in n space dimensions with zero Neumann boundary conditions. Here we present a proof for Theorem 4.2 which, we believe, is simpler than the proof we gave in [9]. However, we see no straighforward analog of our present proof for the case in n space dimensions.

Proof of Theorem 4.2: Given ψ as in our theorem we consider the eigenvalue problem (4.3) and its corresponding eigenvalues λ_j and eigenfunctions $v_j(x)$. By Proposition 4.2 we need only prove that $\lambda_1 > 0$. We argue as follows.

For any real number λ we introduce the ordinary differential equation

$$y''(x) + f'(\psi(x))y(x) = \lambda y(x) \qquad (0 \leq x \leq 1) \tag{4.4}$$

We let $y_1 = y_1(x) = y_1(x;\lambda)$ and $y_2 = y_2(x) = y_2(x;\lambda)$ be the solutions of (4.4) satisfying the initial conditions $y_1(0) = 0$, $y_1'(0) = 1$ and $y_2(0) = 1$, $y_2'(0) = 0$ respectively. Each of these solutions is well defined on the interval $0 \leq x \leq 1$.

Now we consider our given nonhomogeneous equilibrium ψ for (2.1). Comparing (3.4) to (4.4) we see that $\psi'(x)$ is a nonzero solution of (4.4) at $\lambda = 0$ which vanishes at $x = 0$. Therefore, there exists a constant $c_1 \neq 0$ such that $\psi'(x) = c_1 y_1(x,0)$ for all $x \in [0,1]$. It follows that $y_1(x;0)$ vanish both at $x = 0$ and at $x = 1$. By a result well known in the classical theory of oscillations [11, p. 208, Theorem 1.1] $y_2(x;0)$ has at least one zero ξ_0 in the open interval $(0,1)$. By another such result [11, p. 210, Theorem 1.2], for each $\lambda < 0$, the solution $y_2(x;\lambda)$ has at least one zero ξ lying in $(0,\xi_0)$.

Next, we consider the eigenvalue λ_1 and a corresponding eingenfunction $v_1(x)$ coming from (4.3). In connection with (4.3) we have already pointed out that $v_1(x)$ has no zeros lying in the closed interval [0,1]. Comparing (4.3) to (4.4) we find that $v_1(x)$ is a nonzero solution of (4.4) at $\lambda = 0$ with $v_1'(0) = 0$. It follows that there exists a constant $c_2 \neq 0$ such that $v_1(x) = c_2 y_2(x;\lambda_1)$ for all $x \in [0,1]$. Hence, $y_2(x;\lambda_1)$ has no zeros lying in [0,1]. Comparing this last assertion to the statement made at the end of the preceding paragraph, we conclude that $\lambda_1 > 0$. The proof is complete.

5. THE CASE OF CONSTANT VOLTAGE: GLOBAL STABILITY OF A HOMOGENEOUS EQUILIBRIUM

We are now supposing that the quantity V in Eq. (1.1a) is a positive constant. Recalling Eq. (1.3) as well as Eqs. (1.1) we obtain the following system of equations for the temperature distribution $u(x,t)$ in our ballast resistor.

$$u_t(x,t) = u_{xx}(x,t) - g(u(x,t)) + V^2 r(u(x,t))\rho[u(\cdot,t)]^{-2} \quad (0 < x < 1,\ 0 < t < +\infty) \tag{5.1a}$$

$$u_x(0,t) = u_x(1,t) = 0 \qquad (0 < t < +\infty) \tag{5.1b}$$

$$u(x,0) = \phi(x) > 0 \qquad (0 < x < 1). \tag{5.1c}$$

Here, by stipulation,

$$\rho[u(\cdot,t)] \equiv \int_0^t r(u(\xi,t))d\xi. \tag{5.2}$$

In this section and in Section 6 below we shall study Eqs. (5.1).

To begin, Eqs. (5.1) are amenable to the treatment we have rendered in Section 2 for Eqs. (2.1). In particular, we can let X, $\|\ \|$, and Ω be as in that Section, and, henceforth, we can regard ρ in (5.2) as a C^1-smooth functional mapping Ω into R. Indeed, we write

$$\rho[\phi] = \int_0^1 r(\phi(\xi))d\xi \qquad (\phi \in \Omega), \tag{5.3}$$

and we note that, by (H_3) in Section 1, $\rho[\phi] > 0$ for all $\phi \in \Omega$.

In the manner described in Section 2 Eqs. (5.1) determine a nonlinear strongly continuous semigroup $\{U_1(t)\}$ on Ω. For each $\phi \in \Omega$ we can speak of the solution $u(x,t;\phi)$ for (5.1) and the domain of this solution has the form $[0,1] \times [0,s_1(\phi))$ with $0 < s_1(\phi) \leq +\infty$. Furthermore, with respect to (5.1), we can introduce the notions of an orbit $\gamma(\phi)$ and an ω-limit set $\omega(\phi)$. Finally, for Eqs. (5.1) we have an exact analog of Theorem 2.2 with $\{U(t)\}$ replaced by $\{U_1(t)\}$.

Now let $a > 0$ be as in (H_2) in Section 1. By (H_3) in Section 1 there exists a constant $m > 0$ such that

$$r(u) > m \qquad (a \leq u < +\infty). \tag{5.4}$$

Next, let $M > 0$ be as in (H_2). Then, with V fixed and positive, there exist numbers c_1, c_2 such that

$$\begin{aligned} &0 < a < c_1 < c_2 < +\infty \\ &-g(u) + V^2 r(u) M^{-2} > 0 \qquad (0 < u \leq c_1) \\ &-g(u) + V^2 r(u) m^{-2} < 0 \qquad (c_2 \leq u < +\infty). \end{aligned} \tag{5.5}$$

We let Ω^* be the subset of Ω consisting of all those $\phi \in X$ such that $c_1 \leq \phi(x) \leq c_2$ for every $x \in [0,1]$.

Using (5.4), (5.5), and appropriate comparison arguments for (5.1a), one can obtain the following analog of Theorem 3.1.

<u>Theorem 5.1</u>. The set Ω_1^* is positively invariant with respect to (5.1). Furthermore, if ϕ is any element in Ω, then $s_1(\phi) = +\infty$ and the solution $u(x,t;\phi)$ of (5.1) has a nonempty ω-limit set $\omega(\phi) \subseteq \Omega_1^*$. This set $\omega(\phi)$ is compact and connected relative to $\| \ \|$. Also, $\omega(\phi)$ is invariant with respect to (5.1), and $U_1(t)\phi \to \omega(\phi)$ as $t \to +\infty$, the convergence here being relative to $\| \ \|$.

Now we turn our attention to the problem of finding homogeneous equilibria for (5.1). That is, we want to obtain constant solutions $u(x,t) \equiv c$ of (5.1a), where the constants c are to be positive. Following our practice in Sections 3 and 4 we can regard each such equilibrium as a single element $\psi \in \Omega$ with $\psi(x) \equiv c$ on $0 \leq x \leq 1$. Taking into account (5.3) we see that the homogeneous equilibria of (5.1) are given as the roots $\bar{c}$ of the equation

$$0 = -g(c) + V^2 r(c)^{-1}.$$

Using (H_2) and (H_3) in Section 1, one can show that, with V fixed and positive, there exists a unique constant $\bar{c} \in (0,+\infty)$ such that

$$\begin{aligned} -g(c) + V^2 r(c)^{-1} &> 0 \qquad (0 < c < \bar{c}) \\ -g(\bar{c}) + V^2 r(\bar{c})^{-1} &= 0 \\ -g(c) + V^2 r(\bar{c})^{-1} &< 0 \qquad (\bar{c} < c < +\infty) \end{aligned} \tag{5.6}$$

These relations and Theorem 5.1 yield the following proposition.

Proposition 5.2. Given any fixed $V \in (0,+\infty)$, Eqs. (5.1) have a unique homogeneous equilibrium $\bar{\psi} = \bar{\psi}(x) \equiv \bar{c}$. Moreover, $\bar{\psi}$ lies in Ω_1^*. Indeed, $c_2 \geq \bar{c} \geq c_1 > a$, where c_2, c_1, a are as in (5.5).

In a moment we will see that $\bar{\psi}$ has an important stability property. First, however, we must perform some elementary constructions.

From (H_2) of Section 1 we have $g'(u) > 0$ for all $u \in (0,+\infty)$. We define a number $\ell \geq 0$ by setting $\ell = \inf\{g'(u): 0 < u < +\infty\}$. Also, we let m and M' be as in (5.4) and in (H_3) of Section 1 respectively. Then, we define a number $V_* > 0$ by setting

$$V_* = +\{(\pi^2 + \ell)(m^2)(M')^{-1}\}^{1/2}. \tag{5.7}$$

We have the following theorem.

Theorem 5.3. Suppose that V in (5.1) satisfies the inequality $0 < V < V_*$, where V_* is as in (5.7). In accordance with Proposition 5.2, let $\bar{\psi}$ be the unique homogeneous equilibrium of (5.1). Then, for each $\phi \in \Omega$ we have $\|U_1(t)\phi - \bar{\psi}\| \to 0$ as $t \to +\infty$.

Proof: We let V be as in our theorem, we let $\phi \in \Omega$, and we consider the corresponding solution $u(x,t;\phi)$ of (5.1). Where appropriate we shall abbreviate notation and denote this solution by u.

By Theorem 5.1 we have $u(\cdot,t;\phi) \to \Omega_1^*$ as $t \to +\infty$. Hence, given $a > 0$ as in (5.5), there exists $\tau > 0$ such that $u(x,t;\phi) > a$ for all $x \in [0,1]$, $t \in [\tau,+\infty)$. From this and from (5.4) and (5.3) we obtain

$$\rho[u(\cdot,t;\phi)] > m \qquad (\tau \leq t < +\infty). \tag{5.8}$$

Along with (5.8) we note

$$\pi^2\int_0^1 u_x^2 dx \leq \int_0^1 u_{xx}^2 dx \qquad (0 < t < +\infty), \tag{5.9}$$

which is a consequence of Wirtinger's Inequality [28, pp. 141].

With the aid of (5.8) and (5.9) one can establish that

$$\frac{d}{dt}\int_0^1 u_x^2 dx \leq -\alpha\int_0^1 u_x^2 dx \qquad (\tau < t < +\infty),$$

where α is a constant given by $\alpha = \pi^2 + \ell - V^2 m^{-2} M'$. From (5.7) and our assumption that $0 < V < V_*$ we obtain $\alpha > 0$. Therefore,

$$\int_0^1 u_x^2 dx \to 0 \qquad \text{as} \qquad t \to +\infty. \tag{5.10}$$

Now, recalling Theorem 5.1, we consider the ω-limit set $\omega(\phi)$ corresponding to our solution u. From (5.10) it follows that any element $\psi = \psi(x)$ lying in $\omega(\phi)$ must be a constant function $\psi(x) \equiv c$ on $[0,1]$. Moreover, since $\omega(\phi) \subseteq \Omega_1^*$, we have $c_1 \leq c \leq c_2$, where c_1, c_2 are as in (5.5). The same Theorem 5.1 tells us that $\omega(\phi)$ is invariant. Therefore, $\omega(\phi)$ can be represented as the union of orbits $\{\tilde{u}(\cdot,t): -\infty < t < +\infty\}$, where each function $\tilde{u} = \tilde{u}(x,t)$ is a solution of (5.1) on $[0,1] \times (-\infty,+\infty)$. From the preceding paragraph it follows that, for any $t \in (-\infty,+\infty)$, the element $\tilde{u}(\cdot,t)$ lying in $\omega(\phi)$ is constant with respect to x. Thus, we can now speak of $\tilde{u}(t)$ rather than $\tilde{u}(x,t)$. From the preceding paragraph it also follows that $c_1 \leq \tilde{u}(t) \leq c_2$ for every $t \in (-\infty,+\infty)$.

With the aid of (5.1) and (5.2) we see that $\tilde{u}(t)$ is a solution of the ordinary differential equation

$$\dot{u} = -g(u) + V^2 r(u)^{-1}. \tag{5.11}$$

In fact, as we have already shown, $\tilde{u}(t)$ is a solution of (5.11) defined on $(-\infty,+\infty)$ and satisfying $c_1 \leq \tilde{u}(t) \leq c_2$ for every $t \in (-\infty,+\infty)$. But, from (5.6) and the inequality $c_1 \leq \bar{c} \leq c_2$ contained in Proposition 5.2, we see that (5.11) has only one solution of the sort required and that solution is $\tilde{u}(t) \equiv \bar{c}$.

Thus, we have shown that $\omega(\phi)$ consists of the single element $\bar{\psi} = \bar{\psi}(x) \equiv \bar{c}$ described in Proposition 5.2. From Theorem 5.1 we have $U_1(t)\phi \to \omega(\phi)$ as $t \to +\infty$, the convergence here being relative to $\| \; \|$. It follows that $\|U_1(t)\phi - \bar{\psi}\| \to 0$ as $t \to +\infty$, and our proof is complete.

6. THE CASE OF CONSTANT VOLTAGE: APPEARANCE OF STABLE NONHOMOGENEOUS EQUILIBRIA

With V fixed and positive, we let ψ be any equilibrium, homogeneous or nonhomogeneous, for Eqs. (5.1). With respect to ψ we wish to formulate a principle of linearized stability, i.e., an analog of Proposition 4.1. To this end we let D(A) be the domain introduced in Section 2 and we define a linear operator $L: D(A) \to X$ by setting

$$L\phi = \phi'' - g'(\psi)\phi + V^2 r'(\psi)\rho[\psi]^{-2}\phi - 2V^2 r(\psi)\rho[\psi]^{-3}\rho'[\psi](\phi) \qquad (\phi \in D(A)), \tag{6.1}$$

where by $\rho'[\psi]$ we mean the Fréchet derivative given by

$$\rho'[\psi](\phi) \equiv \int_0^1 r'(\psi(\xi))\phi(\xi)d\xi \qquad (\phi \in X).$$

Clearly, L comes from taking the linearization of (5.1) about ψ. The following proposition is our required principle of linearized stability and its proof parallels the proof of Proposition 4.1.

Proposition 6.1. If all the eigenvalues of L have negative real parts, then ψ is asymptotically stable with respect to (5.1). If at least one of the eigenvalues of L has positive real part, then ψ is unstable with respect to (5.1).

Now we want to apply Proposition 6.1 to the study of the single homogeneous equilibrium $\bar{\psi} = \bar{\psi}(x) \equiv \bar{c}$ treated in Section 5.

Setting $\psi(x) \equiv \bar{c}$ in (6.1) we can explicitly calculate all the eigenvalues and eigenfunctions of the corresponding operator L. The result of that calculation is a sequence of eigenvalues $\lambda_0, \lambda_1, \lambda_2, \ldots$ and eigenfunctions $v_0, v_1, v_2, \ldots$ given by the formulas

$$
\begin{aligned}
\lambda_0 &= -g'(\bar{c}) - V^2 r'(\bar{c}) r(\bar{c})^{-2} \\
v_0(x) &= 1 \qquad (0 \le x \le 1) \\
\lambda_n &= -n^2\pi^2 - g'(\bar{c}) + V^2 r'(\bar{c}) r(\bar{c})^{-2} \\
v_n(x) &= \cos n\pi x \qquad (0 \le x \le 1) \qquad (n = 1,2,\ldots).
\end{aligned}
\tag{6.2}
$$

From (6.2) and (H_2), (H_3) in Section 1 we obtain $\lambda_0 < 0$ and $\lambda_1 > \lambda_2 > \lambda_3 > \cdots$. Therefore, to determine the stability properties of $\bar{\psi}$, we ought now to investigate the sign of λ_1.

Recalling Proposition 5.2 and Eqs. (5.6), we see that $\bar{c}$ in (6.2) can be regarded as a function $\bar{c}(V)$ with V varying on $(0,+\infty)$. Indeed, this function is implicitly determined by the relation

$$
g(\bar{c}) r(\bar{c}) = V^2 \qquad (a < \bar{c} < +\infty),\ 0 < V < +\infty), \tag{6.3}
$$

where a is as in (H_2) of Section 1. Using (6.3) one can show that $\bar{c}(V)$ is C^4-smooth with respect to V on $(0,+\infty)$, that $\bar{c}'(V) > 0$ on $(0,+\infty)$, and that $\bar{c}(V)$ maps $(0,+\infty)$ onto $(a,+\infty)$ in a one-to-one fashion.

Thus, we can now regard the eigenvalue λ_1 in (6.2) as a C^4-smooth function $\lambda_1(V)$ depending on V in $(0,+\infty)$. We ought now to ask, how does the sign of $\lambda_1(V)$ vary with V on $(0,+\infty)$?

To begin, with V_* as in (5.7), one can show that $\lambda_1(V) < 0$ for every $V \in (0,V_*)$. Indeed, the reasoning here is an application of straightforward estimates to our formula for λ_1 in

(6.2). From this and from Proposition 6.1 it follows that $\bar{\psi}$ is asymptotically stable for each $V \in (0,V_*)$. This last statement certainly agrees with Theorem 5.3.

Next, we look for a number $V_0 \in [V_*,+\infty)$ such that $\lambda_1(V_0)=0$ and $\lambda_1'(V_0) > 0$. From (6.2) and (6.3) it follows that such a number V_0 exists if and only if there is a number $\bar{c}_0 > a$ for which

$$g(\bar{c}_0)r'(\bar{c}_0)r(\bar{c}_0)^{-1} = \pi^2 + g'(\bar{c}_0) \tag{6.4}$$

$$g(\bar{c}_0)r''(\bar{c}_0)r(\bar{c}_0)^{-1} > \pi^2 r'(\bar{c}_0)r(\bar{c}_0)^{-1} + g''(\bar{c}_0). \tag{6.5}$$

Indeed, under (6.4) and (6.5), V_0 is related to $\bar{c}_0$ through the equation

$$g(\bar{c}_0)r(\bar{c}_0) = V_0^2. \tag{6.6}$$

Henceforth, we shall assume that there do indeed exist numbers $\bar{c}_0 \in (a,+\infty)$ and $V_0 \in (0,+\infty)$ satisfying (6.4), (6.5), and (6.6). Under this assumption we have $\lambda_1(V_0) = 0$ and $\lambda_1'(V_0) > 0$. Without loss of generality we also can assume that $\lambda_1(V) < 0$ for each $V \in (0,V_0)$. Corresponding to $\bar{c}_0$ and V_0 we have the homogeneous equilibrium $\bar{\psi}_0$ for (5.1) given by

$$\bar{\psi}_0(x) = \bar{c}_0 \qquad (0 \leq x \leq 1). \tag{6.7}$$

The significance of the preceding assumptions is evident. As V increases through the value V_0, the equilibrium $\bar{\psi}$, depending on V, ceases to be asymptotically stable and becomes unstable. Indeed the transition takes place at the instant $V = V_0$ and $\bar{\psi} = \bar{\psi}_0$. We can expect that this transition is accompanied by a stationary bifurcation, i.e., a bifurcation of one or more equilibria $\tilde{\psi}$ for (5.1) from the given equilibrium $\bar{\psi}_0$.

Our goal now is to investigate this bifurcation. We must establish that it does indeed take place and we want to determine some of its properties. To do this we shall treat our problem in the framework developed by Crandall and Rabinowitz in [12,13].

Let X and $A: D(A) \subset X \to X$ be as introduced in Section 2.

We set $X_1 = D(A)$ and on X_1 we impose the graph norm $|||\ |||$ given by $|||\phi||| = \|\phi\| + \|A\phi\|$, $\phi \in X_1$. Recalling $v_1 = v_1(x) \equiv \cos \pi x$ from (6.2) we define a projection Q on X by setting

$$(Q\phi)(x) = 2v_1(x)\int_0^1 \phi(\xi)v_1(\xi)d\xi \qquad (\phi \in X,\ x \in [0,1]). \tag{6.8}$$

Next, we let

$$\begin{aligned} Y &= \text{range of } Q \\ Z &= \text{null space of } Q \\ Z_1 &= X_1 \cap Z. \end{aligned} \tag{6.9}$$

We note that Y is a 1-dimensional subspace of X_1 and that Z has co-dimension 1 in X.

On a sufficiently small open neighborhood B of the origin in $R \times X_1$ we can define a function $G\colon B \to X$ by setting

$$G(\sigma,w) = \psi'' - g(\psi) + g(\bar{c})r(\bar{c})r(\psi)\rho[\psi]^{-2} \tag{6.10}$$

with

$$\psi \equiv \bar{c} + w,\ \bar{c} \equiv \bar{c}_0 + \sigma \qquad (\sigma,w) \in B.$$

Here, $\bar{c}_0$ is as in (6.4) and (6.5). By (H_1) the function G is C^3-smooth on B, and clearly $G(\sigma,0) = 0$ for $|\sigma|$ sufficiently small. Also, the problem of finding equilibria ψ of (5.1) with ψ near $\bar{\psi}_0$ and V nea V_0 is equivalent to the problem of finding elements (σ,w) near $(0,0)$ such that $G(\sigma,w) = 0$.

We want to calculate the Fréchet derivative $G_w(0,0)$ of G with respect to w at the origin $(0,0)$ in $R \times X_1$. This derivative is a linear transformation from X_1 into X whose value at a given $h \in X_1$ we can denote by $G_w(0,0)(h)$. The calculation is routine and the result is

$$\begin{aligned} G_w(0,0)(h) &= h'' - g'(\bar{c}_0)h + g(\bar{c}_0)r'(\bar{c}_0)r(\bar{c}_0)^{-1}h \\ &\quad - 2g(\bar{c}_0)r'(\bar{c}_0)r(\bar{c}_0)^{-1}P[h] \qquad (h \in X_1), \end{aligned} \tag{6.11}$$

where

$$P[h] \equiv \int_0^1 h(\xi)\,d\xi \qquad (h \in X_1).$$

Similarly, by direct calculation,

$$G_{w\sigma}(0,0)(v_1) = \{-g''(\bar{c}_0) - \pi^2 r'(\bar{c}_0) r(\bar{c}_0)^{-1} +$$

$$+ g(\bar{c}_0) r''(\bar{c}_0) r(\bar{c}_0)^{-1}\} v_1.$$

By (6.5) we can rewrite this last equation as

$$G_{w\sigma}(0,0)(v_1) = \beta v_1 \qquad \beta > 0. \tag{6.12}$$

By (6.11) the null space and range of $G_w(0,0)$ are the spaces Y and Z respectively defined in (6.9). By (6.12) we have $G_{w\sigma}(0,0)(v_1) \notin Z$. Also, we recall, Y has dimension 1 and Z codimension 1 in X. Hence, as a direct application of a theorem by Crandall and Rabinowitz [12, p. 325, Theorem 1.7] we obtain the following proposition.

Proposition 6.2. Under the assumptions (6.4) and (6.5), let $G: B \subset X \times X_1 \to X$ be as in (6.10). Also, let $Z_1 \subset X_1$ and v_1 be as in (6.9) and (6.2) respectively. Then, there exist an open neighoborhood B_1 of the origin in B, a pair of numbers η_0, $\sigma_0 > 0$, and a pair of functions $\tilde{\sigma}$, $\tilde{z}$ continuously mapping $(-\eta_0, \eta_0)$ into R, Z_1 respectively such that

$$\begin{aligned} B_1 \cap G^{-1}(0) &= \{(\tilde{\sigma}(\eta), \eta v_1 + \eta\tilde{z}(\eta)) : -\eta_0 < \eta < \eta_0\} \\ &\cup \{(\sigma, 0) : -\sigma_0 < \sigma < \sigma_0\}. \end{aligned} \tag{6.13}$$

Moreover, $\tilde{\sigma}(0) = 0$ and $\tilde{z}(0) = 0$.

Thus, we have the required stationary bifurcation. The equilibria $\tilde{\psi}$ obtained in this bifircation are given by the formula

$$\tilde{\psi} = \tilde{\psi}(\eta) = \bar{c}_0 + \tilde{\sigma}(\eta) + \eta v_1 + \eta\tilde{z}(\eta) \quad (-\eta_0 < \eta < \eta_0). \tag{6.14}$$

For each $\eta \in (-\eta_0, \eta_0)$ the element $\tilde{\psi}(\eta)$ is an equilibrium for (5.1) with V^2 given by (6.3) and the formula $\bar{c} = \bar{c}_0 + \tilde{\sigma}(\eta)$.

An important aspect of (6.13) is that, for $|\eta|$ sufficiently small, $\tilde{\psi}(\eta)$ is nonhomogeneous. Thus, we have obtained nonhomogeneous equilibria for (5.1).

Next, we want to determine conditions under which the bifurcation in Proposition 6.1 is supercritical. In other words, we seek conditions quaranteeing that $\tilde{\sigma}$ is C^2-smooth on $(-\eta_0,\eta_0)$ and $\tilde{\sigma}'(0) = 0$, $\tilde{\sigma}''(0) > 0$.

With $\bar{c}_0$ as in (6.4), (6.5), (6.6) we have $g(\bar{c}_0) > 0$ $r''(\bar{c}_0) > 0$. From (H_2u, (H_3) in Section 1 we already have $g'(\bar{c}_0) > 0$, $g''(\bar{c}_0) \geq 0$, $r(\bar{c}_0) > 0$, $r'(\bar{c}_0) > 0$. Hence, we can define positive constants k_1, k_2, k_3, and k by setting

$$\begin{aligned} k_1 &= \tfrac{3}{4}|g'''(\bar{c}_0)| \\ k_2 &= 2g(\bar{c}_0)r'(\bar{c}_0)^2 r(\bar{c}_0)^{-2} \\ k_3 &= 2g(\bar{c}_0)r''(\bar{c}_0)r(\bar{c}_0)^{-1} + g''(\bar{c}_0) \\ k &= \tfrac{4}{3}r(\bar{c}_0)g(\bar{c}_0)^{-1}\{k_1 + k_2k_3 + 2k_3^2\}. \end{aligned} \tag{6.15}$$

Now we state the following proposition.

Proposition 6.3. Under the hypotheses of Proposition 6.2, let $\bar{c}_0$, $\tilde{\sigma}$, $\tilde{z}$, η_0, and k be as in (6.4), (6.5), (6.13), and (6.15). Then, $\tilde{\sigma}$, $\tilde{z}$ are both C^2-smooth on $(-\eta_0,\eta_0)$ and $\tilde{\sigma}'(0) = 0$. Moreover, if

$$r'''(\bar{c}_0) < -k, \tag{6.16}$$

then $\tilde{\sigma}''(0) > 0$.

The complete proof of Proposition 6.3 involves calculations too lengthy to be included here. Therefore, we shall only sketch the proof of that Proposition.

First, to abbreviate notation, we denote the numbers $g(\bar{c}_0)$, $r(\bar{c}_0)$, $g'(\bar{c}_0)$, $r'(\bar{c}_0)$, ... by $g, r, g', r', \ldots$.

Second, in connection with Proposition 6.2, we can invoke a theorem by Crandall and Rabinowitz [12, p. 328, Theorem 1.18] and conclude that $\tilde{\sigma}$, $\tilde{z}$ are each C^2-smooth on $(-\eta_0,\eta_0)$.

Now, from (6.13) we have the relation

$$G(\tilde{\sigma}(\eta),\ \eta v_1 + \eta\tilde{z}(\eta)) = 0 \qquad (-\eta_0 < \eta < \eta_0). \tag{6.17}$$

We differentiate both sides of (6.17) twice with respect to η and then set $\eta = 0$. This gives us an equation involving unknowns $\tilde{\sigma}'(0)$ and $\tilde{z}'(0)$. With that equation and with the aid of Q defined in (6.8) we obtain $\tilde{\sigma}'(0) = 0$ and

$$\tilde{z}'(0) = \alpha_0 + \alpha_2 \cos 2\pi x,$$

where

$$\alpha_0 = \left(-\frac{1}{4}\right)\{g'' + gr''r^{-1}\}\{\pi^2 + 2g'\}^{-1}$$

$$\alpha_2 = \left(\frac{1}{12\pi^2}\right)\{gr''r^{-1} - g''\}.$$

Next, we differentiate (6.17) three times with respect to η and then set $\eta = 0$. This gives us an equation involving the unknowns $\tilde{\sigma}''(0)$ and $\tilde{z}''(0)$. Applying Q in (6.8) to both sides of that equation and letting $\beta > 0$ be as in (6.12), we obtain

$$3\beta\sigma''(0) = \frac{3}{4}\{g''' - gr'''r^{-1}\} + 3gr''r'r^{-2} - 6\alpha_0\{gr''r^{-1} - g'' + \\ - 2gr'^2r^{-2}\} - 3\alpha_2\{2gr''r^{-1} - g''\}, \tag{6.18}$$

where α_0, α_2 are as above. With (6.18) and (6.15) one can establish that, if $r'''(\bar{c}_0) < -k$, then $\tilde{\sigma}''(0) > 0$. Thus, we have all the results required by Proposition 6.3.

Our next task is to determine the stability properties of the bifurcating equilibria $\tilde{\phi}(\eta)$ given by (6.14). The relevant statement is as follows.

<u>Proposition 6.4</u>. Under the hypotheses of Proposition 6.2 and 6.3, including particularly (6.16), the equilibria $\tilde{\phi}(\eta)$ given by (6.14) are asymptotically stable with respect to (5.1) provided that $|\eta|$ is positive and sufficiently small.

Proof: We consider $\tilde{\phi}(\eta)$, $-\eta_0 < \eta < \eta_0$, given by (6.14) and we note that $\tilde{\phi}(0) = \bar{\phi}_0$, where $\bar{\phi}_0$ is as in (6.7).

For each $\eta \in (-\eta_0, \eta_0)$ Eq. (6.1) defines a linear operator L_η on X which corresponds to the linearization of (5.1) about $\tilde{\phi}(\eta)$. In particular, for $\eta = 0$ we have the operator L_0 corresponding to the linearization of (5.1) about $\bar{\psi}_0$. One can easily verify that $L_\eta \to L_0$ as $\eta \to 0$, the convergence here taking place in the operator norm of bounded linear operators from X into itself. Proposition 6.1 suggests that we should investigate the eigenvalues of L_η.

At $\eta = 0$ we already know these eigenvalues. They are the numbers λ_0, λ_1, λ_2, ... appearing in (6.2) with $\bar{c} = \bar{c}_0$. By (6.4) we have $\lambda_1 = 0$. Hence, by our remarks following (6.2), we have $\lambda_0 < 0$ and $0 > \lambda_2 > \lambda_3 > \cdots$.

At this juncture we note that $G_w(0,0)$ in (6.11) is the restriction of L_0 to the space $X_1 \subset X$. Therefore, taking Proposition 6.2 in conjunction with results due to Crandall and Rabinowitz [13, p. 163, Lemma 1.3; p. 165, Theorem 1.16], we obtain the following assertion. There exist a number η_1, $0 < \eta_1 < \eta_0$, and an open neighborhood O_1 about the origin in the complex plane such that, for each $\eta \in (-\eta_1, \eta_1)$, the operator L_η has exactly one eigenvalue $\mu(\eta)$ in O_1, and the real part $\mathrm{Re}(\mu(\eta))$ of $\mu(\eta)$ has the same sign ad does $-\eta\sigma'(\eta)\lambda_1'(0)$. Here, $\lambda_1'(0)$ is the derivative at $V = 0$ of the function $\lambda_1(V)$ we have introduced in our remarks preceding (6.4), (6.5).

From (6.5) we have $\lambda_1'(0) > 0$. From Proposition 6.3 we have $\tilde{\sigma}''(0) > 0$. Hence, for $|\eta|$ sufficiently small but positive, we have $\mathrm{Re}(\mu(\eta)) < 0$. Thus, for $|\eta| > 0$ sufficiently small, we know that one of the eigenvalues of L_η lies in the left half complex plane. Now we must prove that the others have the same property.

Let Σ be the set of all complex numbers λ such that $\mathrm{Re}(\lambda) \geq 0$ and $\lambda \notin O_1$. With the aid of (6.1), (6.2), (6.4), and (6.6), one can show that the resolvent operator $R(\lambda; L_0)$ of L_0 is well defined at every $\lambda \in \Sigma$ and that $\|R(\lambda; L_0)\|$ is uniformly bounded with respect to λ to Σ. Here, $\| \ \|$ denotes the operator norm for bounded linear operators taking X into X.

From the preceding assertions, from the property $\|L_\eta - L_0\| \to 0$ as $\eta \to 0$ noted above, and from a theorem given by Kato [22, p. 214, Theorem 3.17], it follows that, for each η sufficiently small, Σ lies in the resolvent set of L_η. Hence, for such η, any eigenvelue of L_η not lying in O_1 must have negative real part.

Thus, we have established that, for $|\eta|$ sufficiently small and positive, every eigenvalue of L_η has negative real part. This together with Proposition 6.1, yields the conclusion required by Proposition 6.4. Our proof is complete.

This finishes our analysis of the bifurcation brought about by (6.4) and (6.5). Now, with regard to the ballast resistor, we want to comment on two of the assumptions governing that bifurcation, namely, (6.5) and (6.15).

We interpret each of these assumptions as a requirement concerning the local behavior of the electrical resistivity $r(u)$ near the temperature $u = \bar{c}_0$. Specifically, we interpret (6.5) as a requirement that $r''(\bar{c}_0)$ be positive and large relative to $g(\bar{c}_0)^{-1}$, $r(\bar{c}_0)$, $g''(\bar{c}_0)$, and $r'(\bar{c}_0)$. We interpret (6.16) as a requirement that $r'''(\bar{c}_0)$ be negative and that $|r'''(\bar{c}_0)|$ be large relative to $r(\bar{c}_0)\ g(\bar{c}_0)^{-1}$, $r'(\bar{c}_0)\ r(\bar{c}_0)^{-1}$, $g(\bar{c}_0)\ r''(\bar{c}_0)\ r(\bar{c}_0)^{-1}$, $g''(\bar{c}_0)$, and $|g'''(\bar{c}_0)|$.

The following theorem summarizes Proposition 6.2 - 6.4.

<u>Theorem 6.5</u>. Let $\bar{c}_0$, $\bar{\psi}_0$, V_0 be as in (6.4) - (6.7) and (6.16). Then there exist numbers δ, $\mu > 0$ such that, for each V satisfying $V_0 < V < V_0 + \delta$, Eqs. (5.1) have exactly three equilibria in the ball $B(\bar{\psi}_0,\mu)$. One of these is an unstable homogeneous equilibrium $\bar{\psi}$. The other two are asymptotically stable nonhomogeneous equilibria $\tilde{\psi}_1$, $\tilde{\psi}_2$. As $V \to V_0 +$ we have $\tilde{\psi}_1, \tilde{\psi}_2 \to \bar{\psi}_0$, the convergence here being relative to $\|\ \|$ on X.

Indeed, $\tilde{\psi}_1$ and $\tilde{\psi}_2$ are given by (6.14) with appropriate values for η, one positive and one negative.

This concludes our work in Section 6.

REFERENCES

1. N. Alikakos: <u>An Application of the Invariance Principle to Reaction Diffusion Equations</u>, J. Differential Equations <u>33</u>, 201-225, (1979).
2. J. M. Ball: <u>Stability Thoery of an Extensible Beam</u>, J. Differential Equations <u>14</u>, 399-418, (1973).
3. J. M. Ball, L. A. Peletier: <u>Stabilization of Concentration Profiles in Catalyst Particles</u>, J. Differential Equations <u>20</u>, 356-368, (1976).

4. D. Bedeaux: P. Mazur, R. A. Pasmanter, The Ballast Resistor; an Electro-Thermal Instability in a Conducting Wire I; The Nature of the Stationary States, Physica 86 A 355-382, (1977).

5. H. Busch: Über die Erwärmung von Drähten in verdünnten Gasen durch den elektrischen Strom, Ann. Physik 64, 401-450, (1921).

6. R. G. Casten, C. J. Holland: Instability Results for Reaction Diffusion Equations with Neumann Boundary Conditions, J. Differential Equations 27, 266-273, (1978).

7. N. Chafee, E. F. Infante: A Bifurcation Problem for a Nonlinear Partial Differential Equation of Parabolic Type, Applicable Anal. 4, 17-37, (1974).

8. N. Chafee: A Stability Analysis for a Semilinear Parabolic Partial Differential Equation, J. Differential Equations 15, 522-540, (1974).

9. N. Chafee: Asymptotic Behavior for Solutions of a One-Dimensional Parabolic Equation with Homogeneous Neumann Boundary Conditions, J. Differential Equations 18, 111-134, (1975).

10. N. Chafee: Behavior of Solutions Leaving the Neighborhood of a Saddle Point for a Nonlinear Evolution Equation, J. Math. Anal. Appl. 58, 312-325, (1977).

11. E. A. Coddington, N. Levinson: Theory of Ordinary Differential Equations, New York, McGraw-Hill, 1955.

12. M. G. Crandall, P. H. Rabinowitz: Bifurcation from Simple Eigenvalues, J. Funct. Anal. 8, 321-340, (1971).

13. M. G. Crandall, P. H. Rabinowitz: Bifurcation, Perturbation of Simple Eigenvalues and Linearized Stability, Arch. Rational Mech. Anal. 52, 161-180, (1973).

14. C. M. Dafermos: An Invariance Principle for Compact Processes, J. Differential Equations 9, 239-252, (1971).

15. C. M. Dafermos: Applications of the Invariance Principle for Compact Processes. I. Asymptotically Dynamical System, J. Differential Equations 9, 291-299, (1971).

16. C. M. Dafermos: Applications of the Invariance Principle for Compact Processes II. Asymptotic Behavior of Solutions of a Hyperbolic Conservation Law, J. Differential Equations 11, 416-424, (1972).

17. C. M. Dafermos, M. Slemrod: Asymptotic Bahavior of Non-Linear Contraction Semigroups, J. Functional Analysis 13, 97-106, (1973).
18. P. de Mottoni, G. Talenti, A. Tesei: Stability Results for a Class of Non-Linear Parabolic Equations, Ann. Mat. Pura Appl. (4) 115, 295-310, (1977).
19. R. P. Gifford, N. C. Page: Experiments with Ballasts of a Nernst Lamp, B. S. Thesis, MIT, Cambridge, Mass. 1902.
20. J. K. Hale: Dynamical Systems and Stability, J. Math. Anal. Appl. 26, 39-59, (1969).
21. D. Henry: Geometric Theory of Semilinear Parabolic Equations, unpublished lecture notes, University of Kentucky, 1975.
22. T. Kato: Perturbation Theory for Linear Operators, Berlin-Heidelberg, New York, Springer 1966.
23. H. Kielhöfer: On the Lyapunov-Stability of Stationary Solutions of Semilinear Parabolic Differential Equations, J. Differential Equations 22, 193-208, (1976).
24. H. Kielhöfer: Stability and Semilinear Evolution Equations in Hilbert Space, Arch. Rational Mech. Anal. 57, 150-165, (1974).
25. R. Landauer: The Ballast Resistor, Phys. Rev. A (3) 15, 2117-2119, (1977).
26. H. Matano: Convergence of Solutions of One-Dimensional Semilinear Parabolic Equations, J. Math. Kyoto Univ. 18-2 221-227, (1978).
27. P. Mazur and D. Bedeaux: An Electro-Thermal Instability in a Conducting Wire; Homogeneous and Inhomogeneous Stationary States for an Exactly Solvable Model, preprint.
28. D. S. Mitronoviĉ: Analytic Inequalities, Springer-Verlag, New York, 1970.
29. R. A. Pasmanter, D. Bedeaux and P. Mazur: The Ballast Resistor; an Electro-Thermal Instability in a Conducting Wire II; Fluctuations Around Homogeneous Stationary States, Physica 90 A, 151-163, (1978).
30. A. Pazy: A Class of Semi-Linear Equations of Evolution, Israel J. Math. 20, 23-36, (1975).

31. M. Slemrod: Asymptotic Behavior of a Class of Abstract Dynamical Systems, J. Differential Equations 7, 584-600, (1970).

32. T. I. Zelenyak: Stabilization of Solutions of Boundary Value Problems for a Second Order Parabolic Equation With One Space Variable, Differential Equations 4, 17-22, (1968).

EQUILIBRIA OF AN AGE-DEPENDENT POPULATION MODEL

Klaus Deimling

Fachbereich 17
Gesamthochschule Paderborn
Fed. Rep. Germany

Given a population consisting of n species, let $u^i(t,x)$ be the number of individuals of the i-th species at time $t \geq 0$ and age $x \geq 0$. The model is described by

$$u_t^i + u_x^i + d_i(x)u^i + f_i(x,u(t,\cdot))u^i = 0 \tag{1}$$

$$u^i(t,0) = \int_0^\infty b_i(x)u^i(t,x)\,dx \quad \text{and} \quad \text{for } i = 1,\ldots,n, \tag{2}$$

$$u^i(0,x) = u_o^i(x) \tag{3}$$

where $b_i \geq 0$ and $d_i \geq 0$ are birth and death rates in the absence of interaction, while $f_i \geq 0$ describes the death rate due to the interaction of the individuals of different age and species; see [4] for more biological background. We are looking for nontrivial equilibria, i.e. t-independent solutions $\not\equiv 0$ of (1), (2), reporting recent results of J. Prüß (Paderborn) [3]. We consider n = 1 only, for simplicity i.e. we consider (1)-(3) without index i. The main result on equilibria will be a special case of the following theorem on existence of zeros of an equation $Av + Fv = 0$ in a Banach space.

<u>Theorem 1</u> ([3]). Let X be a B-space, $D \subset X$ closed bounded convex, $A : D_A \subset X \to X$ the generator of a C_o-semigroup of

ISBN 0-12-508780-2

linear operators $U(t)$ and $F : D \to X$ continuous and locally Lipschitz such that

$$\beta(F(B)) \leq k_1\beta(B) \text{ for } B \subset D,\ \beta(U(t)) \leq e^{k_2 t}$$
$$\text{for } t \geq 0,\ k_1 + k_2 < 0. \tag{4}$$

$$\lim_{h\to 0^+} h^{-1}\rho(U(h)v + hF(v),D) = 0 \text{ for all } v \in \partial D. \tag{5}$$

Then $Av + Fv = 0$ for some $v \in D \cap D_A$ [Here $\beta(B) = \inf\{r > 0 :$ B can be covered by finitely many balls $B_r(v_i)\}$; $\rho(z,D) =$ = distance from z to D; $\beta(U(t)) := \beta(U(t)\bar{B}_1(0))$].

For the application, it is natural to consider $X = L^1(\mathbb{R}^+)$, a subset D of the cone $K = \{v \in X : v \geq 0 \text{ a.e.}\}$, $Av = -v' + d(x)v$ for $v \in D_A = \{v \in AC_{loc} : Av \in X,\ v(0) = \int_0^\infty b(x)v(x)dx\}$ and $(Fv)(x) = -f(x,v)v(x)$. The main result is

<u>Theorem 2</u> ([3]). <u>Conditions</u>: (i) $b \in L^\infty(\mathbb{R}^+) \cap L^1(\mathbb{R}^+)$, $d \in L^\infty_{loc}$, $b(x) \geq 0$ a.e., $d(x) \geq 0$ a.e. and $\lim_{x\to\infty} d(x) = d_\infty > 0$.

(ii) $\lambda_0 > 0$, where λ_0 is the unique real eigenvalue of A, determined by $\int_0^\infty b(x)\exp(-\lambda x - \int_0^x d(s)ds)dx = 1$.

(iii) $f(x,v) = g(x, \int_0^\infty k(x,y)h(y,v(y))dy) \geq 0$ for $v \in K$; $h(x,0) = g(x,0) \equiv 0$, h Lipschitz; k uniformly continuous and bounded; g locally Lipschitz and $|g(x,\xi)| \leq M(|\xi|)$ with M continuous increasing and $M(0) = 0$.

(iv) There exist $x_0 > 0$ and $R > 0$: $\mu = \text{ess inf}\{d(x) - b(x): x \geq x_0\} > 0$ and $b(x) \leq d(x) + \inf\{f(x,v): v \in K \text{ and } \int_0^{x_0} v(x)dx \geq R\}$ for all $x \leq x_0$.

<u>Claim</u>: $v' + d(x)v + f(x,v)v = 0$, $v(0) = \int_0^\infty b(x)v(x)dx$ has a nontrivial solution in K.

If $\lambda_0 < 0$ then $v = 0$ is the only solution, also in case $\lambda_0 > 0$ if $\sup\{f(x,v): x \in \mathbb{R}^+, v \in K\} < \lambda_0$.

The <u>proof of Theorem 2</u> requires first of all the investigation of the spectrum of A, to see that A generates a semigroup $U(t)$ according to the Hille-Yosida Theorem. Then one needs an appropriate representation of $U(t)$ to see that $\beta(U(t)) \leq e^{-d_0 t}$ on $\mathbb{R}^+$, where $d_0 = \text{ess inf}\{d(x) : x \in \mathbb{R}^+\}$. Next, we observe that $Av + Fv = 0$ is equivalent to $(A - \mu I)v + F_\mu v = 0$ with

$F_\mu = F + \mu I$, that $A - \mu I$ generates $V(t) = U(t)e^{-\mu t}$ satisfying $\beta(V(t)) \le e^{-(\mu + d_o)t}$ and that $\beta(F_\mu B) \le \mu\beta(B)$ for $B \subset D$, if D is the bounded set to be chosen later on and $\mu > \sup\{f(x,v) : x \in \mathbb{R}^+, v \in D\}$. Hence, $V(t)$ and F_μ satisfy (i) of Theorem 1 if $d_o < 0$. Now, the essential problem is to find a closed bounded convex $D \subset K$ such that $0 \notin D$ and the boundary conditition (5) holds. The construction of D is based on the following Lemma which is interesting in itself.

Lemma ([3]). Let X be a Banach space; $D \subset X$ closed convex; $A : D_A \subset X \to X$ a closed densely defined linear operator generating a C_o-semigroup of linear operators $U(t) : X \to X$; $F : D \to X$ continuous and $\beta(FB) \le k\beta(B)$ for some $k \ge 0$ and all bounded $B \subset D$. Then

(i) If $(I - \lambda A)^{-1} : D \to D$ for all small $\lambda > 0$ then $U(t)D \subset D$ on $\mathbb{R}^+$

(ii) If $U(t)D \subset D$ on $\mathbb{R}^+$ and $\lim_{h \to 0^+} h^{-1}\rho(x + hFx, D) = 0$ on D then (5) holds.

(iii) If (ii) holds and $D \cap D_A$ is dense in D then "$(Ax + Fx, x)_- \le 0$ for $x \in D \cap D_A$ and $|x| \ge r$" implies (5) for $D \cap \bar{B}_r(0)$. Here $(\cdot,\cdot)_-$ is the semi-inner-product defined by $(x,y)_- = \min\{y^*(x) : y^* \in X^*, y^*(y) = |y|^2 = |y^*|^2\}$.

(iv) Suppose that (5) holds and consider the half space $H = \{x \in X : x^*(x) \ge \rho\}$ for some $x^* \in D_{A^*}$ such that $A^*x^* \in D_{A^*}$. If $(x, A^*x^*) + x^*(Fx) \ge 0$ on $D \cap \partial H$ then (5) holds for $D \cap H$ too.

Applied to our example, (i) and (ii) give (5) for $D = K$; then (iii) and Theorem 2 (iv) yield (5) for $K_r = K \cap \bar{B}_r(0)$ with large r. Now one can check directly that (5) holds for

$$D_\varphi = \{v \in K_r : \int_0^\infty v(x)\varphi(x)dx \le \sigma\},$$

$$\sigma \text{ large}, \quad \varphi(x) = \exp\left(\int_0^x d(\tau)d\tau - \frac{d_\infty}{2} x\right).$$

Finally, (iv) and the condition $\lambda_o > 0$ give (5) for

$$D = \{v \in D_\varphi : \int_0^\infty v(x)\psi(x)dx \ge \rho\} \text{ for some small } \rho > 0,$$

where ψ is an eigenfunction of the problem adjoint to $Av = \lambda_0 v$, i.e.

$$\psi' = -b(x) + d(x)\psi + \lambda_0\psi, \quad \psi(0) = 1,$$

and as a last step one can weaken $d_0 > 0$ to $d_\infty > 0$, by chosing an appropriate equivalent norm.

REFERENCES

1. G. Di Blasio: Nonlinear Age-Dependent Population Diffusion, J. Math. Biol. 8, 265-284 (1979).
2. M.E. Gurtin, R.C. Mac Camy: Nonlinear Age-Dependent Population Dynamics, Arch. Rat. Mech. Anal. 54, 281-300 (1974).
3. J. Prüß : Equilibrium Solutions of Age-Specific Population Dynamics of Several Species J. Math. Biol., to appear.
4. M. Rotenberg: Equilibrium and Stability in Populations Whose Interactions Are Age-Specific, J. Theor. Biol. 54, 207-224 (1975).

A VARIATION-OF-CONSTANTS FORMULA FOR NONLINEAR VOLTERRA INTEGRAL EQUATIONS OF CONVOLUTION TYPE

Odo Diekmann
Stephan van Gils

Mathematisch Centrum
Amsterdam, The Netherlands

1. INTRODUCTION

There are (at least) two different ways to associate with Volterra integral equations of convolution type a semigroup of operators:

(i) Write the equation in its translation invariant form and prescribe an initial function on an interval of the right length. The semigroup acts on the space of initial functions and it is defined by translation along the solution.

(ii) Consider a space of forcing functions as the state space and define the semigroup by the formula which shows how the equation transforms under translation.

In the linear case, with an appropriate choice of the spaces, one construction is modulo transposition of the matrix-valued kernel the adjoint of the other [2]. In the process of building a qualitative theory this observation, which applies to other delay equations as well [1,3], can be succesfully exploited in the proof of Fredholm alternatives and in the construction of projection operators.

In this note we shall derive an important tool for a geometric theory within the framework of the second construction.

ISBN 0-12-508780-2

It will appear that this somewhat unusual approach has certain advantages. For instance, if

$$\begin{cases} x(t) = \int_0^b B(\tau)g(x(t-\tau))d\tau, & t > 0, \\ x(t) = \phi(t) & -b \le t \le 0, \end{cases}$$

then x is discontinuous in t = 0 unless $\phi \in M$ where by definition

$$M = \{\phi \mid \phi(0) = \int_0^b B(\tau)g(\phi(-\tau))d\tau\}.$$

Of course one can restrict ones attention to the manifold M, but, particularly in perturbation problems where B and g, and hence M as well, may depend on parameters, this leads to technical (though not insuperable) difficulties [4, section 12.3;5]. Such difficulties are less prominent in the theory we are going to sketch.

A first we shall deal with equations where the nonlinearity occurs in the integrand. But in section 5 we shall, by means of an example, indicate how the theory can be extended to equations which contain a nonlinear function of integrals.

2. DEFINITION OF THE SEMIGROUP

In the following B denotes a given $n\times n$-matrix valued function defined and integrable on $\mathbb{R}_+ = [0,\infty)$. We assume that the support of B is contained in the interval [0,b], where b is some positive number. Let $g: \mathbb{R}^n \to \mathbb{R}^n$ be a given uniformly Lipschitz continuous function. We are interested in the equation

$$x = B*g(x) + f \tag{2.1}$$

where, as usual,

$$(B*g(x))(t) = \int_0^t B(\tau)g(x(t-\tau)d\tau.$$

For reasons which are explained in detail in [2] we put rather severe restrictions on the forcing function f. More precisely, we take $f \in X$ where

$$X = \{f \in C(\mathbb{R}_+)\,|\,f(t) = 0 \quad \text{for} \quad t \geq b\}.$$

We provide X with the supremum norm topology.

Let $f \in X$ be arbitrary. Equation (2.1) has a unique continuous solution x defined on $\mathbb{R}_+$. We define, for $s \geq 0$, $S(s)f$ by the relation

$$x_s = B*g(x_s) + S(s)f, \tag{2.2}$$

where $x_s(t) = x(s + t)$. Using the identity

$$(B*g(x_s))(t) = (B*g(x))(t + s) - (B_t*g(x))(s)$$

and (2.1) we obtain

$$(S(s)f)(t) = f(t + s) + (B_t*g(x))(s). \tag{2.3}$$

From the fact that translation is continuous in the L_1-topology we infer that $(B_t*g(x))(s)$ is continuous as a function of t. Moreover, $(B_t*g(x))(s) = 0$ for $t \geq b$. Hence $S(s)$ is a mapping of X into itself. Since $x(t)$ depends continuously on f, uniformly on compact t-intervals, $S(s)$ is continuous.

Theorem 2.1. The mapping $s \to S(s)$ defines a strongly continuous semigroup of continuous (nonlinear) operators on X.

Proof: From (2.2) we deduce that

$$(x_\sigma)_s = B*g(x_\sigma)_s) + S(s)S(\sigma)f,$$

and

$$x_{s+\sigma} = B*g(x_{s+\sigma}) + S(s+\sigma)f.$$

Since $(x_\sigma)_s = x_{s+\sigma}$ this implies that

$$S(s)S(\sigma) = S(s+\sigma).$$

(Note that we use implicitly the uniqueness of the solution of (2.1).) Clearly $S(0) = I$. Finally,

$$(S(s)f)(t) - (S(\sigma)f)(t) = x_s(t) - x_\sigma(t) + (B*(g(x_s) - g(x_\sigma)))(t) \to 0$$

as $s-\sigma \to 0$ uniformly for $t \in [0,b]$. □

3. THE LINEAR CASE

In the special case that $g(x) = x$ the semigroup constructed above consists of linear operators and will be called $T(s)$. Let R denote the resolvent of B, i.e. the unique (matrix -valued) solution of the equation (see [6])

$$R = B*R - B, \tag{3.1}$$

(for later use we note that $B*R = R*B$). The solution of

$$x = B*x + f \tag{3.2}$$

is given explicitly as

$$x = f - R*f. \tag{3.3}$$

Substitution of this expression into (2.3) yields an explicit representation of $T(s)$:

$$(T(s)f)(t) = f(t + s) + (B_t - B_t*R)*f(s). \tag{3.4}$$

The formula (3.4) extends the action of $T(s)$ to integrable functions and hence also to the columns of B. The next result will turn out to be useful.

Lemma 3.1. $(T(s)B)(t) = B_t(s) - B_t*R(s)$.

Proof: By (3.4) and (3.1) we can write

$$(T(s)B)(t) = B_t(s) + (B_t - B_t*R)*B(s) = B_t(s) + B_t*B(s) - B_t*(B + R)(s) = B_t(s) - B_t*R(s). \quad \square$$

One can show that the infinitesimal generator A of T(s) is given by

$$(Af)(t) = f'(t) + B(t)f(0)$$

with

$$\mathcal{D}(A) = \{f \in X \mid f \text{ absolutely continuous and } f'(.) + B(.)f(0) \text{ continuous}\}.$$

Moreover,

$$\sigma(A) = P\sigma(A) = \{\lambda \mid \det[I - \int_0^b e^{-\lambda\tau}B(\tau)d\tau] = 0\},$$

and one can decompose the space X according to the spectrum of A. We refer to [2] for a detailed account of these matters.

4. THE VARIATION-OF-CONSTANTS FORMULA

Suppose now that $g(x) = x + r(x)$. Let for a given $f \in X$ the functions x and y be the solutions of, respectively,

$$x = B*g(x) + f = B*x + B*r(x) + f, \tag{4.1}$$

$$y = B*y + f. \tag{4.2}$$

Lemma 4.1. (Miller [6])

$$x - y = -R*r(x).$$

Proof: Subtracting the equations we obtain

$$x - y = B*(x - y) + B*r(x).$$

Hence, by (3.1),

$$R*(x - y) = R*(x - y) + B*(x - y) + R*r(x) + B*r(x)$$

and so

$$x - y = -R*r(x) - B*r(x) + B*r(x) = -R*r(x). \qquad \square$$

From (2.3) and the corresponding formula for T(s) we deduce, using Lemmas 3.1 and 4.1, that

$$\begin{aligned}(S(s)f)(t) &= (T(s)f)(t) + (B_t*(x + r(x) - y))(s) \\ &= (T(s)f)(t) + ((B_t - B_t*R)*r(x))(s) \\ &= (T(s)f)(t) + \int_0^s (T(s - \tau)B)(t)r(x(\tau))d\tau.\end{aligned}$$

If we define $F: \mathbb{R}_+ \to X$ by $F(s) = S(s)f$ and $\alpha: X \to \mathbb{R}$ by $\alpha(f) = f(0)$ we can rewrite this identity as

$$F(s) = T(s)\ F(0) + \int_0^s (T(s - \tau)B)r(\alpha(F(\tau)))d\tau \tag{4.3}$$

(indeed, note that, by (2.2), $x(s) = \alpha(S(s)f)$). Our main result formulates the "equivalence" between (4.1) and (4.3).

Theorem 4.2.

(i) Let x be the solution of (4.1). Then $F: \mathbb{R}_+ \to X$ defined by $F(s) = x_s - B*g(x_s)$ satisfies (4.3).

(ii) Conversely, let F satisfy (4.3). Then x defined by $x(s) = \alpha(F(s))$ satisfies (4.1) with $f = F(0)$.

Proof: (i) has been proved above, so we concentrate on (ii). Putting $F(0) = f$, $x(s) = \alpha(F(s))$ and applying α to (4.3) we obtain, using Lemma 3.1, (3.4) and (3.1),

$$x = f + (B - B*R)*(f - r(x)) = f - R*f - R*r(x).$$

Hence

$$B^*x = B^*f - B^*f - R^*f - B^*r(x) - R^*r(x)$$
$$= x - f - B^*r(x). \qquad \square$$

Remarks.

(i) For obvious reasons we call (4.3) the variation-of-constants formula.

(ii) If $r(x) = o(x)$, $x \to 0$, then $T(s)$ is the Fréchet derivative of $S(s)$ in $f = 0$.

(iii) Formal differentiation of (4.3) yields the autonomous ordinary differential equation

$$\frac{dF}{ds} = AF + Br(\alpha F) = F' + Bg(\alpha F) \tag{4.4}$$

in the Banach space X. So we have demonstrated the correspondence between solutions of (4.1) and mild solutions of (4.4).

5. A SPECIAL EQUATION

The equation

$$x(t) = \gamma(1 - \int_0^1 x(t - \tau)d\tau) \int_0^1 a(\tau)x(t - \tau)d\tau, \tag{5.1}$$

arises from a model of the spread of a contagious disease, which supplies only temporary immunity, in a closed population. The positive parameter γ is proportional to the population size. The nonnegative kernek $a(\tau)$ describes the infectivity as a function of the time τ elapsed sincc exposure. This infectivity vanishes for $\tau > 1$. Moreover, an infected individual becomes susceptible again after exactly one unit of time. Finally, $x(t)$ is the frequency of those infected at time t. If we define

$$b^1(\tau) = \begin{cases} 1 & \text{if } 0 \leq \tau \leq 1, \\ 0 & \text{otherwise}, \end{cases}$$
$$b^2(\tau) = \gamma a(\tau), \tag{5.2}$$

and if we prescribe x on the interval $-1 \leq t \leq 0$, we can rewrite (5.1) as

$$x = (1 - b^1 * x - f^1)(b^2 * x + f^2), \tag{5.3}$$

where f^1 and f^2 incorporate the influence of the past (the prescribed initial function). We observe that the support of f^1 and f^2 is contained in [0,1]. Motivated by this fact we choose

$$X = \{(f^1, f^2) \mid f^i \in C(\mathbb{R}_+) \quad \text{and} \quad f^i(t) = 0 \quad \text{for} \quad t \geq 1,$$
$$i = 1,2\},$$

proveded with the topology induced by the norm

$$\|(f^1, f^2)\| = \sup_{0 \leq t \leq 1} (|f^1(t)| + |f^2(t)|),$$

as our state space.

Additional properties of f^1 and f^2 will guarantee that (5.3) has a globally defined solution. Here we shall not comment on those properties, but rather we simply assume that they are satisfied.
Let $f = (f^1, f^2)$. The semigroup S(s) is now defined by the formula

$$x_s = (1 - b^1 * x_s - (S(s)f)^1)(b^2 * x_s + (S(s)f)^2), \tag{5.4}$$

or, in other words,

$$(S(s)f)^i)(t) = f^i(t + s) + (b^i_t * x)(s), \quad i = 1,2. \tag{5.5}$$

Introducing $B = (b^1, b^2)$ we can rewrite (5.5) as

$$(S(s)(t) = f(t+s) + (B_t * x)(s). \tag{5.6}$$

The equation (5.1) has two constant solutions. Each of these yields a fixed point of $S(s)$ (for arbitrary s). Here we shall derive the variation of constants formula corresponding to the linearization about $f = 0$, but we remark that a similar formula exists for the other case.
The linearized equation is

$$y = b^2 * y + f^2 \tag{5.7}$$

and the linearized semigroup is

$$(T(s)f)(t) = f(t+s) + (B_t * y)(s), \tag{5.8}$$

(note that, essentially, there is no dependence on f^1 in the linearized problem). Consequently

$$S(s)f = T(s)f + \int_0^s B(. + s - \tau)(x(\tau) - y(\tau))d\tau. \tag{5.9}$$

The following observations are intended to rewrite this identity in a more useful form. We omit the proofs since they are very similar to those of the corresponding results in the foregoing sections.

(i) Let R denote the resolvent corresponding to b^2, i.e. the solution of

$$R = b^2 * R - b^2.$$

Define h by

$$x = b^2 * x + f^2 + h.$$

Then $x - y = h - R*h$ (see Lemma 4.1).

(ii) The definition of h implies

$$h = -(b^1 * x + f^1)(b^2 * x + f^2) = -(S(.)f)^1(0).(S(.)f)^2(0)$$
$$= r(\alpha(S(.)f)),$$

where $\alpha(f) := f(0)$ and $r: \mathbb{R}^2 \to \mathbb{R}$, $r(x_1,x_2) = -x_1x_2$.

(iii)

$$(T(s)B)(t) = B_t(s) - (B_t{*}R)(s).$$

Using (i) - (iii) and (5.9) we obtain the variation-of-constants formula

$$S(s)f = T(s)f + \int_0^s (T(s-\tau)B)r(\alpha(S(\tau)f))d\tau. \tag{5.10}$$

6. CONCLUDING REMARKS

In work in progress we use the variation-of-constants formula for the construction of (local) invariant manifolds (the stable and unstable manifolds of a saddle point as well as the center manifold in the case of critical stability). We intend to apply these results to concrete problems (special equations). In a prelude to Hopf bifurcation R. Montijn has recently obtained rather detailed information about a characteristic equation associated with (5.1). It appears that lots of roots may cross the imaginary axis with nonzero speed. Detailed results will be given in future publications.

REFERENCES

1. J.A. Burns, T.L. Herdman: Adjoint Semigroup for a Class of Functional Differential Equations, SIAM J. Math. Anal. 7 729-745 (1976).
2. O. Diekmann: Volterra Integral Equations and Semigroups of Operators, preprint, Math. Centrum Report TW 197/80, Amsterdam, (1980).
3. O. Diekmann: A Duality Principle for Delay Equations, in preparation.

4. J.K. Hale: *Theory of Functional Differential Equations*, Berlin, Springer, 1977.
5. J.K. Hale: *Behavior near Constant Solutions of Functional Differential Equations*, J. Diff. Equ. *15* 278-294 (1974).
6. R.K. Miller: *Nonlinear Volterra Integral Equations*, New York, Benjamin, 1971.

AN EXAMPLE OF BIFURCATION IN HYDROSTATICS

G. Fusco

Istituto di Matematica Applicata
Università di Roma

1. INTRODUCTION

Let us consider a mass of water contained in a cylindrical vessel with a flat bottom. When the ratio h between the water volume and the bottom area is greater than some critical value h_c, the water forms a layer of constant thickness h in the region away from the walls of the vessel so that the free-surface that separates the water from the environment is a plane. On the other hand, if h is smaller than h_c, the water does not form a layer of constant thickness but - due to cohesion forces - it aggregates in very complicated shapes and there is experimental evidence of the existence of a great (infinite) number of equilibrium configurations.

It seems reasonable to expect that among these infinite equilibria, in the ideal case when the influence of the walls can be neglected, some [equilibria] should exist whose free surface is invariant under some proper subgroup of the group G_0 of the plane rigid transformations. In this paper, we shall refer to an ideal case where the vessel botton can be schematized with a horizontal plane Π and we shall study the existence of equlibria whose free surface Σ is not a horizontal plane and is invariant under the subgroup $G \subset G_0$ generated by the following rigid transformations parallel to Π

ISBN 0-12-508780-2

$$\begin{cases} (x_1,x_2) \to (x_1 + \ell, x_2), & \text{translation by } \ell \text{ along } x_1 \\ (x_1,x_2) \to (x_1, x_2 + \ell), & \text{translation by } \ell \text{ along } x_2 \\ (x_1,x_2) \to (x_2,x_1) & \text{reflection with respect to the plane } x_2 = x_1 \\ (x_1,x_2) \to (-x_2,x_1) & \text{rotation by } \frac{\pi}{2} \text{ around the origin,} \end{cases} \tag{1}$$

where ℓ is a positive number and x_1, x_2 are coordinates parallel to Π. It is easily seen that invariance under G is equivalent to the fact that vertical planes $x_1 = n\ell$; $x_2 = m\ell$, $n, m = 0,1,-1,2,-2,\ldots$ (which divide plane Π into squares of side ℓ) divide Σ into elements which are all equal to one another, each one possessing the property of being symmetric with respect to the vertical planes through the symmetry axis of its square vertical projection onto the plane Π.

In the following we shall prove that under certain conditions on cohesion forces and if the average thickness h [1] of the liquid is less than a certain computable critical value h_c, such equilibria exist. To obtain this result we shall show that the problem of finding equilibria is equivalent to the problem of solving a two parameter (which are h, ℓ) family of equations in a suitable Banach space, and apply bifurcation theory, particularly the technique developed in [1].

It is assumed that cohesion forces internal to the liquid depend on a potential in the sense that if A, B are two elements of liquid of unit volume at a distance a from each other, the potential energy of A in the field generated by B is: $-\omega(a^2)$; with

$$\omega(a^2) = ke^{-\frac{a^2}{\nu^2}}, \qquad (k, \nu > 0). \tag{2}$$

An analogous assumption is made to describe cohesion forces between liquid and supporting plane Π with k replaced by k_0 and the same ν.

---

[1] By h we mean the ratio between the volume of the liquid above each square of side ℓ and the area ℓ^2 of the square.

2. THE ABSTRACT EQUATION DETERMINING EQUILIBRIA

Let Π' be the average free surface of the liquid i.e. the horinzontal plane at a distance h above Π, and let $0\,x_1\,x_2\,z$ be a reference frame with axes x_1, x_2 on Π' and axis z vertical and pointing upward. Let us consider a configuration of the liquid such that the free surface Σ can be described - with respect to $0\,x_1\,x_2\,z$ - by the equation $z = \varphi(x_1,x_2)$, through a function $\varphi: \mathbb{R}^2 \to \mathbb{R}$ which is invariant under (1); we shall assume that φ is continuous and satisfies the incompressibility condition

$$\int_{Q_\ell} \varphi = 0, \tag{3}$$

where Q_ℓ is the square $Q_\ell = \{(x_1,x_2)\,|\,|x_1| < \frac{\ell}{2},\ |x_2| \leq \frac{\ell}{2}\}$. Putting $x = (x_1,x_2)$, $|x| = (x_1^2 + x_2^2)^{\frac{1}{2}}$, the potential $\Omega^i_\varphi(x,z)$ of cohesion forces inside the liquid at point (x,z) is given by

$$\Omega^i_\varphi(x,z) = \int_{\mathbb{R}^2}\int_{-h}^{\varphi(\xi)} \omega(|\xi - x|^2 + (\zeta - z)^2)\,d\xi d\zeta \tag{4}$$

and it is a smooth function of (x,z). The potential $\Omega^e(z)$ due to the gravity and to the cohesion forces between liquid and supporting plane has - on the basis of the previous assumptions - the following expression

$$\Omega^e(z) = -\rho z + \pi k_o \nu^2 \int_{-\infty}^{-h} e^{-\frac{1}{\nu^2}(\zeta - z)^2}\,d\zeta, \tag{5}$$

where ρ is the weight of a unit volume of liquid and the thickness of the support is assumed to be infinite. Let $\Omega_\varphi(x,z)$ be the total potential at point (x,z) i.e. the sum of $\Omega^i_\varphi(x,z)$ and $\Omega^e(z)$ and $C_\varphi = \{(x,z)\,|\,-h \leq z \leq \varphi(x)\}$ the region occupied by the liquid, than we shall say that φ is an equilibrium configuration[2] if and only if there exists a smooth function P_φ : $C_\varphi \to \mathbb{R}^+$ (the pressure field) that satisfies the equation

---

[2]From now on we shall identify the configuration of the liquid with the function φ describing its free surface.

$$\overrightarrow{\text{grad}}(\Omega_\varphi(x,z) - P_\varphi(x,z)) = 0 \quad \text{on } C_\varphi \tag{6}$$

expressing hydrostatic equilibrium, together with the boundary condition

$$P_\varphi(x,\varphi(x)) = p, \tag{7}$$

expressing the coincidence of $P_\varphi(x,z)$ with the environmental pressure p ($p \geq 0$) on Σ.

It follows at once from (6) that, when it exists, P_φ is given by Ω_φ plus a constant c. Therefore (7) implies that a necessary condition for φ to be an equilibrium configuration is to satisfy

$$\Omega_\varphi(x,\varphi(x)) = p - c, \tag{8}$$

i.e. to make Σ a surface of constant potential.

On the other hand, if a configuration φ satisfies (8) for some c, the function $P_\varphi(x,z) \overset{\text{def}}{=} \Omega_\varphi(x,z) + c$ satisfies (6), (7); such a φ is thus an equilibrium configuration if one has

$$P_\varphi(x,z) = \Omega_\varphi(x,z) + c \geq 0 \quad \text{on } C_\varphi \tag{9}$$

It can be proved that (8) implies (9) for small values of $\|\varphi\| = \max|\varphi(x)|$. In fact, we shall prove the existence of a number $\varepsilon > 0$ such that any configuration φ with $\|\varphi\| < \varepsilon$ satisfying (8) satisfies also

$$P_\varphi(x,z) - p = \Omega_\varphi(x,z) - \Omega_\varphi(x,\varphi(x)) \geq 0 \quad \text{on } C_\varphi \tag{10}$$

which is a stronger condition than (9).

To this end, let us consider the following expression for $\Omega^i_\varphi(x,z)$

$$\Omega^i_\varphi(x,z) = \int_{\mathbb{R}^2}\int_{-h}^{\min\varphi} \omega(|\xi - x|^2 + (\zeta - z)^2)\,d\xi d\zeta + \int_{\mathbb{R}^2}\int_{\min\varphi}^{\varphi(\xi)} \omega(|\xi - x|^2 + (\zeta - z)^2)\,d\xi d\zeta \tag{11}$$

and let $J_1(z)$, $J_2(x,z)$ indicate the two integrals on the right hand side of (11). Since $\omega(a^2)$ is a decreasing function of a, $J_1(z)$ attains its maximum for $z = z_o \overset{def}{=} \frac{1}{2}(\min\varphi - h)$ and decreases when $|z - z_o|$ increases. Moreover it is easily seen that $\frac{\partial J_2}{\partial z}(x,z)$ is bounded on C_φ by a number $\beta(\|\varphi\|)$ which goes to zero together with $\|\varphi\|$. Therefore for any $z \in [-h,\varphi(x)]$ one has

$$\Omega_\varphi^i(x,z) = J_1(z) + J_2(x,z) \geq J_1(\varphi(x)) + J_2(x,\varphi(x)) -$$
$$\beta(\|\varphi\|)(\varphi(x) - z) \geq \Omega_\varphi^i(x,\varphi(x)) - \beta(\|\varphi\|)(\varphi(x) - z). \quad (12)$$

On the other hand, since from (5) it follows the existence of a numeber $\beta_o > 0$ such that $\frac{d\Omega^e}{dz}(z) < -\beta_o$, one can also write

$$\Omega^e(z) \geq \Omega^e(\varphi(x)) + \beta_o(\varphi(x) - z). \quad (13)$$

From (12), (13) it follows

$$\Omega_\varphi(x,z) - \Omega_\varphi(x,\varphi(x)) \geq (\beta_o - \beta(\|\varphi\|))(\varphi(x) - z) \text{ on } C_\varphi. (14)$$

Since $\beta(\|\varphi\|)$ goes to zero with $\|\varphi\|$ there exists a number $\varepsilon > 0$ such that $\|\varphi\| < \varepsilon$ implies that the right hand side of (14) is not negative on C_φ. Therefore one has the following theorem:

Theorem 1. There is a number $\varepsilon > 0$ such that any φ with $\|\varphi\| < \varepsilon$ is an equilibrium configuration iff the following condition is fulfilled

$$\Omega_\varphi(x,\varphi(x)) = \text{cost}. \quad (15)$$

Note that as should be expected $\varphi = 0$ is a solution of (15) for both $\Omega_\varphi^i(x,\varphi(x))$ and $\Omega^e(\varphi(x))$ reduce to constants when $\varphi = 0$. Let Φ be the set of continuous functions φ that are invariant under (1) and satisfy (3). To study the existence of non-zero equilibria $\varphi \in \Phi$ it is convenient to rewrite equation (15) as a two parameter (which are h and ℓ) family of equations in a suitable Banach space. To do that we write (15) taking into account the explicit expression (4) of Ω_φ^i and replace x by ℓx and φ by the function $\psi(x) = \varphi(\ell x)$; clearly ψ belongs to the vector space Ψ of continuous functions that are invariant under (1) with $\ell = 1$ and satisfy (3) with $\ell = 1$. When the above substitutions are made, one obtains

$$\ell^2 \int_{\mathbb{R}^2} \int_{-h}^{\psi(\xi)} \omega(\ell^2 |\xi - x|^2 + (\zeta - \psi(x))^2) d\xi d\zeta + \Omega^e(\psi(x)) = \text{cost}, \tag{16}$$

and finding solutions φ of (15) is the same as finding functions $\psi \in \Psi$ that satisfy (16) and the corresponding values of ℓ.

For fixed h, ℓ, ψ the left hand side of (16) can be viewed as the value $\hat{S}(h,\ell,\psi)(x)$ at point x of a function $\hat{S}(h,\ell,\psi)$: $\mathbb{R}^2 \to \mathbb{R}$ defined as

$$\hat{S}(h,\ell,\psi)(x) = \ell^2 \int_{\mathbb{R}^2} \int_{-h}^{\psi(\xi)} \omega(\ell^2 |\xi - x|^2 + (\zeta - \psi(x))^2) d\xi d\zeta + \Omega^e(\psi(x)). \tag{17}$$

The condition that the left hand side of (16) be a constant is the same as the condition that $\hat{S}(h,\ell,\psi)$ be coincident with its mean value $M\hat{S}(h,\ell,\psi)$. Therefore, if we let

$$S(h,\ell,\psi) \overset{\text{def}}{=} (I - M)\hat{S}(h,\ell,\psi), \tag{18}$$

we can write equation (16) in the abstract form

$$S(h,\ell,\psi) = 0. \tag{19}$$

It is a routine to check that $\hat{S}(h,\ell,\psi)$ belongs to the vector space $\hat{\Psi}$ of the continuous functions which are invariant under (1) with $\ell = 1$, thus the same is true for $S(h,\ell,\psi)$ and moreover $S(h,\ell,\psi)$ belongs to Ψ because its mean value is zero by definition. Therefore equation (19) can be viewed as a two parameter family of equations on Ψ. With the norm $\|\psi\| = \max|\psi(x)|$, Ψ is a Banach space and S: $(0,\infty) \times (0,\infty) \times \Psi \to \Psi$ becomes a smooth function of (h,ℓ,ψ); therefore we can apply differential calculus and bifurcation theory for studying the solutions of (19).

3. THE BIFURCATION EQUATION

For any $h, \ell > 0$, $\psi = 0$ is a solution of (19). In order to show the possibility of the existence of solutions $\psi \neq 0$, it will be shown that - under certain conditions on k_o, k, ν - there exist points $(\bar{h},\bar{\ell})$ in the parameter space $(0,\infty) \times (0,\infty)$ that are bifurcation points. A point $(\bar{h},\bar{\ell})$ is defined as being a bifurcation point of the solution set of (19) if given any neighborhood $\mathcal{U}$ of $(\bar{h},\bar{\ell})$ and any neighborhood $\mathcal{V} \subset \Psi$ of $\psi = 0$, a point $(h,\ell) \in \mathcal{U}$ and a non-zero $\psi \in \mathcal{V}$ that satisfy (19) exist. Due to the implicit function theorem, if the Fréchet derivative $D_\psi S(\bar{h},\bar{\ell},0): \Psi \to \Psi$ of S with respect to ψ is a linear homeomorphism at $(\bar{h},\bar{\ell})$, there exist neighborhood $\mathcal{U}$ of $(\bar{h},\bar{\ell})$ and $\mathcal{V}$ of $\psi = 0$ such that $\psi = 0$ is the unique solution of (19) in $\mathcal{V}$ for any $(h,\ell) \in \mathcal{U}$. Therefore, in order that (h,ℓ) be a bifurcation point, $D_\psi S(h,\ell,0)$ has to be singular. The following lemma summarizes the properties of the operator $D_\psi S(h,\ell,0)$ and gives a necessary and sufficient condition for it to be singular.

<u>Lemma 1</u>.

(i) The operator $D_\psi S(h,\ell,0)$ has the following expression

$$D_\psi S(h,\ell,0) = A(\ell) - \sigma(h)I,$$

with $\sigma(h) = \rho + \pi k\nu^2[1 - (1 - \frac{k_o}{k})e^{-\frac{h^2}{\nu^2}}]$ and with the operator $A(\ell)$ defined by

$$[A(\ell)\psi](x) = \ell^2\int_{\mathbb{R}^2} \omega(\ell^2|\xi - x|^2)\psi(\xi)d\xi.$$

(ii) The operator $D_\psi S(h,\ell,0)$ is singular iff $\sigma(h)$ is an eigenvalue of $A(\ell)$

(iii) $\lambda \in \mathbb{R}$ is an eigenvalue of $A(\ell)$ iff there exist integers n, m with $n + m \geq 1$ and such that

$$\lambda = \lambda_{n^2+m^2}(\ell) \overset{\text{def}}{=} \pi k \nu^2 e^{-\pi^2 \frac{\nu^2}{\ell^2}(n^2+m^2)} \quad ;$$

the eigenspace $W(\lambda)$ associated to λ is generated by the eigenvectors

$$w_{n,m}(x_1,x_2) = c_{n,m}(\cos 2\pi n x_1 \cos 2\pi m x_2 + \cos 2\pi m x_1 \cos 2\pi n x_2),$$

$$(c_{n,m} = \sqrt{2} \text{ for } n \neq m;\ c_{n,m} = 1 \text{ for } n = m)$$

with n, m such that $\lambda_{n^2+m^2}(\ell) = \lambda$

(iv) If $D_\psi S(h,\ell,0)$ is singular at (h,ℓ) and N, R are its kernel and its range, one has

a) $N = W(\sigma(h))$,

b) $\psi \in R$ iff $\int_{Q_1} \psi w = 0,\ \forall\ w \in N$.

<u>Proof</u>: If $\psi_1 \ldots \psi_i$ belong to Ψ and $s_1,\ldots,s_i$ are real variable, one has[3]

$$D^i_\psi \hat{S}(h,\ell,\psi)(\psi_1,\ldots,\psi_i) = \frac{\partial^i \hat{S}}{\partial s_1 \ldots \partial s_i}(h,\ell,\psi + \sum_1^i{}_j s_j \psi_j)\Big|_{s_i=\ldots=s_i=0} \tag{20}$$

By applying this relation with i = 1 to (17) and by using (2) and (5), one easily obtains

$$D_\psi \hat{S}(h,\ell,0)(\psi) = A(\ell)\psi - \sigma(h)\psi, \tag{21}$$

---

[3] $D^i_\psi \hat{S}(h,\ell,\psi)$ stands for the i-th order Fréchet derivative of $\hat{S}$ with respect to ψ at (h,ℓ,ψ), while $D^i_\psi \hat{S}(h,\ell,\psi)(\psi_1,\ldots,\psi_i)$ indicates the value at $(\psi_1,\ldots,\psi_i) \in \Psi^i$ of $D^i_\psi \hat{S}(h,\ell,\psi)$ that - as well known, see for example [3] pag. 179 - can be considered as an i-linear simmetric mapping of Ψ^i into Ψ. Equation (20) expresses the connection between Fréchet and Gâteaux derivatives.

Therefore, since $D_\psi S(h,\ell,0) = (I - M)D_\psi \hat{S}(h,\ell,0)$, in order to prove (i), it is only needed to prove that $MA(\ell)\psi$ is zero for any $\psi \in \Psi$. To do this and, at the same time, to prove (iii), let us note that a routine computation shows that $w_{n,m}$ is an eigenvector of $A(\ell)$ corresponding to the eigenvalue

$$\lambda = \ell^2 \int_{\mathbb{R}^2} \omega(\ell^2(\xi_1^2 + \xi_2^2))\cos 2\pi n\xi_1 \cos 2\pi m\xi_2 d\xi_1 d\xi_2 = \pi k \nu^2 e^{-\pi\frac{\nu^2}{\ell^2}(n^2+m^2)} \tag{22}$$

To prove that there is no other eigenvalue of $A(\ell)$ besides the $\lambda_{n^2+m^2}(\ell)$'s let $\mathcal{H} \supset \Psi$ be the Hilbert space of the measurable functions $\psi: \mathbb{R}^2 \to \mathbb{R}$ which are a.e. invariant under with $\ell = 1$, satisfy (3) with $\ell = 1$ and are such that $\int_{Q_1} |\psi|^2 < \infty$; the inner product being defined as

$$\langle \psi_1, \psi_2 \rangle = \int_{Q_1} \psi_1 \psi_2 . \tag{23}$$

$A(\ell)$ can be extended (with the same definition) to an operator $A^{\mathcal{H}}(\ell)$ defined an all $\mathcal{H}$. Since $A^{\mathcal{H}}(\ell)$ is selfadjoint, if $\lambda \neq \lambda_{n^2+m^2}(\ell)$, $n + m \geq 1$, is an eigenvalue of $A^{\mathcal{H}}(\ell)$ and w a corresponding eigenvector, we have: $\langle w, w_{n,m} \rangle = 0$, $n + m \geq 1$, in contraddiction with the fact that $\{w_{n,m}\}$ is an Hilbert basis in $\mathcal{H}$.

This observation prove (iii) and imply: $MA(\ell)\Psi = \{0\}$ because clearly one has: $MA^{\mathcal{H}}(\ell)w_{n,m} = \lambda_{n^2+m^2}(\ell)Mw_{n,m} = 0$, thus by continuity: $MA^{\mathcal{H}}(\ell)\mathcal{H} = \{0\}$.

Since $A(\ell)$ is a compact operator, (ii) follows from the spectral theorem for compact operators and the fact that $\sigma(h)$ does not vanish at any h. (iv)-a is obvious. To show (iv)-b, note that $A^{\mathcal{H}}(\ell)$ maps any $\psi \in \mathcal{H}$ into a function $A^{\mathcal{H}}(\ell)\psi$ which belongs to Ψ. This implies that $(A^{\mathcal{H}}(\ell) - \sigma(h)I)\psi$ is in Ψ if and only if ψ is in Ψ. Therefore R is the intersection of Ψ with $(A^{\mathcal{H}}(\ell) - \sigma(h)I)\mathcal{H}$ and (iv)-b holds because, being $A^{\mathcal{H}}(\ell)$ compact and self-adjoint, $(A^{\mathcal{H}}(\ell) - \sigma(h)I)\mathcal{H}$ is the orthogonal complement of N.

On the basis of Lemma 1, in order that (h,ℓ) be a bifurcation point, it has to be a solution of the equation

$$\sigma(h) = \lambda_{n^2+m^2}(\ell) \tag{24}$$

for some n, m ($n + m \geq 1$). From the expressions of $\sigma(h)$ and $\lambda_{n^2+m^2}(\ell)$ given in Lemma 1 it is immediately seen that (24) admits solutions iff k_o, k, ν satisfy the following condition:

$$\pi k \nu^2 \geq \pi k_o \nu^2 + \rho, \tag{25}$$

that, in mechanical terms, means that cohesion forces inside the liquid must be stronger than gravity forces and cohesion forses between the liquid and its support. Moreover one sees that, if (25) is fullfilled, all solutions of (24) are obtained by associating to any h less than the critical value

$$h_c = \nu[\ln\frac{\pi k \nu^2}{\rho}(1 - \frac{k_o}{k})]^{\frac{1}{2}}, \tag{26}$$

the value ℓ_h of ℓ given by

$$\ell_h = \pi\nu(n^2 + m^2)^{\frac{1}{2}}\{-\ln[1 - (1 - \frac{k_o}{k})e^{-\frac{h^2}{\nu^2}} + \frac{\rho}{\pi k \nu^2}]\}^{-\frac{1}{2}}. \tag{27}$$

From Lemma 1 it follows: dim $N < \infty$; $\Psi = N \oplus R$ moreover the projection P on N and I - P on R are continuous and, since R is a closed subspace of Ψ, the open mapping theorem implies the existence of a bounded linear operator B: $R \to R$ such that $D_\psi S(h,\ell,0)B = I$. These facts make $D_\psi S(h,\ell,0)$ a Fredholm operator and allows us to reduce the problem of finding solutions of (19) to a finite dimensional one, via the classical Liapunov-Schmidt procedure [1] [2]. By writing $\psi = u + y$ with $u = P\psi$: $y = (I - P)\psi$ in (19) and by projecting on N and on R one transforms (19) into the equivalent system:

$$PS(h,\ell, u + y) = 0, \tag{28}$$

$$(I - P)S(h,\ell, u + y) = 0, \tag{29}$$

whose left hand sides, due to continuity of P and I - P, have the same smoothness properties as S; note that $u = 0$, $y = 0$ is a solution of (28) and (29) because $\psi = 0$ is a solution of (19).

Since the Fréchet derivative of the left hand side of (28) with respect to y at $(h,\ell_h,0,0)$ is the same as the restriction $D_\psi S(h,\ell_h,0)|_R$ of $D_\psi S(h,\ell_h,0)$ to R, from the existence of B and the implicit function theorem, it follows the existence of neighborhoods $U \subset (0,\infty) \times (0,\infty) \times N$ of $(h,\ell_h,0)$ and $V \subset R$ of $y = 0$ together with a smooth function $\tilde{y}: U \to V$ such that $\tilde{y}(h,\ell,0) = 0$ and with the property that $\tilde{y}(h,\ell,u)$ is the only solution of (28) in V for any $(h,\ell,u) \in U$.

If one formally replaces y by $\tilde{y}(h,\ell,u)$ in (27), the problem of solving (27), (28) in $U \times V$ is reduced to the problem of solving the following bifurcation equation

$$F(h,\ell,u) \overset{\text{det}}{\equiv} PS(h,\ell,\tilde{y}(h,\ell,u)) = 0, \tag{30}$$

which is equivalent to a system of $r = \dim N$ scalar equation for r scalar unknowns. Note that $F(h,\ell,0) = 0$ because: $S(h,\ell,0) = 0$ and $\tilde{y}(h,\ell,0) = 0$; moreover $D_u F(h,\ell_h,0) = 0$ for obviously: $PD_\psi S(h,\ell_h,0) = 0$. Once that a solution u of (20) for some (h,ℓ) has been found, the corresponding solution ψ of (19) is given by the formula

$$\psi = u + \tilde{y}(h,\ell,u). \tag{31}$$

4. EXISTENCE OF NON-ZERO EQUILIBRIA

In what follows we shall consider solution (h,ℓ_h) of (24) corresponding to the case $n + m = 1$ alone (i.e. we only consider the first eigenvalue of $A(\ell)$) then we have $\dim N = 1$ and, if we write w for w_{1o}, the projection P on N is given by $P_\psi = \langle \psi,w\rangle w$. Therefore if we pose: $\alpha w = u$ and $f(h,\ell,\alpha) = \langle F(h,\ell,\alpha w),w\rangle$ equation (30) is equivalent to

$$f(h,\ell,\alpha) = 0, \tag{32}$$

and it results: $f(h,\ell,0) = 0$; $\frac{\partial f}{\partial \alpha}(h,\ell_h,0) = 0$.
Since $f(h,\ell,0) = 0$, equation (32) is equivalent to

$$\alpha g(h,\ell,\alpha) = 0 \tag{33}$$

where g is defined by

$$\begin{cases} g(h,\ell,0) = \dfrac{\partial f}{\partial \alpha}(h,\ell,0), \\ g(h,\ell,\alpha) = \dfrac{1}{\alpha} f(h,\ell,\alpha), \quad \text{for } \alpha \neq 0. \end{cases} \tag{34}$$

The existence of non-zero solutions of (32) is equivalent to the existence of non-zero solutions of

$$g(h,\ell,\alpha) = 0. \tag{35}$$

Since $\frac{\partial f}{\partial \alpha}(h,\ell_h,0) = 0$, from the definition of g, it follows that $g(h,\ell_h,0) = 0$.

The following Lemma summarizes some properties of partial derivatives of g used to study equation (35) in a neighborhood of $(h,\ell_h,0)$.

Lemma 2.

(i) $\frac{\partial g}{\partial \alpha}(h,\ell_h,0) = 0$.

(ii) There exists $h_o < h_c$ such that

$$\frac{\partial^2 g}{\partial \alpha^2}(h,\ell_h,0) > 0 \quad \text{for } h_o < h < h_c,$$

(iii) $\frac{\partial g}{\partial \ell}(h,\ell_h,0) > 0$.

Proof: The proof is a quite complicated application of differential calculus and is omitted for brevity.

We are now in the position of proving the following theorem:

Theorem 2. If k_o, k, ν satisfy (26), the point $(\bar{h},\ell_{\bar{h}})$ is a bifurcation point of the solution set of (19) for every $\bar{h}$ in the interval (h_o,h_c). More specifically there exist neighborhoods $\mathcal{U}$ of $(\bar{h},\ell_{\bar{h}})$ and $\mathcal{V}$ of $\psi = 0$ such that if $(h,\ell) \in \mathcal{U}$ then

(i) if $\ell \geq \ell_h$, $\psi = 0$ is the only solutions of (19) in $\mathcal{V}$

(ii) if $\ell < \ell_h$, there exist exactly three solutions of (19) in $\mathcal{V}$: the solution $\psi = 0$ and two solutions $\psi_1 \neq \psi_2 \neq 0$. To a first approximation ψ_1, ψ_2 are given by

$$\psi_{1,2} = \pm\left[2\frac{\frac{\partial g}{\partial \ell}(h,\ell_h,0)}{\frac{\partial^2 g}{\partial \alpha^2}(h,\ell_h,0)}(\ell - \ell_h)\right]^{\frac{1}{2}} w.$$

Proof: The proof is essentially the same as for theorem 8-2 in [1]. From Lemma 2 we have:

$$\begin{cases} \frac{\partial g}{\partial \alpha}(h,\ell_h,0) = 0, \\ \frac{\partial^2 g}{\partial \alpha^2}(h,\ell_h,0) > 0, \end{cases} \tag{36}$$

for every $h \in (h_o,h_c)$, thus the implicit function theorem implies the existence of a neighborhood U of $(\bar{h},\ell_{\bar{h}})$, of a number $\delta > 0$, and of a smooth function $\tilde{\alpha}$: $U \to (-\delta,\delta)$ which is uniquely defined and such that

$$\begin{aligned} &\tilde{\alpha}(h,\ell_h) = 0, \\ &\frac{\partial g}{\partial \alpha}(h,\ell,\tilde{\alpha}(h,\ell)) = 0, \quad (h,\ell) \in U. \end{aligned} \tag{37}$$

Therefore if U and δ are chosen small enough then the function $g(h,\ell,\cdot)$ possesses a unique minimum in $(-\delta,\delta)$ for every $(h,\ell) \in U$ and equation (35) has exactly two solutions in $(-\delta,\delta)$ if the minimum $g(h,\ell,\tilde{\alpha}(h,\ell)) \overset{\text{def}}{=} m(h,\ell)$ is < 0, one solution if $m(h,\ell) = 0$, no solution if $m(h,\ell) > 0$. Since we have $\tilde{\alpha}(h,\ell_h) = 0$ and $g(h,\ell_h,0) = 0$, the minimum $m(h,\ell)$ vanishes on the curve $\ell = \ell_h$, moreover from $\frac{\partial g}{\partial \alpha}(h,\ell_h,0) = 0$ and Lemma 2 (iii) it follows: $\frac{\partial m}{\partial \ell}(h,\ell_h) = \frac{\partial g}{\partial \alpha}(h,\ell_h,0) = 0$. Therefore $\ell \gtrless \ell_h$ is equivalent to $m(h,\ell) \lesseqgtr 0$. On the basis of this results the proof may be easily completed by taking into account the equivalence between equations (19), (32) and by observing also that, when taking lower order terms only, equation (35) becomes

$$g(h,\ell,\alpha) \cong \frac{\partial g}{\partial \ell}(h,\ell_h,0)(\ell - \ell_h) + \frac{1}{2}\frac{\partial^2 g}{\partial \alpha^2}(h,\ell_h,0)\alpha^2 = 0, \tag{38}$$

while to a first approximation equation (31) may be written as

$$\psi \cong \alpha w \tag{39}$$

because $\tilde{y}(h,\ell,0) = 0$ and $D_u\tilde{y}(h,\ell_h,0) = 0$.

5. CONCLUSION

In agreement with physical intuition, Theorem 2 states that a layer of liquid laying on a horizontal plane possesses - under certain conditions for k_o, k, ν, h - equilibrium configurations whose free surfaces are invariant under G and are not horizontal planes. On the basis of Theorem 2 these equilibria exists in pairs; we remark that this is a consequence of the fact that, as one sees immediately, if $\varphi \in \Phi$ is an equilibrium also $\varphi' \in \Phi$ defined by $\varphi'(x_1,x_2) = \varphi(x_1+\frac{\ell}{2},x_2+\frac{\ell}{2})$ is an equilibrium.

REFERENCES

1. S.N. Chow, J.K. Hale, J. Malet-Paret: Application of Generic Bifurcation I, Archive for Rational Mechanics and Analysis, 62, n. 3, 209-235, (1976).
2. J.K. Hale: Application of Alternative Problems, Lefschetz Centre for Dynamical Systems, Lecture Notes, Div. Appl. Math. Brown Univ. Providence, R.I., 1971.
3. J. Dieudonné: Fondations of Modern Analysis, Academic Press, New York, 1968.
4. M.G. Crandall, P.H. Rabinowitz: Bifurcation from Simple Eigenvalues, J. Funct. Anal., 8, 321-340, (1971).
5. M.A. Krasnoselskii: Topological Methods in the Theory of Nonlinear Integral Equations, New York, MacMillan, 1964.
6. D.H. Sattinger: Pattern Formation in Convective Phenomena, In: Turbulence and Navier-Stokes Equation, Lecture Notes in Mathematics, n. 565, 159-173, Berlin, Heidelberg, New York, Springer Verlag, 1976.

7. A.N. Tikhonov, V.B. Glasko: Use of the Regularization Method in Non-Linear Problems, Zh. vychisl. Mat. mat. Fiz 5, 3, 463-473, (1965).

SOME EXISTENCE AND STABILITY RESULTS FOR SOLUTIONS OF REACTION-DIFFUSION SYSTEMS WITH NONLINEAR BOUNDARY CONDITIONS

Jesús Hernández

Departamento de Matemáticas
Universidad Autónoma, Madrid, Spain

1. INTRODUCTION

The purpose of this paper is to give a brief survey of some results obtained by the author on existence and stability for equations of reaction-diffusion type. More detailed statements and proofs will be found in [12].

We study the parabolic system

$$\frac{\partial u}{\partial t} - \Delta u = f(x,u,v) \qquad \text{in } \Omega \times (0,T) \tag{1.1}$$

$$\frac{\partial v}{\partial t} - \Delta v = g(x,u,v) \qquad \text{in } \Omega \times (0,T) \tag{1.2}$$

with the nonlinear boundary conditions

$$\frac{\partial u}{\partial n} + b(u) = 0 \qquad \text{on } \partial\Omega \times (0,T) \tag{1.3}$$

$$\frac{\partial v}{\partial n} + c(v) = 0 \qquad \text{on } \partial\Omega \times (0,T) \tag{1.4}$$

and the initial conditions

ISBN 0-12-508780-2

$$u(x,0) = \hat{u}_o(x) \quad v(x,0) = \hat{v}_o(x) \quad \text{in } \bar{\Omega}, \tag{1.5}$$

where Ω is an smooth bounded domain in R^N, $T > 0$, f,g,b, and c are smooth and b and c are increasing. In fact, our results are also valid in much more general situations, including the case of unilateral constraints (cf. Remarks below).

A considerable amount of work has been devoted recently to these questions, in particular to the asymptotic behaviour of solutions of (1.1)-(1.5). It is interesting in this context to study existence and stability of stationary solutions of (1.1)-(1.5), i.e., solutions of the associated elliptic problem.

$$-\Delta u = f(x,u,v) \quad \text{in } \Omega \tag{1.6}$$

$$-\Delta v = g(x,u,v) \quad \text{in } \Omega \tag{1.7}$$

where the boundary conditions (1.3) (1.4) are satisfied on $\partial\Omega$.

Many papers have been written concerning existence and uniqueness for solutions of (1.1)-(1.5), including the case where the nonlinear terms f and g depend on the gradients ∇u and ∇v. Cf., e.g., [4],[5],[8],[15],[17], and [22]. Literature dealing with the elliptic problem (1.6) (1.7) is not so extensive. Existence theorems were obtained by Amann in [4] as a corollary of his results for the parabolic case. Moreover, some people, including the author, have given direct existence proofs by using topological tools and *a priori* estimates (cf. [13] and its references).

On the other hand, it seems reasonable to try to extend to systems comparison methods based on sub and supersolutions, which are very well-known in the case of one equation (cf. [2],[20]). As pointed out in [20], the above system can be treated in exactly the same way as that of a single equation provided f and g are both increasing in u and v. But, if f is not increasing in v (or if g is not increasing in u), then a simple counterexample shows that, with the natural definition of sub and supersolutions given in [20], the corresponding existence theorem fails to hold.

In this paper we show that if sub and supersolutions satisfy stronger assumptions (namely, Müller type conditions), then it is possible to obtain existence results. This kind of assumptions, which include the case of an invariant rectangle, were used for the parabolic problem in, e.g. [4],[5],[15],[17]. We point out that we only assume that f and g are smooth, without any additional growth assumption.

Section 2 is devoted to state and prove some existence results. For the sake of brevity we only consider the elliptic problem. Our method is very simple: by using a result of H. Brézis and an elementary comparison argument, the problem is reduced to an application of Schauder's fixed point theorem. This gives the existence of a weak solution and a regularity argument shows that it is in fact a classical solution. We also give an example arising in combustion theory. In Section 3 we employ the same kind of comparison techniques to prove uniqueness and global stability for solutions of the elliptic system if, roughly speaking, the Lipschitz constants of f and g are "small". This gives an elementary proof of results obtained by Amann [4] in a more general setting.

After the completion of this work, the author learned about some similar results obtained independently by other researchers.Existence for the elliptic problem was obtained by D. Clark [9] and L.Y. Tsai [21] by using differential inequalities: the paper [21] treats the case of nonlinearities depending on the gradient and consequently its proof is much more involved than ours. C.V. Pao [19] proves existence of maximal and minimal solutions for the combustion system by finding a monotone scheme which involves the growth properties of the nonlinear terms. Concerning stability, the same idea is used in [18] and [19] to handle particular examples but, contrarily to [4] and the present paper, growth assumptions are largely exploited. On the other hand, the results in [18] are more difficult to obtain, because Lipschitz constants are not supposed to be small.

Finally, the author is indebted to H. Brézis for some useful remarks.

2. EXISTENCE RESULTS

We shall give our existence theorem in the case of the system

$$-\Delta u = f(u,v) \quad \text{in } \Omega \tag{2.1}$$

$$-\Delta v = g(u,v) \quad \text{in } \Omega \tag{2.2}$$

with the nonlinear boundary conditions

$$Bu = \frac{\partial u}{\partial n} + b(u) = 0 \quad \text{on } \Gamma \tag{2.3}$$

$$Cv = \frac{\partial v}{\partial n} + c(v) = 0 \quad \text{on } \Gamma, \tag{2.4}$$

where Ω is a bounded domain in R^N ($N \geq 2$) with a very smooth boundary $\Gamma = \partial\Omega$ (in particular Γ will be $C^{2,\alpha}$ for $0 < \alpha < 1$) and n denotes the outer normal.

We assume that

$$f,g \text{ are } C^1 \tag{2.5}$$

$$b,c \text{ are } C^2, \text{ increasing, and } b(0) = c(0) = 0. \tag{2.6}$$

We state and prove our existence theorem only for the simple case (2.1)-(2.4) with assumptions (2.5) (2.6). Cf. Remarks 2.3 - 2.6 below for other possible extensions.

A couple of pairs (u_o,v_o) - (u^o,v^o) is called a sub-super-solution for problem (2.1) - (2.4) if $u_o \leq u^o$, $v_o \leq v^o$, and moreover

$$u_o,v_o,u^o,v^o \in H^2(\Omega) \cap L^\infty(\Omega) \tag{2.7}$$

$$-\Delta u_o - f(u_o,v) \leq 0 \leq -\Delta u^o - f(u^o,v) \quad \text{for any } v \in [v_o,v^o] \tag{2.8}$$

$$Bu_o \leq 0 \leq Bu^o \quad \text{on } \Gamma, \tag{2.9}$$

$$-\Delta v_o - g(u,v_o) \leq 0 \leq -\Delta v^o - g(u,v^o) \quad \text{for any } u \in [u_o,u^o] \; , \tag{2.10}$$

$$Cv_o \leq 0 \leq Cv^o \qquad \text{on} \quad \Gamma. \tag{2.11}$$

In all the above inequalities $u \leq v$ means $u(x) \leq v(x)$ a.e. on Ω (or on Γ) and $[u,v] = \{z \in L^2(\Omega)\colon u(x) \leq z(x) \leq v(x)$ a.e. on $\Omega\}$.

Remark 2.1. It is clear that this definition is much more stringent than the natural generalization given in [20] where, e.g., (2.8) are only satisfied for $v = v_o$ and $v = v^o$, respectively. On the other hand, if f and g are increasing in u and v, both definitions coincide. Definitions given in [18] and [19] are particular cases of it where some growth conditions on f and g are assumed.

Remark 2.2. As was pointed out in the introduction, conditions of type (2.8) - (2.11) were introduced by Müller in a classical paper. Moreover, if all four u_o,v_o,u^o,v^o are constant functions, then (2.8) - (2.11) is equivalent to the fact that $[u_o,u^o] \times [v_o,v^o]$ is an invariant rectangle.

Our main existence theorem is the following.

Theorem 2.1. Suppose that $(u_o,u^o) - (v_o,v^o)$ is a sub-super solution for problem (2.1) - (2.4) with assumptions (2.5)(2.6). Then there exists at least one (classical) solution (u,v) of (2.1) - (2.4) satisfying $u_o \leq u \leq u^o$ and $v_o \leq v \leq v^o$.

Let $E = [L^2(\Omega)]^2$ and let $K = [u_o,u^o] \times [v_o,v^o]$. By (2.7), K is a closed convex bounded subset of E. Next define a nonlinear operator $T\colon K \longrightarrow E$ in the following way: for $(\bar{u},\bar{v}) \in K$, $(w,z) = T(\bar{u},\bar{v})$ is the (unique) solution of the system

$$-\Delta w + Mw = f(\bar{u},\bar{v}) + M\bar{u} \qquad \text{in} \quad \Omega \tag{2.12}$$

$$Bw = 0 \qquad \text{on} \quad \Gamma \tag{2.13}$$

$$-\Delta z + Mz = g(\bar{u},\bar{v}) + M\bar{v} \qquad \text{in} \quad \Omega \tag{2.14}$$

$$Cz = 0 \qquad \text{on} \quad \Gamma, \tag{2.15}$$

where $M > 0$ is a constant such that the right-hand side in (2.12) (resp. in (2.14)) is increasing in $\bar{u}$ (resp. in $\bar{v}$). We can find such M by (2.5) and (2.7).

Indeed, we first remark that $f(\bar{u},\bar{v}) + M\bar{u} \in L^{\infty}(\Omega)$. Hence, as a consequence of, e.g., Coroll. 13 in [6], there exists a unique solution $w \in H^2(\Omega)$ of (2.12), (2.13). The same argument works for z. Thus, T is well-defined and has the following properties.

<u>Lemma 2.1.</u> T is compact and $T(K) \subset K$.

<u>Proof</u>: The proof of the fact that T is compact is straight forward and will be omitted. Concerning the inclusion $T(K) \subset K$, we only prove $u_0 \leq w$, because the other inequalities can be obtained in a completely similar way by using (2.8) - (2.11). It is obvious that $u_0 \leq w$ is equivalent to $(u_0 - w)^+ = 0$, where $u^+ = \max(u,0)$. In view of (2.8) with $v = \bar{v}$ and (2.12) we have

$$0 \geq -\Delta(u_0 - w) + f(\bar{u},\bar{v}) - f(u_0,\bar{v}) + M(\bar{u} - w) \quad \text{in } \Omega.$$

Multiplying this inequality by $(u_0 - w)^+$ and integrating over Ω by using Green's formula and (2.9), we get easily $(u_0 - w)^+ = 0$.

Hence, Lemma 2.1 and Schauder's fixed point theorem give the existence of at least one fixed point of T or, equivalently, one weak (i.e., in $H^2(\Omega)$) solution of (2.1) - (2.4). The following lemma, which can be proved by employing the regularity results in [1] and [7] together with Schauder theory, shows that weak solutions of (2.1) - (2.4) are classical solutions.

<u>Lemma 2.2.</u> The pair $(u,v) \in K$ is a classical solution of (2.1) - (2.4) if and only if it is a fixed point of T.

Theorem 2.1 can be generalized in different directions. First, it is clear that it extends to any finite number of weakly coupled equations and that x-dependence in f and g is allowed if it is sufficiently smooth. More relevant extensions are the object of the following remarks (see [12] for more precise information).

Remark 2.3. We point out that Theorem 2.1 is also valid, with only very minor modifications in the proof, for the case of unilateral boundary conditions, i.e., unilateral constraints given by maximal monotone graphs (cf. [6],[7]). In this case sub and supersolutions are defined by combining (2.7) - (2.11) with the definition in [11]. Indeed, existence and regularity results are still applicable and the comparison technique is the same as in [7] or [11]. It is clear that now we only get weak solutions.

Remark 2.4. It is possible to replace the operator $-\Delta$ by second order elliptic linear differential operators of the form

$$L_k u = -\sum_{i,j=1}^{N} \frac{\partial}{\partial x_i}\left(a_{ij}^k(x)\frac{\partial u}{\partial x_j}\right) + c^k(x)u \quad k = 1,2, \tag{2.16}$$

where $a_{ij}^k = a_{ji}^k$, $a_{ij}^k \in C^\infty(\bar{\Omega})$, $c^k \in C(\bar{\Omega})$, $c^k \geq 0$, $c^k \not\equiv 0$. We also assume that L_k is uniformly elliptic. This is a corollary of Theorems 1 and 2 in [14].

Remark 2.5. An alternative proof of our results can also be given following the lines of [3]. This allows weaker regularity assumptions on the coefficients and more general (e.g., coupled) nonlinear boundary conditions.

Remark 2.6. It is clear that an analogous existence and uniqueness theorem can be proved for the parabolic problem (1.1) - (1.5) (cf. [4],[5],[15] for the case of linear boundary conditions). The proof uses the classical results in [10] and [16] and the existence and regularity results in [6] and [7].

Example 2.1. Consider the system

$$-\Delta u = -\alpha(u + 1)^p \exp\left(\frac{\gamma v}{v + 1}\right) \quad \text{in } \Omega \tag{2.17}$$

$$-\Delta v = \alpha\beta(u + 1)^p \exp\left(\frac{\gamma v}{v + 1}\right) \quad \text{in } \Omega \tag{2.18}$$

$$Bu = Cv = 0 \quad \text{on } \Gamma, \tag{2.19}$$

where $\alpha,\beta,\gamma > 0$ are certain physical constants and $p \geq 2$ is an integer (cf. [4],[5],[19]). It is not difficult to see that

$(u_o,v_o) - (u^o,v^o)$ is a sub-supersolution if

$$u_o = -1 \qquad u^o = 0 \qquad v_o = 0$$

and v^o is the unique solution of the nonlinear problem (we assume $c \not\equiv 0$)

$$-\Delta v^o = \alpha\beta e^{\gamma} \qquad \text{in } \Omega$$

$$Cv^o = 0 \qquad \text{on } \Gamma.$$

(Observe that $v^o > 0$ on $\bar{\Omega}$). This extends the existence results of [4] to nonlinear boundary conditions.

Remark 2.7. We point out that the right-hand side in (2.17) (resp. in (2.18)) is decreasing (resp. increasing) in v if $u + 1 \geq 0$. It is not difficult to replace $e^{\frac{\gamma v}{v+1}}$ by different functions of v, without satisfying growth conditions, in such a way that a similar result is obtained.

3. GLOBAL STABILITY RESULTS

The aim of this section is to show how the same kind of comparison techniques employed to prove existence in Section 2 can be used to obtain global stability results.

Now, for convenience, we replace both Laplacians in (1.1) (1.2) by two different uniformly elliptic second order linear differential operators L_1 and L_2 of the form (2.16) as in Remark 2.4. Consider the parabolic problem

$$\frac{\partial u}{\partial t} + L_1 u = f(u,v) \qquad \text{in } \Omega \times (0,T) \tag{3.1}$$

$$\frac{\partial v}{\partial t} + L_2 v = g(u,v) \qquad \text{in } \Omega \times (0,T) \tag{3.2}$$

with, to simplify the problem, the linear boundary conditions

$$B_1 u = \frac{\partial u}{\partial n} + bu = 0 \qquad \text{on } \Gamma \times (0,T) \qquad (3.3)$$

$$C_1 v = \frac{\partial v}{\partial n} + cv = 0 \qquad \text{on } \Gamma \times (0,T) \qquad (3.4)$$

where b and c satisfy

$$b,c \in C^{1,\alpha}(\Gamma), b,c \geq 0,\ b \not\equiv 0 \text{ on } \Gamma. \qquad (3.5)$$

We also have the initial conditions

$$u(x,0) = \hat{u}_o(x) \qquad v(x,0) = \hat{v}_o(x) \qquad \text{on } \bar{\Omega}. \qquad (3.6)$$

Suppose that there exists a sub-supersolution (u_o,v_o) - (u^o,v^o) for the elliptic system associated with (3.1)-(3.4). Then by Theorem 2.1 there exists at least one solution $(\bar{u},\bar{v})$ in the corresponding interval. Moreover, it is clear that (u_o,v_o)-(u^o,v^o) is also a sub-supersolution for the parabolic problem (3.1) - (3.6) if the inequalities for the initial values are satisfied. Hence we have

$$L_1\bar{u} = f(\bar{u},\bar{v}) \qquad \text{in } \Omega \qquad (3.7)$$

$$L_2\bar{v} = g(\bar{u},\bar{v}) \qquad \text{in } \Omega \qquad (3.8)$$

$$B_1\bar{u} = C_1\bar{v} = 0 \qquad \text{on } \Gamma. \qquad (3.9)$$

Recall that this solution $(\bar{u},\bar{v})$ is not necessarily unique. But we shall prove, roughly speaking, that if the Lipschitz constants for f and g are "small", then this solution is unique and, moreover, it is globally asymptotically stable.

It is well-known that the linear eigenvalue problem

$$L_1 w = \lambda w \qquad \text{in } \Omega$$

$$B_1 w = 0 \qquad \text{on } \Gamma,$$

has a least simple positive eigenvalue $\lambda_1 > 0$, and the same thing happens with

$$L_2 w = \lambda w \quad \text{in } \Omega$$

$$C_1 w = 0 \quad \text{on } \Gamma.$$

Let $\lambda_2 > 0$ the corresponding least eigenvalue. Choose λ satisfying

$$0 < \lambda < \min\{\lambda_1, \lambda_2\}. \tag{3.10}$$

Hence it is also well-known that problem

$$L_1 \varphi = \lambda\varphi + 1 \quad \text{in } \Omega \tag{3.11}$$

$$B_1 \varphi = 0 \quad \text{on } \Gamma \tag{3.12}$$

has a unique positive solution φ such that $\varphi > 0$ on $\bar{\Omega}$, and that problem

$$L_2 \psi = \lambda\psi + 1 \quad \text{in } \Omega \tag{3.13}$$

$$C_1 \psi = 0 \quad \text{on } \Gamma \tag{3.14}$$

has a unique positive solution $\psi > 0$ on $\bar{\Omega}$.

Then it is possible to prove the following result (cf. [12]).

Lemma 3.1. Suppose that f and g satisfy (2.5)(2.6) and that they are globally Lipschitz continuous, with Lipschitz constants K_1 and K_2 satisfying

$$K_i(\|\varphi\|_{L^\infty(\Omega)} + \|\psi\|_{L^\infty(\Omega)}) \leq 1 \qquad i = 1,2. \tag*{$(3.14)_i$}$$

If we define

$$\tilde{u}_o(x,t) = \bar{u}(x) - p(t)\varphi(x) \qquad \tilde{u}^o(x,t) = \bar{u}(x) + p(t)\varphi(x)$$

$$\tilde{v}_o(x,t) = \bar{v}(x) - p(t)\psi(x) \qquad \tilde{v}^o(x,t) = \bar{v}(x) + p(t)\psi(x),$$

where $p(t) = ce^{-\lambda t}$, satisfying (3.10), c real, $c > 0$, then

$(\tilde{u}_o, \tilde{v}_o) - (\tilde{u}^o, \tilde{v}^o)$ is a sub-supersolution for the parabolic problem (3.1) - (3.6) if the inequalities

$$\bar{u}(x) - c\varphi(x) \leq \hat{u}_o(x) \leq \bar{u}(x) + c\varphi(x)$$

$$\bar{v}(x) - c\psi(x) \leq \hat{v}_o(x) \leq \bar{v}(x) + c\psi(x)$$

are satisfied on $\bar{\Omega}$.

Now, we can state our stability theorem. The proof can be carried out by using Lemma 3.1, the existence and uniqueness theorem for the parabolic problem (3.1) - (3.6), and the fact that φ and ψ are > 0 on $\bar{\Omega}$.

Theorem 3.1. Suppose that all the hypotheses of Lemma 3.1 are satisfied. Then there exists a unique solution $(\bar{u}, \bar{v})$ of the elliptic problem associated with (3.1),(3.2),(3.3),(3.4) and this solution is globally asymptotically stable in the sense that for every solution of the parabolic problem (3.1) - (3.6) we have

$$\lim_{t \to \infty} u(x,t) = \bar{u}(x) \qquad \lim_{t \to \infty} v(x,t) = \bar{v}(x)$$

uniformly in x.

Remark 3.1. The case of nonlinear boundary conditions is a little more complicated and is considered in [12].

Remark 3.2. Applications of Theorem 3.1 to some concrete situations, including Example 2.1, can also be found in [12].

REFERENCES

1. S. Agmon, A. Douglis, and L. Nirenberg: Estimates Near the Boundary for Solutions of Elliptic Partial Differential Equations Satisfying General Boundary Conditions, I. Comm. Pure Appl. Math. 12, 623-727,(1959).
2. H. Amann: Fixed Point Equations and Nonlinear Eigenvalue Problems in Ordered Banach Spaces, SIAM Review 18, 620-709, (1976).

3. H. Amann: Nonlinear Elliptic Equations with Nonlinear Boundary Conditions. Proc. of the Second Schveningen Conference on Differential Equations, ed. by W. Eckhaus, Amsterdam, North Holland, 1976.
4. H. Amann: Existence and Stability of Solutions for Semi-Linear Parabolic Systems and Applications to Some Diffusion-Reaction Equations. Proc. Roy. Soc. Edinburgh 81A, 35-47, (1978).
5. J. Bebernes, K.N. Chueh, and W. Fulks: Some Applications of Invariance to Parabolic Systems, Indiana Univ. Math. J. 28, 269-277, (1979).
6. H. Brézis: Monotonicity Methods in Hilbert Spaces and Some Applications to Nonlinear Partial Differential Equations. In Contributions to Nonlinear Functional Analysis, ed. by E.H. Zarantonnello, New York, Academic Press, 1971, 101-156.
7. H. Brézis: Problèmes unilatéraux, J. Math. Pures Appl. 51, 1-168, (1972).
8. K.N. Chueh, C. Conley, and J. Smoller: Positively Invariant Regions for Systems of Nonlinear Diffusion Equations, Indiana Univ. Math. J. 26, 373-391 (1977).
9. D. Clark: On Differential Inequalities for Systems of Diffusion-Reaction Type. To appear.
10. A. Friedman: Partial Differential Equations of Parabolic Type. Englewood Cliffs, Prentice-Hall, 1964.
11. J. Hernández: Bifurcación y soluciones positivas para ciertos problemas de tipo unilateral. Thesis. Madrid, Universidad Autónoma, 1977.
12. J. Hernández: Existence and Global Stability Results for Reaction-Diffusion Systems with Nonlinear Boundary Conditions. To appear in Nonlinear Anal.
13. J. Hernández: Positive Solutions of Reaction-Diffusion Systems with Nonlinear Boundary Conditions and The Fixed Point Index. To appear in the proceedings of the conference Nonlinear Phenomena in The Mathematical Sciences, Arlington (Texas), June 1980, New York, Academic Press.
14. H. Kawohl: Coerciveness for Second-Order Elliptic Differential Equations with Unilateral Constraints, Nonlinear Analysis, 2, 189-196 (1978).

15. H.J. Kuiper: Existence and Comparison Theorems for Nonlinear Diffusion Systems. J. Math. Anal. Appl. 60, 166--181 (1977).
16. O. Ladyzenskaia, V. Solonnikov, and N. Uraltseva: Linear and Quasi-Linear Equations of Parabolic Type. Providence, 1968, Amer. Math. Soc. Translations.
17. V. Lakshmikantham: Comparison Results for Reaction-Diffusion Equations in a Banach Space. Conferenze del Seminario di Matematica, Bari, 1979, 121-156, and this volume.
18. A. Leung and D. Clark: Bifurcations and Large-Time Asymtotic Behavior For Prey-Predator Reaction-Diffusion Equations with Dirichlet Boundary Data. J. Diff. Equa. 35, 113-127 (1980).
19. C.V. Pao: Asymptotic Stability of Reaction-Diffusion Systems in Chemical Reactor and Combustion Theory, J. Math. Anal. Appl., to appear.
20. D.H. Sattinger: Monotone Methods in Nonlinear Elliptic and Parabolic Equations. Indiana Univ. Math. J. 21, 979-1000 (1972).
21. Long-Yi Tsai: Existence of Solutions of Nonlinear Elliptic Systems, Bull. Inst. Math. Acad. Sinica 8, 111-126 (1980).
22. H.F. Weinberger: Invariant Sets for Weakly Coupled Parabolic and Elliptic Systems, Rend. di Matematica 8, 295--310 (1975).

ON THE ASYMPTOTIC BEHAVIOR OF THE SOLUTIONS OF THE NONLINEAR EQUATION $\ddot{x} + h(t,x)\dot{x} + p^2(t)f(x) = 0$.

N. Ianiro

Istituto di Meccanica e Macchine della
Facoltà di Ingegneria, Università dell'Aquila,
L'Aquila, Italy.

C. Maffei

Istituto di Matematica dell'Università
di Camerino, Camerino, Italy.

1. INTRODUCTION

The non linear equation

$$\ddot{x} + h(t,x,\dot{x})\dot{x} + p^2(t)f(x) = 0, \ x \in R, f(0) = 0, \ xf(x) > 0 \text{ for } x \neq 0, \tag{1.1}$$

is a basic mathematical model for the representation of damped oscillatory phenomena: it is of interest to investigate the asymptotic properties of the rest point of this equation. This problem has drawn the attention of many investigators during the last years: for example in the case of $p(t)$ = const. Z. Artstein [1], using techniques inspired by the invariance principle of LaSalle, shows that the rest point of (1.1) is asym-

ISBN 0-12-508780-2

ptotically stable if $\int_0^s h(\tau,x,\dot{x})d\tau$ is a uniformly continuous function. In the same hypothesis of $p(t) = \text{const.}$, the authors [2], have studied the asymptotic properties of the null solution of (1.1) by using several auxiliary functions.

The results presented in [4], [5] are obtained in the case $p(t) > a > 0$. In [5] is considered the motion of a holonomic system with n degrees of freedom subject to a force arising from a potential $p^2(t)U(q_1,\dots,q_n)$ and to a bounded full dissipation. Supposing that $\dot{p}(t)/p^2(t)$ is bounded, the x-asymptotic stability is proved through a new general method known as "method of families of auxiliary functions".

In the reference [4] the same result is obtained, for the linear case, under suitable conditions involving $h(t)$ and $p(t)$. This result is more general than that obtained in [5] with respect to the hypothesis on the damping forces, while Salvadori's hypotheses are weaker for unbounded forces.

In this paper we consider the case in which the damping coefficient is independent on $\dot{x}$, $h(t,x,\dot{x}) \equiv h(t,x)$; our purpose is to give a general condition on $h(t,x)$ and $p(t)$ such that the x-asymptotic stability of the rest point is ensured.

More precisely we suppose that the function $m(t) = \int_0^t h(\tau,x(\tau))d\tau$ is uniformly continuous from a given t on, and that $h(t,x)/p(t) + \dot{p}(t)/p^2(t)$ is bounded from below from a suitable function $\psi_\nu(t)$ (see hypothesis (b) in the Theorem 2.1). These conditions, as one can easily verify, are more general, for the bidimensional case, than the hypotheses in [4] and can be applied also if the equation is not linear.

The technique used here, inspired by Matrosov's method [3], requires the existence of a second auxiliary function in those sets in which the derivative of Liapunov function is no more sign defined. Finally note that this method allows us to show the attractivity of the origin without verifying invariance properties.

2. THE ASYMPTOTIC STABILITY

We shall make use of the following notations and definitions: let G be an open set of R^2 containing the origin, for $z \in R^2$ and any $\gamma \in (0,\rho(z,\partial G))$ let be $B_\gamma = \{z \in G: \|z\| \leq \gamma\}$; moreover for any ν and $\gamma (0 < \nu < \gamma)$ let be $C_{\nu,\gamma} = \{z \in G: \nu \leq \|z\| \leq \gamma\}$. Denote by $z(t) = (x(t,t_o,x_o,\dot{x}_o), \dot{x}(t,t_o,x_o,\dot{x}_o))$ a noncontinuable solution of the equation (1.1) passing through $(t_o,x_o,\dot{x}_o)$.

Definition 2.1. The solution $z(t) = (0,0)$ is said:

(i) x-(uniformly) stable if for any $\varepsilon \in (0,\gamma]$ and $t_o \in R^+$ there exists a $\delta(\varepsilon,t_o) > 0$ $(\delta(\varepsilon) > 0)$ such that $\|z(t_o)\| < \delta$ implies that $\|x(t,t_o,x_o,\dot{x}_o)\| < \varepsilon$ for all $t \geq t_o$;

(ii) x-(equi) attractive if for any $t_o \in R^+$ there exists a $\sigma > 0$ such that $\|z(t_o)\| < \sigma$ imply that $z(t)$ exists for all $t \geq t_o$, and for any $\nu > 0$ there exists a $T = T(\nu,t_o,z(t_o))$ $(T = T(\nu,t_o))$ such that $\|x(t,t_o,x_o,\dot{x}_o)\| < \nu$ for any $t \geq t_o + T$;

(iii) x-(equi) asymptotically stable if it is x-stable and x-(equi) attractive.

Definition 2.2. A function $W \in C^1([R^+ \times B_\gamma],R)$ is definitely divergent on a set $E \subset R^2$ if for every $\nu \in (0,\gamma)$, for every $t_o \geq 0$ and for every continuous function $\varphi: R^+ \to B_\gamma$, there exist two positive constants $\varepsilon = \varepsilon(\nu,\gamma,\varphi)$, $A = A(\nu,\gamma,\varphi)$ such that if $\nu \leq \|\varphi(t)\| \leq \gamma$, $\rho(\varphi(t),E) < \varepsilon$ for all $t \geq t_o$ one has

$$\left|\int_{t_o}^{t} \dot{W}(\tau,\varphi(\tau))d\tau\right| \geq A(t - t_o),$$

where $\dot{W}(t,x)$ is the derivative of W along the solutions of (1.1). (see Def. 2.1 in [2]).

We now consider the equation

$$\ddot{x} + h(t,x)\dot{x} + p^2(t)f(x) = 0, \tag{2.1}$$

and prove the following theorem

Theorem 2.1. Assume for the equation (2.1) the following conditions:

(a) $p(t)$ is differentiable, $p(t) > a > 0$, $\dot{p}(t)/p^2(t)$ is bounded;

(b) $h(t,x)$ is continuous and such that for every compact set $K \subset B_\gamma$, for every continuous function $\varphi: R^+ \to K$ and for every $\alpha > 0$ there exist $T = T(\alpha,K,\varphi) > 0$, $\xi = \xi(\alpha,K,\varphi) > 0$ such that if

$$\left|\int_t^{t+s} h(\tau,\varphi(\tau))d\tau\right| > \alpha \quad \text{for } t > T, \text{ then } s > \xi;$$

(c) for every $\nu \in (0,\gamma)$, $(x,\dot{x}) \in C_{\nu,\gamma}$, let be

$$\frac{h(t,x)}{p(t)} + \frac{\dot{p}(t)}{p^2(t)} \geq \psi_\nu(t) > 0,$$

where $\psi_\nu(t)$ is a continuous nonincreasing function such that

$$\int_0^\infty \psi_\nu(t)p(t)dt = \infty;$$

(d) f is continuous, $xf(x) > 0$ for $x \neq 0$;

then the solution $x = 0$, $\dot{x} = 0$ of (2.1) is x-equiasymptotically stable.

Proof: The change of time variable

$$s(t) = \int_0^t p(\tau)d\tau \tag{2.2}$$

trasforms the equation (2.1) into the equation

$$x'' + \frac{h(t(s),x)}{p(t(s))} + \frac{\dot{p}(t(s))}{p^2(t(s))} \cdot x' + f(x) = 0 \tag{2.3}$$

where $' = d/ds$, or into the equivalent system

$$\begin{cases} x' = y \\ y' = -\tilde{h}(s,x)y - f(x) \end{cases} \tag{2.4}$$

where $\tilde{h}(s,x) = \dfrac{h(t(s),x)}{p(t(s))} + \dfrac{\dot{p}(t(s))}{p^2(t(s))}$.

We prove that the null solution of (2.4) is equiasymptotically stable.
In fact, set $V(x,y) = y^2/2 + \int_0^x f(\xi)d\xi$, along the solutions of (2.4) one has

$$V'_{(2,4)} = -y^2\tilde{h}(s,x) \leq -y^2\psi_\nu(s) \leq 0.$$

Therefore the function $V(x,y)$ verifies the hypotheses of the Liapunov stability theorem and the null solution of (2.4) is uniformly stable, i.e. for any $\beta \in (0,\gamma]$ there exists a $\delta(\beta) > 0$ such that $s_o \in R^+$, $\|z(s_o)\| < \delta(\beta)$ imply that $\|z(s)\| < \beta$ for all $s \geq s_o$; here $z = (x,y)$.

Let $\sigma = \delta(\gamma)$, $s_o \in R^+$ and $\|z(s_o)\| < \sigma$. We shall prove that for any $\nu \in (0,\gamma)$ there exists a $s'_o \geq s_o$ such that $\|z(s'_o)\| < \delta(\nu)$. This is enough to say that $\|z(s)\| < \nu$ for all $s \geq s'_o$: the origin will be attractive.
We suppose, by contradiction, that there exists a $z(s_o) \in B_\sigma$, such that $z(s) \in C_{\delta(\nu),\gamma}$ for $s \geq s_o$. The proof is obtained by steps.

Step 1. We show that there exist an $\varepsilon > 0$ and a $T > 0$ such that for any $s' \geq T$ the solution $z(s)$ cannot stay in the set $N_\varepsilon = \{z \in B_\gamma : \delta(\nu) \leq \|z\| \leq \gamma, \rho(z,E) < \varepsilon\}$ for all $s \geq s'$, where $E = \{z \in B_\gamma : y = 0\}$.
In fact consider the function $W(z) = xy$; one has

$$W'_{(2,4)} = y^2 - \tilde{h}(s,x)xy - xf(x)$$

Let α be a fixed number, from the condition (b) and (2.2) it follows that there exist a $\bar{\xi} \geq a\xi$ and an $S \geq aT$ such that

$$\left|\int_s^{s+\bar{\xi}} \frac{h(\sigma,x(\sigma))}{p(t(\sigma))} d\sigma\right| \leq \alpha \qquad \text{for } s \geq S;$$

consider $s' \geq S$ and $\sigma \in (s',s)$; then

$$\int_{s'}^{s} W'(\sigma,z(\sigma))d\sigma \leq (m/c - m + \gamma\sqrt{m/c}\,L)(s - s') + \gamma\sqrt{m/c}\,n\alpha$$

for $y^2(\sigma) \leq m/c$ where $m = \min\{xf(x),\ x \in C_{\delta(\nu),\gamma}\}$, c is a positive constant, $|\dot{p}/p^2| \leq L$ and $n < (s - s')/a\xi$.

If c is large enough, the function W is definitely divergent on the set E, where $A = A(\delta(\nu),\gamma,\varphi)$ of the Def. 2.2 is equal to $m/c - m + \gamma\sqrt{m/c}\,L + \gamma\sqrt{m/c}\cdot\alpha/a\xi$ and $\varepsilon = \varepsilon(\delta(\nu),\gamma,\varphi)$ is equal to $\sqrt{m/c}$.
This fact and the boundedness of the function W - say $|W| \leq B$- imply that there exists an $s \geq s'$ such that $\rho(z(s),E) \geq \varepsilon$; to be more explicitely one has

$$s - s' \leq 2B/A. \tag{2.5}$$

Step 2. We first observe that for every $\chi > 0$, there exists a sequence $\{s_n\}$, such that $\rho(z(s_n),E) < \chi$ for $n \to \infty$; this easily follows from hypothesis (c).
Choose now $\chi = \varepsilon/2$; then because of the step 1 and the previous observation the two divergent sequences $\{s'_n\}$ and $\{s''_n\}$ must exist such that for $k = 1,2,\ldots,n$, $s'_k < s''_k$ and

$$\rho(y(s'_k),E) = \varepsilon/2, \quad \rho(y(s''_k),E) = \varepsilon, \quad \varepsilon/2 < \rho(y(s),E) < \varepsilon$$

$$\text{for } s \in (s'_k,s''_k).$$

One has

$$V(s''_k,z(s''_k)) - V(s'_k,z(s'_k)) \leq -\frac{\varepsilon^2}{4}\int_{s'_k}^{s''_k} \psi_{\delta(\nu)}(s)ds.$$

But $\rho(z(s''_k),z(s'_k)) \geq \varepsilon/2$, then using the condition (b) it follows that there must exist a $\bar{k} = \bar{k}(\varepsilon,B_\gamma,(z(\cdot))$ and an $\bar{\eta} = \bar{\eta}(\varepsilon,B_\gamma,z(\cdot))$ such that $s''_k - s'_k > \bar{\eta}$ for $k > \bar{k}$.
Then

$$V(s''_k,z(s''_k)) - V(s'_k,z(s'_k)) \leq -\frac{\varepsilon^2}{4}\,\psi_{\delta(\nu)}(s''_k)\bar{\eta}. \tag{2.6}$$

Step 3. We show that a complete oscillation around E takes a finite interval of time. In fact, call $\{\sigma_n\}$ the sequence of all those instants such that for $k = 1,2,\ldots,n$ $s'_k < s''_k < \sigma_k \leq$ $\leq s'_{k+1} < \ldots$ and $\rho(y(\sigma_k),E) = \varepsilon/2$, $\rho(y(s),E) > \varepsilon/2$ for $s \in (s''_k,\sigma_k)$.
It is obvious that

$$s'_{k+1} - s'_k = (s''_k - s'_k) + (\sigma_k - s''_k) + (s'_{k+1} - \sigma_k);$$

taking into account (2.5) and that one can have $\delta(\nu) \leq$ $\leq |x(s)| < \gamma$ and $|y(s)| \geq \frac{\varepsilon}{2}$ for a time shorter or equal than $4\gamma/\varepsilon$, it follows

$$s'_{k+1} - s'_k \leq 2B/A + 4\gamma/\varepsilon + 2B/A = \tilde{T}.$$

This implies also

$$s''_k < k\tilde{T} \tag{2.7}$$

Step 4. Finally we can exclude infinite oscillation of the solution z(s) around the set E. Using (2.6) and (2.7) and taking into account that the function $\Psi_{\delta(\nu)}(s)$ is not increasing, one has that in every interval $s''_k - s'_k$ the function V decreases at least by $\varepsilon^2/4\ \Psi_{\delta(\nu)}(k\tilde{T})\bar{\eta}$.
Then

$$\lim_{s\to\infty} V(s,z(s)) \leq V(s_o,z(s_o)) - \varepsilon^2/4\ \bar{\eta} \sum_{1}^{\infty}{}_k\ \Psi_{\delta(\nu)}(k\tilde{T}) \leq$$
$$V(s_o,z(s_o)) - \varepsilon^2/4\ \bar{\eta} \cdot \frac{1}{\tilde{T}} \cdot \int_0^{\infty} \Psi_{\delta(\nu)}(s)ds;$$

from hypothesis (c) and the condition that V is a positive definite function one has a contradiction: it must exist an instant $s'_o \geq s_o$ for which $\|z(s'_o)\| < \delta(\nu)$. This proves the attractivity of the solution x = 0, y = 0 of the system (2.4). Since the uniform stability and the attractivity imply the equiasymptotic stability, we have proved that the null solution of (2.4) is equiasymptotically stable.
Consequently the solution $x = 0$, $\dot{x} = 0$ of (2.1) is x-equiasymptotically stable and the proof is complete. We remark that the solution $x = 0$, $\dot{x} = 0$ of (2.1) is not in general equiasymptotically stable (with respect to x, $\dot{x}$), unless some additional conditions are satisfied (for example if p(t) is bounded).

For the equation

$$\ddot{x} + p^2(t)f(x) = 0 \qquad (2.8)$$

theorem 2.1 specializes as follows.

Corollary 2.1. Suppose that

(a) $p(t)$ is differentiable, $p(t) > a > 0$, $\dot{p}(t) > 0$, $\dot{p}(t)/p^2(t)$ bounded;

(b) for any $\nu \in (0,\gamma)$, $\dot{p}(t)/p^2(t) \geq \psi_\nu(t) > 0$, where $\psi_\nu(t)$ is a nonincreasing function such that

$$\int_0^\infty \psi_\nu(t)p(t)dt = \infty;$$

(c) $xf(x) > 0$ for $x \neq 0$.

Then the null solution of (2.8) is x-uniformly asymptotically stable.

REFERENCES

1. Z. Artstein: Topological Dynamics of Ordinary Differential Equations and Kurzweil Equations, J. Diff. Eqns. 23, 224-243, (1977).
2. N. Ianiro, C. Maffei: Nonautonomous Differential Systems: Several Auxiliary Functions in The Stability Problem, Applied Math. and Computation, to appear.
3. V.M. Matrosov: On the Stability of Motion, J. Appl. Math. Mech. 26, 1337-1353, (1962).
4. R.J. Ballieu, K. Peiffer: Attractivity of the Origin for the Equation $\ddot{x} + f(t,x,\dot{x})|\dot{x}|^\alpha\dot{x} + q(x) = 0$, J. of Math. Analysis and Appl. 65, 321-332, (1978).
5. L. Salvadori: Famiglie ad un parametro di funzioni di Liapunov nello studio della stabilità, Symposia Mat. Ist. Naz. Alta Mat. Bologna 4, 309-330, (1971).

NUMERICAL METHODS FOR NONLINEAR BOUNDARY VALUE PROBLEMS AT RESONANCE

R. Kannan[1]

The University of Texas at Arlington
Arlington, Texas

1. INTRODUCTION

We consider here the problem of numerical solution of nonlinear boundary value problems at resonance. Briefly stated, a nonlinear boundary value problem at resonance is an operator equation of the type Eu = Nu where E is a linear differential operator with a non-trivial kernel and $\mathcal{D}(E)$ contained in a real Banach space S and N is a nonlinear operator over S. As examples of such problems we present the following:

(i) $x'' = f(t,x,x')$, $x(0) = x(2\pi)$, $x'(0) = x'(2\pi)$ (the case of forced oscillations);

(ii) $\Delta u = 0$ in Ω, $\frac{\partial u}{\partial n} = \sigma(u^4 - u_o^4)$ on $\partial\Omega$, $\sigma \geq 0$ (heat loss due to radiation);

(iii) $u_t - u_{xx} = f(t,x,u)$, u is 2π-periodic in x and t (ring shaped heat conductor);

(iv) $u_{tt} - u_{xx} = f(t,x,u)$, $u(0,t) = u(\pi,t) = 0$, $u(t,x) = u(t + 2\pi,x)$.

In the recent years considerable progress has been made on

---

[1]Research partially supported by U.S. Army Reseach Grant DAAG29-80-C-0060.

ISBN 0-12-508780-2

the question of existence of solutions of the above class of problems. Our attention in this report is restricted strictly to the question of finding numerical solutions to the above class of problems. For the sake of simplicity, we will use the forced oscillation problem (i) as our model. However, most of the discussions are true for the other problems as well.

2. OPERATOR-THEORETIC FORMULATION

Let X and Y be Banach spaces and let $E: \mathcal{D}(E) \subset X \to Y$ be a linear differential operator and $N: \mathcal{D}(N) \subset X \to Y$ be a nonlinear operator. We assume below that the null-space of E is nontrivial.

We assume that there are decompositions $X = X_o + X_1$, $Y = Y_o + Y_1$ of X and Y into complementary closed linear subspaces and let $P: X \to X$, $Q: Y \to Y$ be linear bounded projection operators such that

$$X_o = PX, \quad X_1 = (I - P)X, \quad Y_o = QY, \quad Y_1 = (I - Q)Y,$$
$$\ker E \subset X_o, \text{ range } E = Y_1.$$

Then, $E: \mathcal{D}(E) \cap X_1 \to Y_1$ is onto and one-to-one and thus the inverse linear operator $H: Y_1 \to \mathcal{D}(E) \cap X_1$ is defined. We now assume that (a) $H(I - Q)E = I - P$, (b) $QE = EP$, (c) $EH(I - Q) = I - Q$. Under these assumptions the equation

$$Eu = Nu \tag{1}$$

is equivalent to the system of operator equations

$$u = Pu + H(I - Q)Nu, \tag{2}$$

$$0 = Q(Eu - Nu). \tag{3}$$

Thus for the problem of forced oscillations (i) we may choose $X = Y = S = L_2[0,2\pi]$ a real Hilbert space and $Ex = x''$ with $\mathcal{D}(E) = \{x(t) \in S: x(t), x'(t)$ absolutely continuous,

$x''(t) \in L_2$, $x(0) = x(2\pi)$, $x'(0) = x'(2\pi)\}$. N: $\mathcal{D}(N) \subset S \to S$ is then the nonlinear Nemitsky operator generated by $f(t,x,x')$. If, for example, we choose $X_o = \ker E$, then $P = Q$ is the averaging operator and $H(I - Q): Y_1 \to X_1$ is the linear operator genrated by the Generalized Green's function corresponding to E.

3. PRELIMINARY REMARKS ON APPROXIMATE SOLUTIONS

In this section we discuss some methods to find an approximate solution to the nonlinear operator equation (1). If $v = Pu$ and T is the map defined by $Tu = v + H(I - Q)Nu$ then auxiliary equation (2) reduces to a fixed point problem. We assume that the restriction of E to X_o is bounded. The method of Cesari [3] may now be outlined as follows: let B be the bounded closed convex set in S defined by

$$B = \{u = v + w \in S,\ v \in X_o,\ w \in X_1;\ \|v - v_o\| \leq \delta;\ \|w\| \leq \eta\},$$

where v_o is a fixed element of $S \cap X_o$. If, for every $v_1 \in X_o$ such that $\|v_1 - v_o\| \leq \delta$, the operator $\tau\colon v_1 + w \to v_1 + H(I - Q)N(v_1 + w)$ is a strict contraction where $v_1 + w \in B$ then equation (2) has a unique solution $u = \tau(v_1) = v_1 + w$ on the set $C = \{u_1 = v_1 + w,\ \|w\| \leq \eta\} \subset B$.

Moreover the solution $\tau(v)$ is continuously dependent on v. Thus solving equation (3), the bifurcation equation, now reduces to solving the operator equation

$$Q(E\tau(v) - N\tau(v)) = 0. \tag{4}$$

If $\dim X_o < \infty$, then Cesari [3] obtains sufficient conditions for the solvability of (4). In the process we have shown that (1) has a solution $u = v + w$ such that $\|v - v_o\| \leq \delta$ and $\|w\| \leq \eta$. Thus v_o may be considered an approximate solution of (1) with these error estimates.

Note that in the above process only an estimate is found on the error and there is no attempt at improving the estimate. Clearly for the purpose of establishing existence of solutions the above will suffice. However the above method has, in the recent years, been adapted to obtain a method of successive approximations by a number of authors [2], [10], [8], [4].

We now outline this scheme of successive approximations. With the above definition of B, we now assume the following:

$$\mathcal{D}(N) \supset B; \ \|Nu\| \leq J \text{ for } x \in B \text{ and } \|Nu_1 - Nu_2\| \leq L\|u_1 - u_2\| \text{ for } u_1, u_2 \in B \tag{5}$$

$$K = \|I - DQE\| + \|D\|\|Q\|L < 1 \tag{6}$$

$$\|H(I - Q)\| \leq \ell, \ \ell J \leq \eta, \ K + \ell L + \|D\|\|Q\|\ell L^2 < 1 \tag{7}$$

$$\|DQEv_0 - N(v_0 + w)\| \leq (1 - K)\delta \text{ for all } \|w\| \leq S \tag{8}$$

where $D: Y_0 \to X_0$ is a bounded linear operator with trivial kernel. Consider now the method of successive approximations

$$u_k = v_k + w_k, \ u_k \in X_0, \ w_k \in X_1, \ Pu_k = v_k, \ w_0 = 0$$

$$w_k = H(I - Q)N(v_{k-1} + w_{k-1})$$

$$v_k = v_{k-1} - D[QEv_{k-1} - QN(v_{k-1} + w_k)].$$

Under the above hypotheses it can be shown [4] that $u_k \in B$ for all k and converges to a solution u of the system (2) and (3), i.e., a solution u of (1).

We conclude this section with some remarks on the work of Urabe [11,12] for finding approximate solutions of the problem of forced oscillations. Thus we consider the problem

$$\begin{aligned} x'' &= f(t,x,x') \\ x(0) &= x(2\pi), \quad x'(0) = x'(2\pi) \end{aligned} \tag{9}$$

for the existence of 2π-periodic solutions. We seek a solution of the type $x(t) = a_o + \sum_1^\infty a_n \cos nt + \sum_1^\infty b_n \sin nt$. The method used by Urabe is a Galerkin approach to the above problem. Thus, for any x(t) expressed in a Fourier series as above, let $R_m x = a_o + \sum_1^m a_n \cos nt + \sum_1^m b_n \sin nt$. We now consider the problem

$$x''_m = R_m f(t, x_m, x'_m)$$
$$x_m(0) = x_m(2\pi), \quad x'_m(0) = x'_m(2\pi) \tag{10}$$

for the existence and approximation of solutions. Under sufficient hypotheses on f, Urabe [11] proves that the existence of an isolated solution of (9) implies the existence of an isolated solution of (10) for all m (sufficiently large) and conversely. The approximate solution of (10) is then obtained as follows. Clearly if $x_m = a_o + \sum_1^m a_n \cos nt + \sum_1^m b_n \sin nt$, then we can rewrite (10) as a system of algebraic equations by using the fact that sin nt, cos nt, n = 0,1,... form an orthogonal system. Thus we get the system of equations

$$\int_0^{2\pi} f(t, a_o + \sum_1^m a_n \cos nt + \sum_1^m b_n \sin nt, x'_m) = 0$$

$$\int_0^{2\pi} f(t, a_o + \sum_1^m a_n \cos nt + \sum_1^m b_n \sin nt, x'_m) \cos nt = -n^2 a_n$$

$$\int_0^{2\pi} f(t, a_o + \sum_1^m a_n \cos nt + \sum_1^m b_n \sin nt, x'_m) \sin nt = -n^2 b_n.$$

This system is now solved for approximate values of a_o, a_n, b_n, n = 1,...,m thereby giving an approximate solution $x(t) \cong a_o + \sum_1^m a_n \cos nt + \sum_1^m b_n \sin nt$.

4. OUTLINE OF NUMERICAL METHODS

A. Galerkin Approach

Before we discuss the problem of numerical methods we recall that $\dim X_o < \infty$. Let $p: Y_o \to X_o$ be a continuous operator such that $p^{-1}(0) = 0$ and further let p map bounded subsets of Y_o into bounded subsets of X_o. The system of equations (2) and (3) can be written as a fixed point problem in the following two ways.

(I) Let $T_1: (v,w) \to (\bar{v},\bar{w})$ be defined by

$$\begin{aligned} \bar{w} &= H(I - Q)N(v + w) \\ \bar{v} &= v + pQ(E - N)(v + w) \end{aligned} \tag{11}$$

Then solving the system (2) and (3) is equivalent to finding a fixed point of T_1.

(II) Let $T: X \to X$ be defined by

$$Tu = Pu + H(I - Q)Nu + pQ(E - N)u. \tag{12}$$

Problem (2), (3) is now equivalent to finding a fixed point of T.

We now concentrate on case II. Thus we have reduced the nonlinear operator equation (1) to a single nonlinear operator equation (12). For the sake of simplicity, in the following discussion we assume X = Y = a real Hilbert space S, $X_o = Y_o$, $X_1 = Y_1$ and P = Q (as in the case of E self-adjoint). The operator p then reduces to the identity operator and equation (12) can then be reduced to

$$u = Pu + N(I - P)Nu + P(Eu - Nu). \tag{13}$$

Let S_m be a sequence of closed subspaces of S and let $B_m = S_m \cap B$. Let R_m be the corresponding sequence of projection operators of S onto S_m. We can then consider the corresponding sequence of approximate operator equations

$$u_m = T_m u_m = R_m[Pu_m + H(I - P)Nu_m + P(Eu_m - Nu_m)] \text{ over the space } S_m. \tag{14}$$

If we assume that the nonlinear operator N is Fréchet-differentiable, then so are the operators T and T_m. It can then be shown that if $I - T'(u_o)$ for $u_o \in B$ is continuously invertible, then so is $I - T_m'(R_m u_o)$ for all sufficiently large m and conversely, where u_o is the solution (isolated) of (1). Under assumptions leading to the closeness of T_m and T, we can then show that the approximate equations $u_m = T_m u_m$ are uniquely solvable in a neighborhood of u_o and the sequence of solutions u_m converges to a solution of (1). It must be noted that in the above discussions we do not require that u_o be the solution of (1), but close to a solution of (1).

Note that in the above discussion we could have instead assumed that P^m is a sequence of operators such that $u - P^m u \to 0$ for every $u \in S$ and then equation (13) gives rise to a sequence of problems, each of which is equivalent to (1), given by

$$u = P^m u + H(I - P^m)Nu + P^m(Eu - Nu).$$

or

$$0 = (I - P^m)u + H(I - P^m)Nu + P^m(Eu - Nu). \tag{15}$$

The assumption $u - P^m u \to 0$ implies that the operator equation

$$0 = P^m(Eu - Nu) \tag{16}$$

can then be treated as an approximation to (15) or equivalently (1). In the case of the problem of forced oscillations (16) is precisely the set of algebraic equations of Urabe referred to in the previous section. For a detailed analysis of the sufficient conditions for the convergence of the solutions of the approximate equations to a solution of the problem (1) and the corresponding error estimates we refer to [7].

B. Modified Successive Approximations

As seen before, the nonlinear operator equation (1) is equivalent to the system of operator equations

$$w = H(I - Q)N(v + w)$$
$$0 = Q(E(v + w) - N(v + w)).$$

The iterative scheme discussed in the earlier section rests on a one step iteration at each stage. However, the hypothesis $K + \ell L + \|D\|\|Q\|\ell L^2 < 1$ implies that the operator $H(I - Q)N$ with a Lipschitz constant of ℓL is a strict contraction for each v_{k-1}. Thus we can define w_{k+1} as the nth-iterate of the map $H(I - Q)N$ acting on $v_k + w_k$ where n is large enough to attain a desired accuracy. Then we are replacing $H(I-Q)N(v_k+w_k)$ by $F_m(v_k,w_k)$ where $F_m(v_k,w_k)$ converges (in m) to the solution of $w = H(I - Q)N(v_k + w)$. Clearly the choice of $F_m(v_k,w_k)$ is not unique. Note that the compactness of $H(I - Q)N$ enables us to choose, instead of the iterates of $H(I - Q)N$, any collectively compact sequence of operators [1] which converges to $H(I - Q)N$. Thus if we consider the problem of forced oscillations then $H(I - Q)$ has a Generalized Green's function representation in terms of its kernel $K(t,s)$ and the auxiliary equation can be written as an integral equation

$$w(t) = \int_0^{2\pi} K(t,s)f(s,u(s),u'(s))ds.$$

One could then use an appropriate numerical integration procedure so that for $F_m(v_k,w_k)$ we have $\sum^m w_j K(s_j,t)f(s_j,u(s_j), u'(s_j))$. A third choice for $F_m(v_k,w_k)$ would be to use a finite difference procedure to solve the auxiliary equation. As can be seen easily, the auxiliary equation can be also written as

$$Ew = (I - Q)N(v + w)$$

so that for $F_m(v_k,w_k)$ we could use the approximate solution of the equation

$$E_m w = (I - Q)N(v_k + w_k)$$

where E_m is an approximation to E by the method of finite differences.

For a detailed analysis of these various choices, the convergence rates etc. and application to ordinary differential equations of the type (i) (cf. Section 1) and parabolic equations of the type (iii) (cf. Section 1) one is referred to [7].

C. Error Analysis

We conclude this section with an analysis of the bifurcation equations. As was remarked in Section 3, the auxiliary equation may be solved uniquely in the set B for each v such that $\|v - v_o\| \leq \delta$. Thus the bifurcation equation reduces to

$$Q(E\tau(v) - N\tau(v)) = 0 \tag{17}$$

where $\tau(v)$ is the unique solution corresponding to v. Comparing this with the equation

$$Q(Ev - Nv) = 0 \tag{18}$$

it is clear that solvability of (18) implies the solvability of the implicit equation (17) when "$\tau(v)$ is very close to v" for every v such that $\|v - v_o\| \leq \delta$. But $\tau(v) = v + w$ where $w = H(I - Q)N(v + w)$ and thus we require that "w be small". Thus, assuming sufficient continuity hypotheses on N, the problem of existence and error estimates reduces to a study of the equations (17) and (18). The two major questions thus are:

(i) how small should $w = \tau(v) - v$ be in order that solvability of (18) implies solvability of (17) and conversely;

(ii) rate of convergence of solutions of equations close to (17) by any of the methods of approximating $\tau(v)$.

In fact for the example

$$x'' = -x^3 + \sin t$$

$$x(0) = x(2\pi), \quad x'(0) = x'(2\pi) = 0$$

considered by Cesari [5], where we look for odd periodic solutions, it can be shown by virtue of the abstract theorem in [7] that if we choose dim $X_0 = 2$, $X_0 = \langle \sin t, \sin 3t \rangle$, then $\tau(v)$ is "small" enough and it is sufficient to find only the first two terms by Galerkin's method for v_0, as opposed to Urabe's result where higher order Galerkin approximations were found. Applications of these ideas to other classes of periodic boundary value problems may be seen in [7].

D. Remarks on Partial Differential Equations

Unlike the case of ordinary differential equations in the case of elliptic partial differential equations as in (ii) (cf. Section 1) the operator $H(I - Q)$ is often not explicitly known in terms of its Green's function. However looking at the auxiliary equation (2) as

$$Ew = (I - Q)N(v + w)$$

we can consider the corresponding step of the iterative process as

$$Ew_{k+1} = (I - Q)N(v_k + w_k).$$

Note that the operator E has a nontrivial kernel. However since we are looking for a w_{k+1} belonging to Y_0 the discrete analogue of the above system will be consistent. Thus in order to find a w_{k+1} satisfying the above linear problem we have to find an iterative procedure for a singular system which is consistent. For an analysis of these classes of problems we refer to [9].

Finally we consider the problem of periodic solutions of hyperbolic equations (iv). Noting the fact that a solution $u(t,x)$ of (iv) can be ecpressed in the form $\sum a_k(t) \sin kx$ we can now write the approximate problem by looking at an approximate Galerkin solution of the type $u_m(t,x) = \sum_1^m a_k(t)\sin kx$. For each m, $u_m(t,x)$ now satisfies a coupled nonliner system of ordinary differential equations given by

$$a_k''(t) + k^2 a_k(t) = f(t,x,\sum_1^m a_k(t)\sin kx).$$

we are now back in the setting of the earlier discussions and thus one can study the convergence of the $u_m(t)$ to a solution of (iv) and the corresponding error estimates [6].

REFERENCES

1. P.M. Anselone: Collectively Compact Operator Approximation Theory, Englewood Cliffs, Prentice Hall, 1971.
2. C. Banfi, G. Casadei: Calcolo di soluzioni periodiche di equazioni differenziali nonlineari, Calcolo, 5, 1-10, (1968).
3. L. Cesari: Functional Analysis and Galerkin's Method, Michigan Math. J., 11, 385-414, (1964).
4. L. Cesari: Alternative Method, Finite Elements and Analysis in the Large, Proceedings Int. Conf. Diff. Equations, Uppsala, 11-25, (1977).
5. L. Cesari: Functional Analysis and Periodic Solutions of Nonlinear Differential Equations, in Contributions to Differential Equations, New York, John Wiley, 149-187, 1963.
6. M. Countryman: Numerical Methods for Periodic Solutions of Nonlinear Hyperbolic Equations, Ph.D. dissertation, Univ. of Texas at Arlington.
7. R. Kannan, K.J. Morel: Numerical Methods for Periodic Solutions of Nonlinear Ordinary Differential Equations, to appear.
8. D. Ku: Boundary Value Problems and Numerical Estimates, Ph.D. dissertation, Univ. of Michigan, (1976).

9. M. Ray: Numerical Analysis of Elleptic Boundary Value Problems at Resonance, Ph.D. dissertation, Univ. of Texas at Arlington.

10. D.A. Sanchez: An Iteration Scheme for Boundary Value Alternative Problems, Numer. Math., 23, 223-230, (1975).

11. M. Urabe: Galerkin's Procedure for Nonlinear Periodic Systems, Arch. Rat. Mech. Anal., 20, 120-152, (1965).

12. M. Urabe, A. Reiter: Numerical Computation of Nonlinear Forced Oscillations by Galerkin's Procedure, J. Math. Anal. Appl., 14, 107-140, (1966).

ON ORBITAL STABILITY AND CENTER MANIFOLDS

Nicholas D. Kazarinoff

S.U.N.Y. at Buffalo, N.Y.

This talk is an exposition of two "folk theorems". It represents joint work of Y.H. Wan, Brian Hassard, and myself. Why give a talk on folk theorems? I can only answer that it is comforting to both speaker and audience to talk about results we already know: nearly everyone can nod knowingly.

The Setting. The background assumed is Hopf Bifurcation. I assume that we are in the case of Hopf Bifurcation in a real Hilbert space H. The operator equation is

$$\dot{x} = f(x,\mu) = L_\mu x + h_\mu(x), \quad (x \in U \subset H, \quad \dot{} = \frac{d}{dt}), \tag{1}$$

where I assume

U is a neighborhood of $0 \in H$,
$h_\mu(0) = Dh_\mu(0) = 0$ for $-\varepsilon < \mu < \varepsilon$,
$\sigma(L_0) = \pm i\omega_0 \bigcup_{j=3}^{\infty} \lambda_j(0) \quad (\text{Re } \lambda_j(0) \leq -d < 0)$,
$\lambda_1(\mu) = \bar{\lambda}_2(\mu) = i\omega_0 \quad (\omega_0 > 0)$ at $\mu = 0$,
$d \text{ Re } \lambda_1(\mu)/d\mu$ at $\mu = 0$ is not zero (transversality),
L_μ is the infinitesimal generator of an analytic semigroup $\exp(L_\mu t)$,
h_μ is an H-valued C^r-function ($r \geq 5$) defined on U, and there exists exactly one bifurcated periodic solution of (1) in U for each μ in either $(-\varepsilon,0)$ or $(0,\varepsilon)$ for some $\varepsilon > 0$.

ISBN 0-12-508780-2

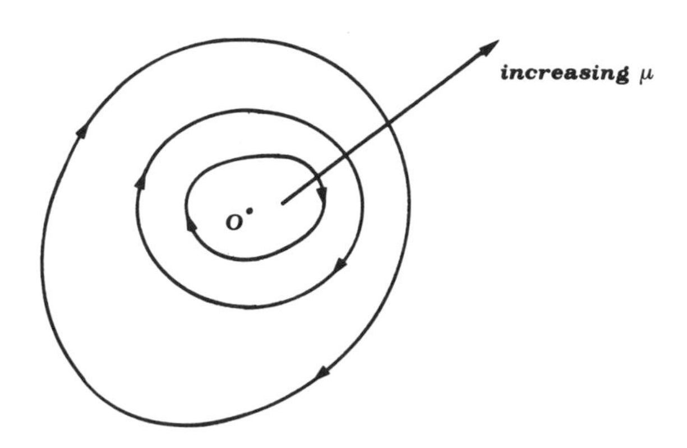

Fig. 1. A family of periodic orbits bifurcating from $0 \in H$, one for each $\mu \in (0,\varepsilon)$, say.

At the time (1) was studied by Hopf in the case $H = R^n$ ($n \geq 2$) it was not understood that:

I. The bifurcating periodic solutions of (1) lie on an invariant manifold, the center manifold;

and

II. If the bifurcating periodic solutions of (1) are asymptotically, orbitally stable within the center manifold, then they are asymptotically, orbitally stable in H if $\operatorname{Re} \lambda_j(0) \leq -d < 0$ $(j \leq 3)$.

For $H = R^n$ Hopf settled (II). The understanding of (I), even for $H = R^n$, took twenty five more years until A. Kelley proved the center manifold theorem in R^n [4]. In the case of delay-differential equations (H is infinite-dimensional) N. Chafee gave a complete account of both (I) and (II) [1]. In the case of partial differential equations of parabolic type D. Henry developed a complete theory for (I) and (II) [3], and J. Marsden and M. McCracken gave a complete theory in their book [5]. Recently, P. Negrini and A. Tesei have also studied (I) and (II) for the case of (1) in a Banach space [7]. The clearest account of which I know, in the case $H = R^n$, is given by P. Hartman in his book [2].

Since 1976, after reading the book by Marsden and McCracken

carefully, B. Hassard, Y.H. Wan, and I have been concerned about (I) and (II) in the infinite-dimensional case. While believing the results, we felt that the proofs in Marsden-McCracken were incomplete. Upon visiting Europe and, in particular, Italy in 1978, I became aware that this concern was shared by others.

P. Negrini and A. Tesei have settled (I) and (II) in the infinite-dimensional case by an approach different from that in Marsden-McCracken or Hartman. My goal in this talk is to outline proofs of the Center Manifold Theorem and (II) for (1) based upon the work of Marsden-McCracken and Hartman.

Theorem I. (The Center Manifold Theorem). Under the above hypotheses, (1) equation (1) has a C^{r-1} center manifold M, which is locally invariant, and (2) there exists an open neighborhood N of the origin in H such that if $\phi_\mu^t(x_o)$ is a solution of equation (1) with initial point $x_o \in N$ and $\phi_\mu^t(x_o) \in N$ for all $t \geq 0$, then $\|\phi_\mu^t(x_o) - M\| \to 0$ at an exponential rate, i.e., M is locally attracting.

Theorem II. The Hopf bifurcating solutions of (1) are asymptotically, orbitally stable in H if the corresponding periodic solutions of (1) restricted to its center manifold M are asymptotically, orbitally stable and provided $\mathrm{Re}\, \lambda_j(0) \leq -d < 0$ for $j \geq 3$.

I thank Professor L. Salvadori for steadily encouraging me to find an elementary proof of Theorem II.

Outline of Proofs. It is convenient to restate the spectral hypotheses on equation (1) at this point, introducing some notation.

Hypotheses. At $\mu = 0$, $H = V_c \oplus V_s$ (c for center and s for stable); and

(i) $\mathrm{Re}\{\sigma(L_o|V_s)\} \leq -d < 0$ for some $d > 0$,

(ii) $\mathrm{Re}\{\sigma(L_o|V_c)\} = 0$ and $\dim V_c = 2$,

(iii) $\sigma(e^{L_\mu t}) = e^{\sigma(L_\mu)t} \cup \{0\}$ if $t > 0$, and

(iv) transversality holds for $\sigma(L_\mu|V_c)$.

Note. In the case of Theorem I, (iv) is unnecessary and in

stead of (ii) one need only assume that dim $(L_0|V_c)$ is finite and positive. Also, elements of $\sigma(L_0)$ may be allowed to have positive real parts. Of course, if some do, then Theorem II cannot hold.

Under these hypotheses each $x \in H$ can be written as

$$x = (x_c, x_s), \text{ where } x_c \in V_c \text{ and } x_s \in V_s.$$

Thus equation (1) becomes (we suspend the equation $\dot{\mu} = 0$ to (1))

$$\begin{aligned} \dot{x}_s &= Ax_s + X(x_c, x_s) \\ \dot{x}_c &= Cx_c + Y(x_c, x_s) \\ \dot{\mu} &= 0 \end{aligned} \qquad \left\{\begin{aligned} &\text{Re } \sigma(A) \leq -d < 0, \\ &\text{Re } \sigma(C) = 0 \end{aligned}\right\} \tag{2}$$

Here X and Y are C^r in U and contain no linear terms.

I first outline the proof of Theorem II, assuming Theorem I. Let $\phi_\mu^t(x_0)$ be the solution of the suspended system (2) with initial data (x_0, μ). Then by Theorem I, on M, the center manifold for (2),

$$x_s = g_\mu(x_c),$$

where $g_\mu(\cdot)$ is a C^{r-1} function. Further, by the local attractivity of M, there exists a neighborhood N of 0 in H, a $\mu_0 > 0$, and a $K > 0$ with $0 < K < 1$ such that if we write

$$\phi_\mu^t(x_0) = (x_c(t,\mu), x_s(t,\mu), \mu)$$

and

$$w_\mu(t) = x_s(t,\mu) - g_\mu(x_c(t,\mu)),$$

then

$$\|w_\mu(t)\|_H \leq K^t \|w_\mu(0)\|_H$$

provided $|\mu| \leq \mu_0$

$$\bigcup_{\tau=0}^{t} (x_c(\tau,\mu), x_s(\tau,\mu)) \subset N.$$

Consider μ fixed with $|\mu|$ so small that (α) $|\mu| \le \mu_0$ and (β) the bifurcated periodic solution γ_μ of (2) lies in N. (Note that if $\gamma_\mu^* = (x_c(\cdot,\mu),\mu)$ is a bifurcated periodic solution of (2) restricted to M, namely, of

$$\begin{aligned} \dot{x}_c &= Cx_c + Y(x_c, g_\mu(x_c)) \\ \dot{\mu} &= 0 \end{aligned} \tag{3}$$

then $\gamma_\mu = (x_c(\cdot,\mu), g_\mu(x_c(\cdot,\mu)),\mu)$ is a bifurcated solution of (2). Now, if γ_μ^* is asymptotically, orbitally stable in M, then, by a theorem of Massera [6], γ_μ^* has a Lyapunov function $V_\mu(x_c)$ that is positive definite in some neighborhood of γ_μ^* in M and whose derivative $\dot{V}_\mu(x_c)$ taken along trajectories of (3), is negative definite in that neighborhood. We next extend the definition of $V_\mu(\cdot)$; namely if $\Phi_\mu^t(x_0)$ is a solution of (2), we write $V_\mu(x_c)$ to mean the Lyapunov function $V_\mu(\cdot)$ for γ_μ^* evaluated at the finite-dimensional component x_c of $(x_c,x_s,\mu) = \Phi_\mu^t(x_0)$.

The idea behind the proof is to take advantage of the strict attraction exerted on any trajectory of (1), with initial point sufficiently near to γ_μ, by the center manifold and by γ_μ in the x_c-coordinates. To combine these two kinds of attraction effectively, for $\eta > 0$ and $\delta > 0$ (η and δ small), we define

$$B_{\delta,\eta} = \{(x_c,x_s,\mu) \mid \|w_\mu\| \le \eta \text{ and } V_\mu(x_c) \le \delta\}.$$

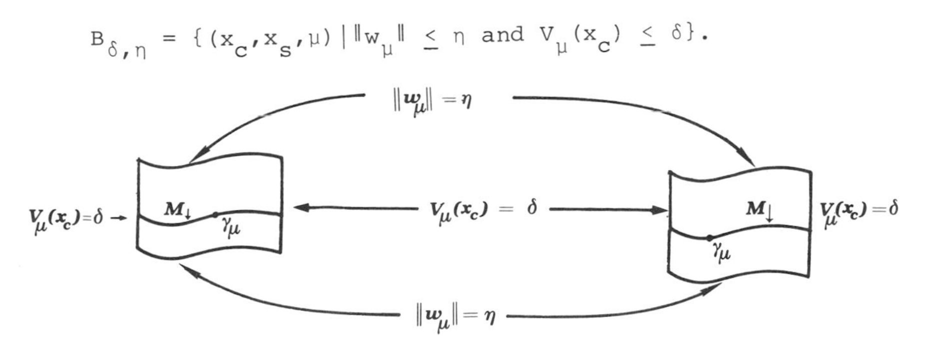

Fig. 2. $B_{\delta,\eta}$

Now, for δ small enough, (a) $B_{\delta,0} \subset N$ and (b) $\dot{V}_\mu(x_c) \leq 0$ on $B_{\delta,0}$ (= 0 only on the periodic orbit γ^*_μ). Having fixed $\delta > 0$ satisfying (a) and (b), we next show that if η is chosen to be small enough, then $B_{\delta,\eta}$ is positively invariant with respect to (2). First, on that portion of $\partial B_{\delta,\mu}$ where $\|w_\mu\| = \eta$, each trajectory of (2) enters $B_{\delta,\eta}$ since M is locally attracting, provided η is small enough. Second, let S be the remaining portion of $\partial B_{\delta,\eta}$, where $\|w_\mu\| < \eta$ but $V_\mu(x_c) = \delta$. Then $\dot{V}_\mu(x_c,x_s)<0$ on S. To see this, recall that in N

$$\dot{x}_c = Cx_c + Y(x_c,x_s).$$

By direct computation,

$$\dot{V}_\mu(x_c,x_s) = \operatorname{grad} V_\mu(x_c)\cdot\dot{x}_c = \operatorname{grad} V_\mu(x_c)\cdot[Cx_c + Y(x_c,x_s)]. \tag{4}$$

But

$$Y(x_c,x_s) = Y(x_c,g_\mu(x_c)) + [Y(x_c,x_s) - Y(x_c,g_\mu(x_c))].$$

Hence (4) becomes

$$\dot{V}_\mu(x_c,x_s) = \dot{V}_\mu(x_c,x_s)\Big|_{M\cap S} + \operatorname{grad} V_\mu(x_c)\cdot[Y(x_c,x_s) - Y(x_c,g_\mu(x_c))].$$

But on $M \cap S$, $\dot{V}_\mu$ is strictly negative independently of η:

$$\dot{V}_\mu(x_c,x_s)\Big|_{M\cap S} \leq -c(\delta) < 0,$$

where $c(\delta)\downarrow 0$ as $\delta\downarrow 0$, since on S, $\|x_c - \gamma_\mu\| > k_1 > 0$ for some $k_1 > 0$. Furthermore, since $Y \in C^r$ $(r \geq 5)$, the following inequality holds on S for some $K > 0$:

$$\|Y(x_c,x_s) - Y(x_c,g_\mu(x_c))\| \leq K\|x_s - g_\mu(x_c)\| < K\eta.$$

If now η is chosen to be $c(\delta)/(2K)$, then $-c(\delta) + K\eta < 0$. Hence $\dot{V}_\mu(x_c,x_s) < 0$ on S; and since M is strictly attracting on $\partial B_{\delta,\eta(\delta)}\setminus S$, here $\eta(\delta) = c(\delta)/(2K)$, $B_{\delta,\eta(\delta)}$ is positively invariant for (2).

Pick a $\delta = \delta_1$ for which $B_{\delta_1,\eta(\delta_1)}$ is positively invariant. Suppose there exists a trajectory $\Phi_\mu^t(x_o)$ with $x_o \in \partial B_{\delta_1,\eta(\delta_1)}$ that does not enter each $B_{\delta,\eta(\delta)}$ with $0 < \delta < \delta_1$. In particular, suppose this trajectory does not enter $B_{\delta_2,\eta(\delta_2)}$. But since M is exponentially attracting, this trajectory can only remain a finite time in the portion of $D = B_{\delta_1,\eta(\delta_1)} \setminus B_{\delta_2,\eta(\delta_2)}$ where $\|w\| > \eta(\delta_2)$. However, this trajectory cannot remain an infinite time in the remaining portion of D since then V_μ would approach $-\infty$ along this trajectory because of the existence of a negative upper bound on $\dot{V}_\mu$ in this portion of D. Thus we have a contradiction, and every trajectory of (2) that enters $B_{\delta_1,\eta(\delta_1)}$ enters each $B_{\delta,\eta(\delta)}$ with $0 < \delta < \delta_1$. Consequently, γ_μ is asymptotically, orbitally stable.

I now turn to the proof of Theorem I. The parameter μ will be suppressed throughout. This proof was constructed by Yieh-Hei Wan. Choose a δ with $0 < \delta < d$. One can find a norm on H, equivalent to the original norm, and such that in the new norm

$$\|e^{At}\| \le e^{-dt} \quad \text{for} \quad t > 0 \quad \text{and} \quad e^{-\delta} \le \|e^{-C}\| \le e^{+\delta}$$

Further, following Marsden-McCracken and Hartman, one can "cut off" the semiflow generated by (1) outside an ε-neighborhood of $0 \in H$ and obtain a system

$$\begin{aligned} \dot{y} &= Cy + \bar{Y}(y,z) \\ \dot{z} &= Az + \bar{Z}(y,z) \end{aligned} \tag{5}$$

with $\varepsilon y = x_c$ and $\varepsilon z = x_s$, that is globally defined for $(y,z) \in H$, which agrees with (1) in some (small) neighborhood of 0, which is linear outside of some larger neighborhood of 0 ($\bar{Y} \equiv \bar{Z} \equiv 0$), and which generates a C^r-semiflow $\Phi_t(y_o,z_o)$ with the following properties. If we define

$$\bar{Y}_{t,\varepsilon}(y_o,z_o) = y_t - e^{Ct}y_o$$

and

$$\bar{Z}_{t,\varepsilon}(y_o,z_o) = z_t - e^{At}z_o,$$

then

$$\bar{Y}_{t,\varepsilon}(y_o,z_o) \quad \text{and} \quad \bar{Z}_{t,\varepsilon}(y_o,z_o) = 0 \quad \text{for} \quad \|y_o\| > 1$$

uniformly for $0 \leq t \leq 1$; and

$$\lambda(\bar{Y}_{t,\varepsilon},\bar{Z}_{t,\varepsilon}) \stackrel{d}{=} \sup\{\|D_z^{j_1} D_y^{j_2} \bar{Y}_{t,\varepsilon}(y_o,z_o)\|,$$
$$\|D_z^{j_1} D_y^{j_2} \bar{Z}_{t,\varepsilon}(y_o,z_o)\|; (y_o,z_o) \in H; 0 \leq j_1 + j_2 \leq r\} \to 0$$

as $\varepsilon \to 0$, <u>uniformly</u> in t for $t \in [0,1]$.

The time-one map is $\Phi_1(y_o,z_o) = (y_1,z_1)$. The existence of an invariant manifold for this map is guaranteed by the following lemma proved by Marsden-McCracken (Lemma 2.4, p. 32 of their book).

<u>Lemma</u>. If δ is close enough to 0, then for sufficiently small ε, there exists an invariant manifold for Φ_1 that is defined on X_c by a C^{r-1} function $z = g_\varepsilon(y)$ with $g_\varepsilon(0) = 0$, and $\|Dg_\varepsilon(y)\| < 1$. Furthermore,

$$\|z_n - g_\varepsilon(y_n)\| \to 0 \quad \text{as} \quad n \to \infty,$$

where

$$(y_n,z_n) = \Phi_1^n(y_o,z_o) = \Phi_n(y_o,z_o).$$

Fix any such small ε and δ, say $\varepsilon = \varepsilon_o$ and $\delta = \delta_o$ with

$$c - a > 4\lambda,$$

where

$$a = \|e^A\| \quad \text{and} \quad 1/c = \|e^{-C}\|.$$

Following Hartman (Chapter IX), one can show that

(α) If $z_o = g_{\varepsilon_o}(y_o)$, then, $\|y_m\| \geq (c - 2\lambda)^m \|y_o\|$ for $m = 1,2,\ldots$

(β) $\|v_m\| \leq (a + 2\lambda)^m \|v_o\|$ for $m = 1,2,\ldots$, where $v_t = z_t - g_{\varepsilon_o}(y_t)$.

Furthermore, one can show that

(γ) For $c > 2\lambda$ the restriction of Φ_1 to the invariant manifold $z = g_{\varepsilon_o}(y)$ is a diffeomorphism (onto).

The result (γ), which is critical to this proof, follows from

<u>Proposition</u>. Let $\varphi: R^n \to R^n$ be a C^1-map of the form

$$\varphi(y) = By + G(y), \tag{6}$$

where B is a nonsingular constant matrix. If

$$\|B^{-1}\| \sup_y \|DG(y)\| < 1,$$

then φ is a diffeomorphism onto.

To establish (γ) one applies this Proposition to the time-one map

$$\Phi_1(y) = e^C y + \bar{Y}_{1,\varepsilon_o}(y, g_{\varepsilon_o}),$$

which has the form (6).

The point in Marsden-McCracken's proof of Theorem I that appears to need elaboration is how to get the existence of a unique invariant manifold for the semiflow Φ_t on a <u>continuous interval of time</u>, say for all $t_o \in (0,1)$. The lemma (γ) of Wan and a clever backward time argument by Wan do the job.

To outline a proof of part (1) of Theorem I it suffices to show that the manifold given by $z = g_{\varepsilon_o}(y)$ is invariant for all the maps Φ_{t_o} for t_o between 0 and 1. Let $z_o = g_{\varepsilon_o}(y_o)$. By (γ), one can find (y_{-n}, z_{-n}) $(n = 1,2,\ldots)$ such that

$$z_{-n} = g_{\varepsilon_o}(y_{-n}) \quad \text{and} \quad \Phi_n(y_{-n}, z_{-n}) = (y_o, z_o).$$

By the facts that $\|Dg_{\varepsilon_o}(y)\| < 1$ and $g_{\varepsilon_o}(0) = 0$, and (β), one obtains

$$\|z_{-n+t_0}\| + \|y_{-n+t_0}\| \geq \|z_{-n+t_0}\| + \|g_{\varepsilon_0}(y_{-n+t_0})\|$$

$$\geq \|v_{-n+t_0}\| \quad \geq (a + 2\lambda)^{-n}\|v_{t_0}\|. \tag{7}$$

Since

$$\|y_{-n+t_0} - e^{At_0}z_{-n}\| \leq \lambda(\|y_{-n}\| + \|z_{-n}\|),$$

$$\|y_{-n+t_0}\| + \|z_{-n+t_0}\| \leq 2[\|y_{-n}\|e^d + \lambda(\|y_{-n}\| + \|y_{-n}\|)]$$

$$\leq 2\|y_{-n}\|[e^d + 2\lambda]. \tag{8}$$

By (7), (8), and (α),

$$2(e^d + 2\lambda)(c - 2\lambda)^{-n}\|y_0\| \geq (a + 2\lambda)^{-n}\|v_{t_0}\|,$$

or

$$2(e^d + 2\lambda)(a + 2\lambda)^n(c - 2\lambda)^{-n}\|y_0\| \geq \|v_{t_0}\|,$$

Let $n \to \infty$. Since $c - a > 4\lambda$, one obtains $v_{t_0} = 0$. Thus $z = g_{\varepsilon_0}(y)$ is invariant under the map Φ_{t_0} <u>for any</u> t_0 <u>between</u> 0 <u>and</u> 1.
The proof of part (2) of Theorem 1 is straightforward. Let

$$\theta = \sup_{(y,z)\in H} \{\|D_y\bar{Y}(y,z)\|, \|D_y\bar{Z}(y,z)\|; \|D_z\bar{Y}(y,z)\|, \|D_z\bar{Z}(y,z)\|\}. \tag{9}$$

Clearly, $\theta \to 0$ as $\varepsilon \to 0$. Thus we may assume that $-d + 2\theta < 0$. We now write $v(t)$ and $g(y)$ for v_t and $g_{\varepsilon_0}(y)$, respectively, for convenience. We shall prove that

$$\|v(t)\| \leq \|v(0)\|e^{(-d+2\theta)t} \qquad \text{for } t \geq 0.$$

The set of initial points of C^1-solutions of (5) is dense in H. Thus one can assume (y_t, z_t) is C^1 for $t \geq 0$. The operator equation for $v = z - g(y)$ is

$$\dot{v} = Av + V(y,v),$$

where $V(y,0) = 0$ and

$$V(y,v) = Ag(y) + \bar{Z}(y,v+g(y)) - (Dg)(Cy + \bar{Y}(y,v+g(y))).$$

Now

$$V(y,v) \leq \left\{ \sup_{(y,z)\in H} \|D_v V\| \right\} \|v - 0\|. \tag{10}$$

Moreover, since $\|Dg\| < 1$ (by Marsden-McCracken's Lemma),

$$\|D_v V\| = \|D_z \bar{Z}(y,v+g(y)) - (Dg) D_z \bar{Y}(y,v+g(y))\| \tag{11}$$

$$\leq 2\theta,$$

by (9).

Now let

$$h(t) = e^{dt} v(t)$$

Then

$$v(t) = e^{At} v(0) + \int_0^t e^{A(t-s)} V(y(s),v(s))ds;$$

and, by (10) and (11),

$$h(t) \leq h(0) + \int_0^t (2\theta) h(s) ds. \tag{12}$$

We apply Gronwall's Lemma to (12), and we obtain for $t \geq 0$

$$h(t) \leq h(0)e^{2\theta t} \quad \text{or} \quad \|v(t)\| \leq \|v(0)\| e^{(-d+2\theta)t}.$$

This completes our outline of the proof of Theorem I.

I close with two remarks.

Remark 1. The Negrini-Tesei proof of Theorem I is cleaner and much shorter than the proof we have outlined; while the proof of Theorem II given by Negrini-Tesei appears to be both much more complicated and longer than ours.

Remark 2. If one allows the nonlinear part of $h_\mu(x)$ in (1) to vary subject to the constraint that the norms of X and Y in (2), as well as the norms of all their partial derivatives with respect to x_c and x_s of orders less than or equal to r, are uniformly bounded in U, say, by 1, then there exists a single neighborhood of 0 in which all of the center manifolds for (1), corresponding to the allowed functions h_μ, exist.

REFERENCES

1. N. Chafee: A Bifurcation Problem for Functional Differential Equations of Finitely Retarded Type, J. Math. Anal. Appl. 35, 312-348 (1971).
2. P. Hartman: Ordinary Differential Equations, Baltimore, (1973).
3. D. Henry: Geometric Theory of Semilinear Parabolic Equations, Lecture Notes, University of Kentucky, (1974).
4. A. Kelley: The Stable, Center-Stable, Center, Center-Unstable Manifolds, Appendix C in R. Abraham, J. Robbin: Transversal Mapping and Flows, New York, Benjamin, (1968).
5. J. Marsden, M. McCracken: The Hopf Bifurcation and Its Applications, Berlin-Heidelberg-New York, Springer, (1976).
6. J.L. Massera: On Liapunoff's Conditions of Stability, Ann. Math., 50, 705-721, (1949).
7. P. Negrini, A. Tesei: Attractivity and Hopf Bifurcation in Banach Spaces, J. Math. Anal. Appl., to appear.

A BUNCH OF STATIONARY OR PERIODIC SOLUTIONS NEAR AN EQUILIBRIUM BY A SLOW EXCHANGE OF STABILITY

Hansjörg Kielhöfer

Institut für Angewandte Mathematik und Statistik
Universität Würzburg
Fed. Rep. Germany

We consider the nonlinear evolution equation

$$\frac{du}{dt} + G(\lambda,u) = 0 \tag{1}$$

depending on a real parameter λ in some real Banach space E. If E is finite dimensional this equation represents an ordinary dynamical system, if E is infinite dimensional it is the abstract version of some class of nonlinear parabolic partial differential equations. The assumptions imposed admit, for instance, the Navier Stokes system in its Hilbert space formulation such that the class of possible applications is large enough.

We assume that for some λ-interval Λ equation (1) has a stationary solution $v(\lambda)$, i.e. $G(\lambda,v(\lambda)) = 0$ for $\lambda \in \Lambda$. Without loss of generality (see, [1]) we can set $v(\lambda) = 0$ and

$$G(\lambda,0) = 0 \text{ for all } \lambda \in \Lambda \subset \mathbb{R} . \tag{2}$$

Bifurcation from that "trivial solution" is observed in applications if it loses its stability at some critical value of λ which we normalize at $\lambda_0 = 0$. By the "Principle of Linearized Stability" (see [6],[7], e.g.) this means, for instance,

NONLINEAR DIFFERENTIAL EQUATIONS:
INVARIANCE, STABILITY, AND BIFURCATION

ISBN 0-12-508780-2

that some eigenvalue of the linearization at $v = 0$ crosses the imaginary axis. We render our assumptions a little more precise:

The mapping

$$G : \Lambda \times D \to E, \tag{3}$$

where D is a continuously embedded subspace of E, is decomposed into a linear part and a nonlinear remainder satisfying a "little-o-condition" at $u = 0$ with respect to the norm in D:

$$G(\lambda,u) = A(\lambda)u + F(\lambda,u). \tag{4}$$

If we confine ourselves to stationary bifurcation we refer to the assumptions in [10], if bifurcation of periodic solutions is considered we need the assumptions in [3] or [8]. For practical use this means the following: if $A(\lambda)$ is some elliptic differential operator of order 2p with constant domain of definition D then in order to study periodic bifurcation the differential operator F is of order at most $2p - 1$, whereas in order to investigate stationary bifurcation the nonlinearity can be of order 2p, too. The space D must be dense in E only in the first case.

Throughout this paper we assume that

$$\begin{cases} \mu(\lambda) \text{ is a simple eigenvalue of } A(\lambda) \text{ with} \\ \operatorname{Re}\mu(0) = 0 \text{ and } ik\operatorname{Im}\mu(0) \text{ is not in the} \\ \text{spectrum of } A(0) \text{ for } k \in \mathbf{Z} \setminus \{1,-1\}. \end{cases} \tag{5}$$

We have to distinguish two cases which entail different bifurcation behaviour:

1) $\operatorname{Im}\mu(0) = 0$,
2) $\operatorname{Im}\mu(0) \neq 0$.

As is well known the first case leads to stationary bifurcation, whereas the second case leads to Hopf bifurcation of periodic solutions of equation (1).

Closely connected to bifurcation at a simple eigenvalue is a "Principle of Exchange of Stability" (PES) which means that the stability lost by the trivial solution is taken over by the bifurcating solution which may be stationary or periodic. A good reference for bifurcation as well as for the PES are the papers of Crandall and Rabinowitz [1],[2],[3].

The next examples show, however, that bifurcation at simple eigenvalues is not as simple as it seems to be. That is the reason why we went beyond the cases considered in the papers mentioned above and why we established a theorem on degenerate bifurcation.

Example a): In $E = C[0,\pi]$ the Sturm-Liouville operator

$$G(\lambda,u) = -u'' - u + \lambda u' + \lambda^2 u + u^3 \tag{6}$$

is defined on $D = C^2[0,\pi] \cap \{u, u(0) = u(\pi) = 0\}$. Obviously zero is a simple eigenvalue of $A(0)$ but $u = 0$ is the only solution of the equation $G(\lambda,u) = 0$, $u \in D$ (Multiply by u and integrate over $[0,\pi]$).

Example b): If in $E = \mathbb{R}^2$ the operator G is defined by

$$G(\lambda,u) = \begin{pmatrix} 0 & -1 \\ 1+\lambda & \lambda^2 \end{pmatrix}\begin{pmatrix} u_1 \\ u_2 \end{pmatrix} + \begin{pmatrix} 0 \\ u_2^3 \end{pmatrix}, \quad u = \begin{pmatrix} u_1 \\ u_2 \end{pmatrix}, \tag{7}$$

then the equation $\frac{du}{dt} + G(\lambda,u) = 0$ is equivalent to the non-linear oscillation problem

$$\ddot{x} + x + \lambda x + \lambda^2 \dot{x} + \dot{x}^3 = 0. \tag{8}$$

The numbers $\pm i$ are simple eigenvalues of $A(0)$ but $u = 0$ (or $x = 0$) is the only periodic solution of (1) (or (8)) (Multiply (8) by $\dot{x}$ and integrate over one period).

Example c): In $E = \mathbb{R}$ we consider

$$\begin{aligned} G(\lambda,u) &= (\lambda - u)(\lambda - u^2)(\lambda - u^3)u \\ &= \lambda^3 u + F(\lambda,u) = 0. \end{aligned} \tag{9}$$

The solution set is shown in Fig. 1. Obviously $(\lambda,u) = (0,0)$ is a bifurcation point, but in contrast to the result in [1]

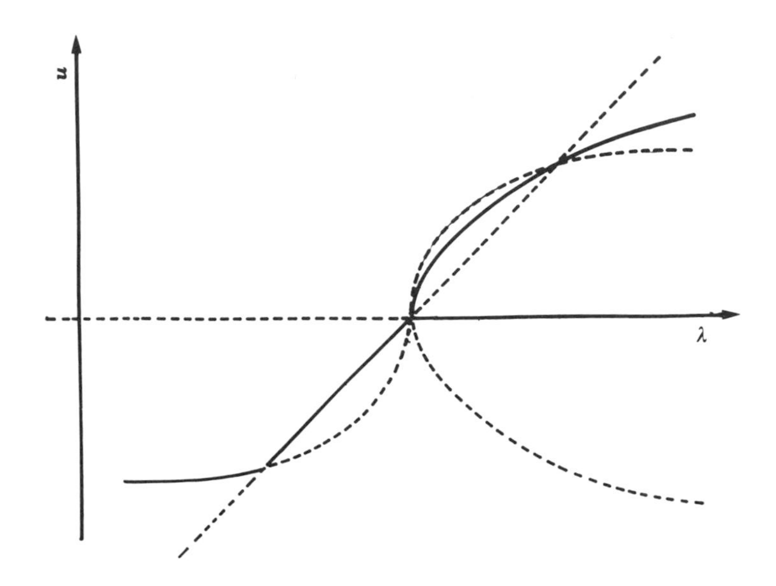

Fig. 1.

the bifurcation branch is not unique. What happens to the PES in this case? When considering the ordinary differential equation $\frac{du}{dt} + G(\lambda,u) = 0$ the stability is given by the sign of the derivative of G with respect to u at the stationary solutions themselves. These are simple zeros of G and therefore the stability property of two consecutive solutions for fixed λ changes. In Fig. 1 the broken lines are unstable whereas the others are stable.

The reason why our examples a) to c) do not fit into the framework of the bifurcation theorems which guarantee a unique branch emanating at a simple eigenvalue is that we did not require the essential assumption

$$\mathrm{Re}\mu'(0) \neq 0. \qquad (10)$$

This nondegeneracy condition means that the curve of eigenvalues $\mu(\lambda)$ crosses the imaginary axis with nonvanishing velocity. Whereas E. Hopf gave this condition explicitly, it is hidden in the conditions on stationary bifurcation in [1], [2].

Our examples do not fulfill condition (10) since $\mu(\lambda) = \frac{5}{4}\lambda^2$ in case a), $\mathrm{Re}\mu(\lambda) = \frac{1}{2}\lambda^2$ in case b), and $\mu(\lambda) = \lambda^3$ in case c).

The last example shows, however, that bifurcation may occur if the nondegeneracy condition (10) is not fulfilled. (Actually, if in the expressions (6) and (8) the cubic term has a negative sign, branches of stationary and periodic solutions emanate sub-and supercritically from the trivial solution in examples a) and b) respectively).

Before stating our main results we mention the work of Flockerzi [4] and Weinberger [12] who considered bifurcation of solutions and their linearized stability in the degenerate case, too.

<u>Theorem</u>. (see [9],[10]): Let $G(\lambda,u)$ be analytic with respect to λ and u, $G(\lambda,0) = 0$, and let $\mu(\lambda)$ be a simple eigenvalue of $A(\lambda) = G_u(\lambda,0)$. We assume

$$\mathrm{Re}\mu^{(j)}(0) = 0,\ j = 0,\ldots,m-1,\ \mathrm{Re}\mu^{(m)}(0) \neq 0 \qquad (11)$$

for some $m \geq 1$. Then the following holds:

1) If m is odd, $(\lambda,u) = (0,0)$ is a bifurcation point of at least one branch of stationary (periodic) solutions of (1), provided $\mathrm{Im}\mu(0) = 0$ ($\mathrm{Im}\mu(0) \neq 0$).
   If m is even, bifurcation may or may not occur. Sufficient conditions, for instance, can be given in terms of the second and third derivatives of G with respect to u at $(0,0)$ (see [9],[10]).
2) At most m nontrivial branches can exist.
3) A general "Principle of Exchange of Stability" says that adjiacent branches do not have the same stability properties.
4) The linearized stability property of the bifurcating branches can be determined from the bifurcation equation that gives the branches themselves. Newton's diagram gives the direction of bifurcation of all solutions shown to exist together with their stability property.
5) The number m and $\mathrm{Re}\mu^{(m)}(0)$ have not to be known a priori.

They are given by the bifurcation equation. On the other hand, if they are known they can directly be put into the bifurcation equation.

The last point can be considered to be the most important. It turned out that the number $\mathrm{Re}\mu^{(m)}(0)$ is the leading coefficient in the one dimensional bifurcation equation which is obtained by a reduction procedure due to Lyapunov and Schmidt. (In case of $\mathrm{Im}\mu(0) \neq 0$ this reduction is more involved since the unknown period as a second parameter has to be eliminated by an additional reduction step). The unexpected double meaning of the number $\mathrm{Re}\mu^{(m)}(0)$ yields all points 1) to 4) of our Theorem.

We give some examples:

I. Stationary bifurcation from the trivial solution of

$$G(\lambda,u) = -\Delta u - \lambda_1 u - \lambda^7 u + u^7 - \lambda^2 u^4 + \lambda^4 u^2 = 0 \qquad u|_{\partial\Omega} = 0, \tag{12}$$

where Ω is some bounded domain in $\mathbb{R}^n$ and λ_1 is the smallest (simple) eigenvalue of $-\Delta$ together with homogeneous Dirichlet boundary conditions. The bifurcation diagram is shown in Fig.2 where the broken lines represent unstable solutions whereas the others are stable. (The stability refers to stationary solutions of the corresponding parabolic initial-boundary value problem $u_t + G(\lambda,u) = 0$. As was shown in [7] we have asymptotic stability of all derivatives of u up to order 2 in any L^p-or C^α-norm or instability of u in the L^1-norm.)

II. Periodic bifurcation from the trivial solution of the nonlinear oscillation

$$\ddot{x} + x + \lambda^7\dot{x} - \dot{x}^9 + \lambda\dot{x}^5 - \lambda^3\dot{x}^3 = 0. \tag{13}$$

The stability indicated in Fig. 3 means asymptotic orbital stability. (All diagrams show the bifurcation directions given by the first term in the respective expansions).

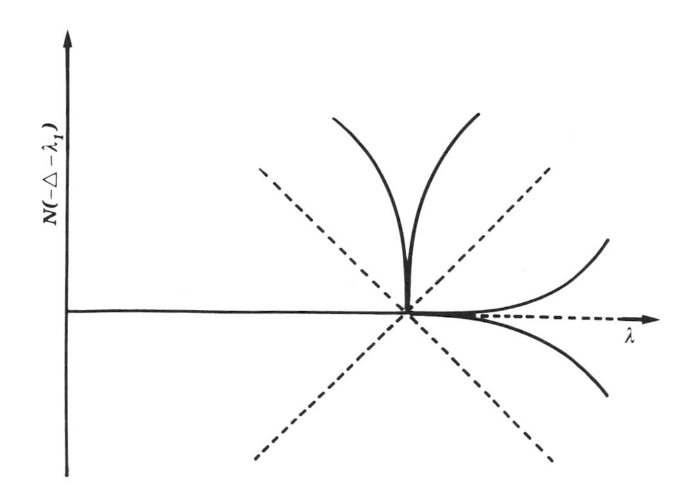

Fig. 2.

III. The "Brusselator" suggested by Lefever and Prigogine:

$$\frac{dX}{dt} = a - (1 + b)X + X^2Y$$
$$\frac{dY}{dt} = bX - X^2Y , \qquad a,b > 0 , \tag{14}$$

where X, Y are chemical concentrations which are spatially constant. The stationary solution $X = a$, $Y = \frac{b}{a}$ is stable if $b < 1 + a^2$ and unstable if $b > 1 + a^2$. If b crosses $1 + a^2$ the loss of stability is through two complex conjugate points $\pm$ ia. Therefore point 1) of our Theorem implies a bifurcation of periodic solutions provided the crossing of b through $1 + a^2$ is parameterized by any analytic function $h : b(\lambda) = 1 + a^2 + h(0), h(\lambda) = 0$. A closer analysis shows that due to the nonlinearity a unique stable periodic branch bifurcates supercritically, whatever the choice of h might be ($a > \sqrt{10/19}$, see Fig. 4).

Obviously the nondegeneracy condition (10) ($m = 1$) is generic whereas $m > 1$ in (11) is a degenerate or exceptional case. By a small perturbation any case $m > 1$ can be carried over into a generic case $m = 1$.

A natural question which arises is the following: What

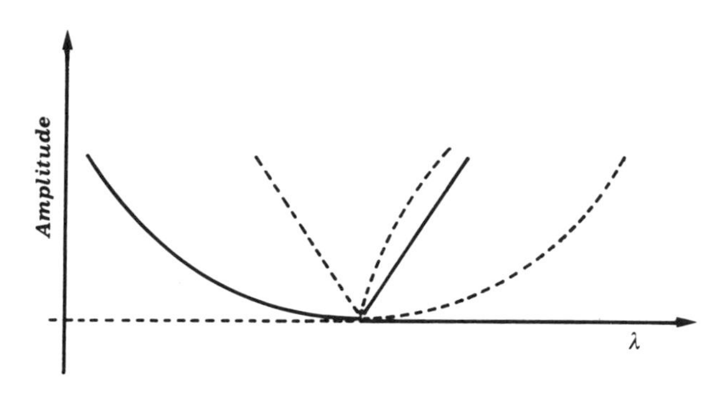

Fig. 3.

happens to the bifurcation picture under that perturbation if many branches emanate at the bifurcation point for $m > 1$?

Obviously all branches but one have to detach from the bifurcation point.

This may happen in many different ways. We give an example due to Poston and Stewart which we found in a recent paper of Golubitsky and Schaeffer [5] but which we present from a different point of view:

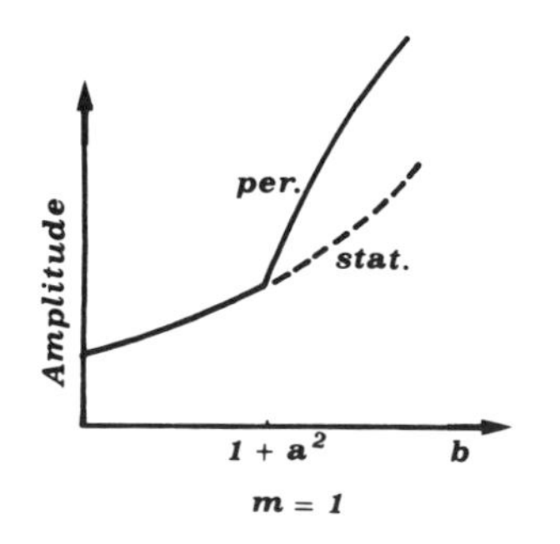

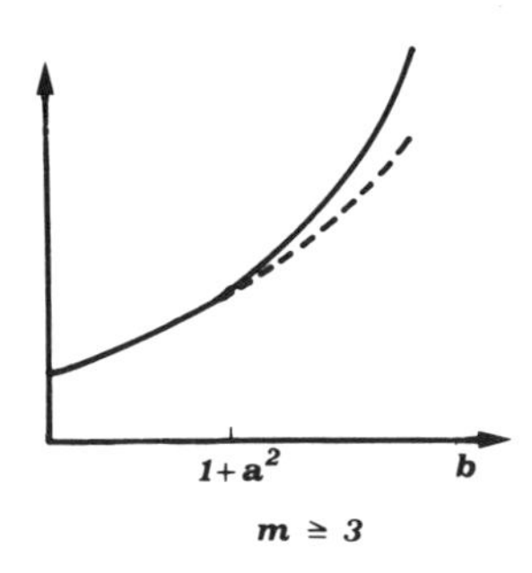

Fig. 4.

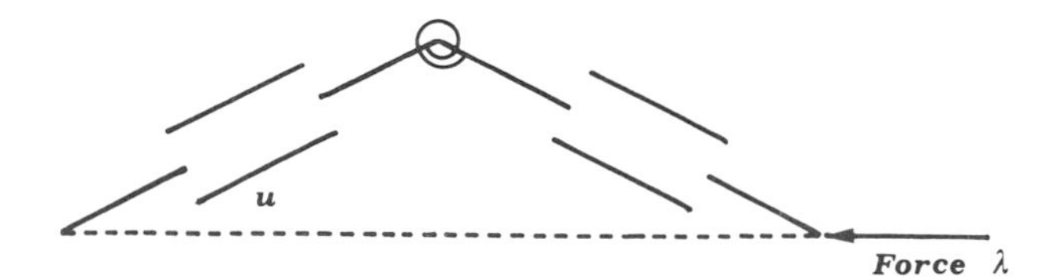

Fig. 5.

IV. We buckle a strut with compressible links of unit length with spring constant k. At the hinge we attach a torsional spring of unit strength. If k is small the trivial solution $u = 0$ (where u is the angle between the links and the trivial solution) is stable for all external forces $\lambda > 0$ since stronger forces are needed to rotate the torsional spring than to compress the two linear springs.

If k is large the system is approximately rigid and we expect a perturbation of the well known Euler buckling.

How is the transition from small k to large k? Considering the equations (see [5]).

$$\begin{aligned} &u - 2\lambda X \sin u = 0 \\ &2k(X - 1) + 2\lambda \cos u = 0 \ , \end{aligned} \tag{15}$$

where X is the common length of the two springs, we find for $k_0 = 2$ a degenerate bifurcation at $\lambda_0 = 1$ with $m = 2$. The sub-and supercritically emanating branches are unstable. For $k < 2$ these branches detach from the trivial solution which is stable throughout. For $k > 2$ two generic bifurcations occur at $\lambda_{1,2} = (k \pm (k^2 - 2k)^{\frac{1}{2}})/2$. The branches for k near $k_0 = 2$ are perturbations of the degenerates case but the detachment is different for $k < 2$ and $k > 2$ (see Fig. 6). From a physical point of view diagram 3) can only be understood if it is considered globally.

Finally, for $k > \frac{8}{3}$ we find the expected supercritical

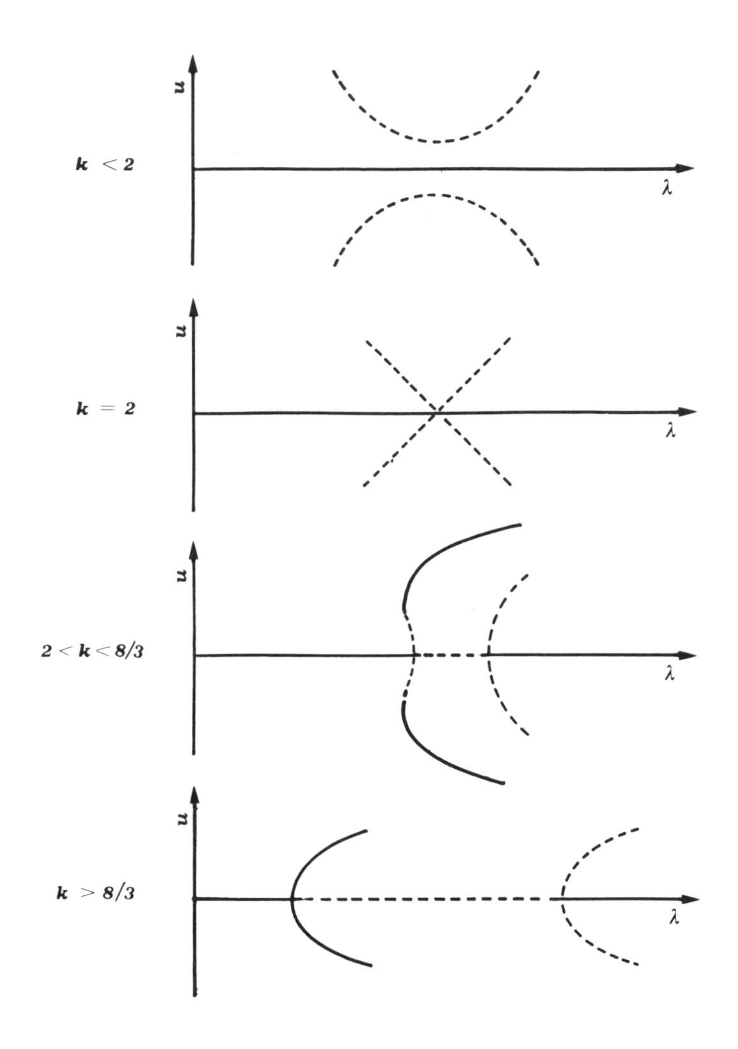

Fig. 6.

stable bifurcation. There is still a second bifurcation which requires compressing the springs to almost zero length. Beyond that bifurcation point the trivial solution is stable again.

This example shows that degenerate bifurcation with even m may occur as a limiting case between two completely different solution sets, namely when isolated branches meet each other and thus bifurcate. The problem of perturbing a degenerate bifurcation might be complicated, in general, since the class of possible perturbations may involve more than one parameter k. In any case it gives rise to the question how all perturbed bifurcation diagrams may look like. Thus "imperfect" bifurcation or "secondary" bifurcation may be discovered which might be overlooked when only generic bifurcation is studied.

A simple example shows what may happen:

V. We consider the Sturm-Liouville problem

$$\begin{aligned} G_o(\lambda,u) &= -u'' - u - \lambda^3 u + \lambda u^3 = 0 \\ u(0) &= u(\pi) = 0 \end{aligned} \tag{16}$$

together with its perturbations

$$\begin{aligned} G_1(\lambda,u) &= -u'' - u - \varepsilon_1\lambda u - \lambda^3 u + \varepsilon_1 u^2 + \lambda u^3 = 0 \\ G_2(\lambda,u) &= -u'' - u - \varepsilon_2\lambda u - \lambda^3 u + \lambda u^3 = 0 \\ G_3(\lambda,u) &= -u'' - u + \varepsilon_3\lambda u - \lambda^3 u + \lambda u^3 = 0 \end{aligned}$$

with the same boundary conditions and small numbers $\varepsilon_1,\varepsilon_2,\varepsilon_3 > 0$. The bifurcation diagrams are given in Fig. 7. Note that the perturbed branches have the same stability properties as the unperturbed branches as long as these are not vertical (see Theorems 6.2 and 6.4 in [10]). The linearized stability gives the asymptotic stability or instability for all small perturbations in these cases. This last example shows how imperfect and secondary bifurcation may occur near a simple eigenvalue. We think that in order to study secondary bifurcation a perturbation of a degenerate simple eigenvalue bifurcation is as adequate as a perturbation of a multiple eigenvalue bifurcation (see [11]).

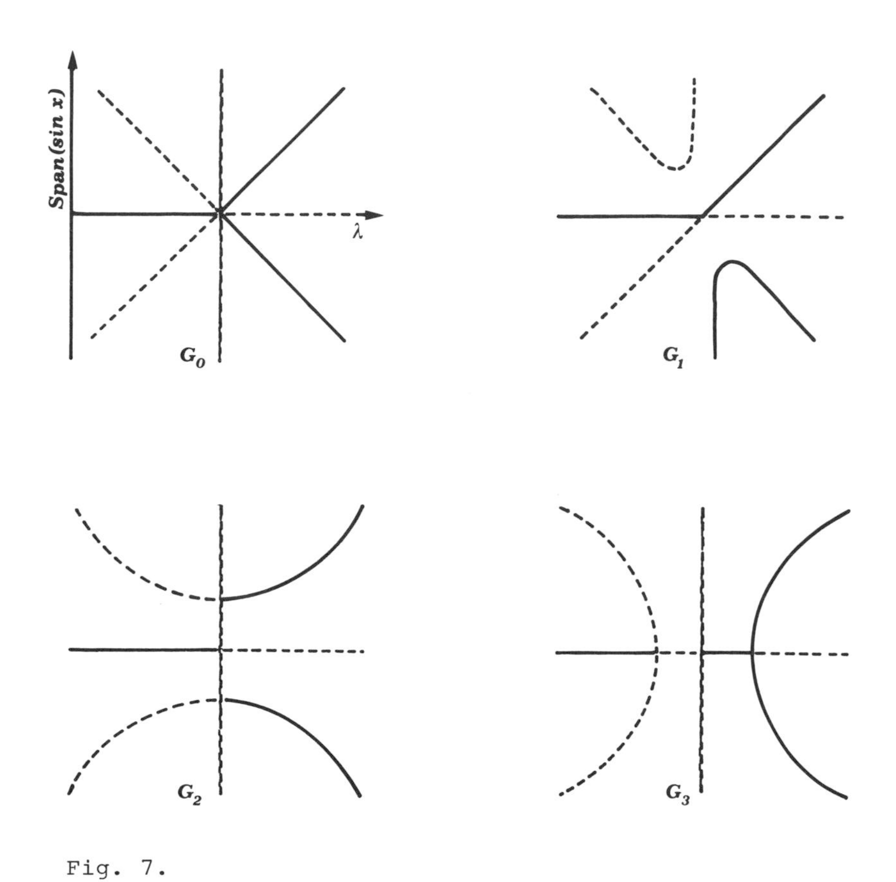

Fig. 7.

REFERENCES

1. M. Crandall, P. Rabinowitz: Bifurcation from Simple Eigenvalues, J. Funct. Anal. 8, 321-340, (1971).
2. M. Crandall, P. Rabinowitz: Bifurcation, Perturbation of Simple Eigenvalues, and Linearized Stability, Arch. Rat. Mech. Anal. 52, 161-180, (1973).
3. M. Crandall, P. Rabinowitz: The Hopf Bifurcation Theorem in Infinite Dimensions, Arch. Rat. Mech. Anal. 67, 53-72, (1977/78).
4. D. Flockerzi: Existence of Small Periodic Solutions of Ordinary Differential Equations in $\mathbb{R}^2$, Arch. d. Math. 33, 263-278, (1979).

5. M. Golubitsky, D. Schaeffer: A Theory of Imperfect Bifurcation via Singularity Theory, Comm. Pure Appl. Math. 32, 21-98, (1979).
6. H. Kielhöfer: Stability and Semilinear Evolution Equations in Hilbert Space, Arch. Rat. Mech. Anal. 57 , 150-165, (1974).
7. H. Kielhöfer: On the Lyapunov Stability of Stationary Solutions of Semilinear Parabolic Differential Equations, J. Diff. Equat. 22, 193-208, (1976).
8. H. Kielhöfer: Hopf Bifurcation at Multiple Eigenvalues. Arch. Rat. Mech. Anal. 69, 53-83, (1979).
9. H. Kielhöfer: Generalized Hopf Bifurcation in Hilbert Space, Math. Meth. in the Appl. Sci., 1, 498-513, (1979).
10. H. Kielhöfer: Degenerate Bifurcation at Simple Eigenvalues And Stability of Bifurcating Solutions, to appear in: J. Funct. Anal.
11. M. Shearer: Secondary Bifurcation Near A Double Eigenvalue, Siam J. Math. Anal. 11, 365-389, (1980).
12. H. Weinberger: On The Stability of Bifurcating Solutions in: Nonlinear Analysis. A Collection of Papers in Honor of Erich Rothe, ed. by L. Cesari and R. Kannan. New York, Academic Press, 219-233, (1978).

PERIODIC AND NONPERIODIC SOLUTIONS OF REVERSIBLE SYSTEMS

Klaus Kirchgässner[1]

Mathematisches Institut, Universität Stuttgart
West Germany

0. INTRODUCTION

In this contribution a survey is given on recent results about the existence of bounded solutions for finite systems of ordinary differential equations which are reversible in the sense of G.D. Birkhoff. Generally speaking, these solutions appear under various contidions when, for some value of an external parameter, an equilibrium point exhibits bifurcation. To be more specific, let us consider the system

$$\frac{du}{dt} = A(\lambda)u + f(\lambda,u) \tag{0.1}$$

in $\mathbb{R}^n$, $A(\lambda)$ being an $n \times n$ matrix depending twice continuously differentiable on the real parameter λ, f is a C^2 function of its arguments which, together with its gradient vanishes for $u = 0$.

The system (0.1) is called *reversible* if there exists a reflection R on $\mathbb{R}^n$ ($R^2 = \text{id}$) such that

$$A(\lambda)R = -RA(\lambda), \quad f(\lambda,Ru) = -Rf(\lambda,u) \tag{0.2}$$

---

[1] Currently visiting Departement of Mathematics, Colorado State University, Fort Collins, Colorado.

ISBN 0-12-508780-2

for all λ and u; R is assumed to be independent of λ. Obviously $u = 0$ is a solution for every λ and thus the really interesting question is for bounded nontrivial solutions of (0.1). In particular, we are interested in solutions bifurcating from the trivial one. A first choice is the case of periodic solutions where, in view of the autonomy of (0.1) the period chosen is arbitrary. A necessary condition for bifurcation of periodic solutions is the existence of purely imaginary eigenvalues of $A(\lambda)$. However, Hopf bifurcation cannot occur since, as the reader may convince himself easily, simple eigevalues remain sitting on the imaginary axis if A is reversible.

For the case $n = 2$ the local and global existence of periodic solutions has been proved by Wolkowisky [12], the global aspects depending strongly on the validity of nodal properties and the restriction to two dimensions, thus following closely the remarkable papers of Crandall and Rabinowitz [2,3]. The extension of the local bifurcation result to arbitrary dimensions is straightforward if one follows the general lines of Theorem 1 in [1]. However, the global result is by no means trivial. Using a new uniqueness result, global existence of periodic solutions for arbitrary dimensions could be established [6].

Leaving the relatively easy case of periodic solutions aside one could ask for bounded nonperiodic solutions bifurcating from the equilibrium. For $n = 2$, one easily constructs those solutions as transient orbits connecting equilibrium points having the form of pulses or waves by phase-plane methods. Again, a general analysis of such phenomena in higher dimensions is lacking, although methods to handle specific situations with asymptotic methods are available (c.f. [7], [10]). From the point of view of bifurcation theory, these transient or singular solutions appear at the lowest point of the continuous spectrum as envelopes of periodic solutions of indefinitely increasing period.

A particularly deep existence problem arises when A has k pairs of single eigenvalues on the imaginary axis. As related questions for Hamiltonian systems indicate, one external parameter generally does not suffice to make the bifurcation of the expected quasiperiodic solutions of $m \leq k$ indipendent frequencies possible [8], [11]. The natural question therefore is to ask for the bifurcation of m-dimensional tori on which the flow is quasiperiodic (in any integer with $m \leq k$) for system having at least $k - 1$ external parameters. Recently quite satisfactory results have been given in this direction [11].

A last question which arises concerns the validity of these results in infinite-dimensional spaces, i.e., for partial differential equations. Indeed, those results have been proved for periodic and singular solutions in the local case in [4], whereas the global case for periodic solutions and the existence of bifurcating quasiperiodic solutions is still open.

We shall organize this contribution in four parts. The first will cover preliminary material, in particular the new uniqueness theorem which is at the heart of almost all of the following results. The second part will be concerned with local and global aspects of bifurcating periodic solutions, and at its end we shall briefly formulate the theorems on quasiperiodic bifurcation, the proofs of which, even only sketched, would go far beyond the frame of this contribution. In the third part we discuss the existence of singular solutions bifurcating at the lowest point of the continuous spectrum and finally, in the fourth part we show generalizations to partial differential equations and some of their applications.

1. PRELIMINARIES

Let $I(\lambda) = \frac{d}{dt} - A(\lambda)$ then every "small" solution of (0.1) is uniquely determined by its projection into the kernel of $I(\lambda)$ (smallness and the type of projection is specified below). Such a result only superficially seems to be a consequence of a center manifold theorem since, even if we knew uniqueness of this manifold, solutions with noncompact orbits may not belong to it.

Throughout this paper we shall denote by Σ_λ the spectrum of $A(\lambda)$, the corresponding eigenprojection by $P(\lambda)$. The spaces used are defined such that immediate generalizations to partial differential equations are possible.

$$Y^k = \{u \in \mathbb{H}^k_{\ell oc}(\mathbb{R})/E_k(u) < \infty\},\ k = 0,\ 1 \tag{1.1}$$

where $\mathbb{H}^k$ denotes the n-fold product of the real Sobolev spaces H^k and where

$$E_k(u) = \sup_{j \in \mathbb{Z}^1} \|u\|_{k,j},\ \mathbb{Z}^1 = \mathbb{Z}\setminus\{0\} \tag{1.2}$$

and where we set

$$\|u\|_{k,j} = \|u\|_{\mathbb{H}^k(K_j)} \tag{1.3}$$

for the sequence of compacta

$$K_j = [j - 1,\ j],\ K_{-j} = -K_j,\ j \in \mathbb{N}.$$

Solutions bounded in the sense used here are elements of Y^o or Y^1. Of course, those in Y^1 are bounded in the classical sense, whereas those in Y^o are, if f grows sublinearly in u (see Theorem 1.2).

<u>Theorem 1.1</u>. (Local uniqueness modulo ker $I(\lambda)$) Fix $\lambda \in \mathbb{R}$ and assume that all imaginary eigenvalues of $A(\lambda)$ are simple. Then there exists a positive number ε (dependent on λ) such that any two solutions u and v of (0.1) in Y^1 do coincide, if

$$P(\lambda)u(0) = P(\lambda)v(0),\ E_1(u) < \varepsilon,\ E_1(v) < \varepsilon$$

is satisfied. In particular, 0 is E_1-locally unique if $A(\lambda)$ has only eigenvalues with nonvanishing real parts.

A proof of this theorem can be found in [5] and in an infinite-dimensional case in [4]. For the global existence of periodic solutions one needs a version which is stronger in the sense that E_1 is replaced by E_o and ε is independent of λ.

<u>Theorem 1.2</u>. Assume, in addition to Theorem 1.1, that $A(\lambda) = g(\lambda)C(\lambda)$ where g is some smooth real-valued function of λ satisfying $g(\lambda) \geq 1$ and where

$$\lim_{\lambda\to\infty} C(\lambda) = C(\infty),$$

Moreover, assume that there are positive numbers, γ_o, δ, η, and β with $\beta < 1$, such that

$$|f(\lambda,u)| \leq \gamma_o(1 + |u|^{\beta}), \quad u \in \mathbb{R}^n$$

$$|\nabla f(\lambda,u)| \leq \delta|u|, \qquad |u| < \eta$$

holds for all $\lambda\in[\lambda_o,\infty)$. Then the conclusions of Theorem 1.1 are valid in $[\lambda_o,\infty)$ if E_1 is replaced by E_o and ε being independent of λ.

The proof of this theorem can be found in [6]. Observe that $A(\lambda) = \lambda A$, A fixed, fulfills the requirements of Theorem 1.2 for $\lambda_o = 1$. Remark, moreover, that reversibility is not needed in either one of the above theorems.

2. PERIODIC AND QUASIPERIODIC SOLUTIONS

Periodic solutions of any period can be shown to exist if $A(\lambda)$ has two simple imaginary eigenvalues satisfying a generic nondegeneracy condition and if (0,1) is reversible. The proof of the local bifurcation result is a direct consequence of Theorem 1 in [1] if one considers the spaces for some $p > 0$

$$Z_p^k = \{u \in \mathbb{H}^k_{\ell oc}(\mathbb{R})/u(t+p) = u(t),\ Ru(t) = u(-t) \text{ a.e.}\}$$

with the norm

$$|u|_p = (u,u)_p^{1/2}, \quad (u,v) = \frac{1}{p}\int_0^p u(t)v(t)dt \quad \text{for } k = 0$$

and

$$\|u\|_p = ((u,u))_p^{1/2}, \quad ((u,v))_p = (u,v)_p + (\dot{u},\dot{v})_p \quad \text{for } k = 1.$$

Theorem 2.1. Let system (0.1) be reversible. Assume that $A(\lambda)$, for $\lambda > \lambda_o$, has exactly one pair of imaginary eigenvalues - denoted by $\pm i\omega(\lambda)$ - which are simple and satisfy $\omega'(\lambda) \neq 0$ for all λ. Then, for every λ_k, with $\omega(\lambda_k) = 2\pi k/p$, $(\lambda_k, 0)$ is a point of bifurcation in $\mathbb{R} \times Z_p^1$ of p-periodic solutions. In some neighborhood of this point the nontrivial solutions determine a $\mathbb{R} \times Z_p^1$ -valued C^1-curve $(\lambda(\varepsilon), u(\varepsilon))$ with

$$u(\varepsilon) = \varepsilon(u_o + z(\varepsilon)), \quad z(0) = 0$$

$$\lambda(\varepsilon) = \lambda_k + \tau(\varepsilon), \quad \tau(0) = 0$$

and

$$x_o = \phi_r \cos\frac{2\pi k}{p} - \phi_i \sin\frac{2k\pi}{p} t$$

where $\phi = \phi_r + i\phi_i$ denotes a normalized eigenvector of $A(\lambda_k)$ to $i\omega(\lambda_k)$.

To obtain a global version of this theorem observe that the results of Rabinowitz apply [9]. The component C_p^k of nontrivial solutions (λ, u) containing $(\lambda_k, 0)$ either meets the boundary of $(\lambda_o, \infty) \times Z_p^1$ or another bifurcation point. Using Theorem 1.2 we are able to exclude the second alternative. Observe that the index with respect to 0 is defined for the curve given by $P(\lambda)u(\lambda)(t)$, $t \in [0,p]$ if $0 \notin Pu(\lambda)(t)$ for every t. Theorem 2.1 yields, that this index is k near $(\lambda_k, 0)$. If C_p^k would meet another bifurcation point there exist sequences t_n, $(\lambda_n, u_n) \in C_p^k$, such that $P(\lambda_n)u_n(t_n) = 0$. One can show that, for small periods p, this yields u_u eventually satisfying $E_o(u_n) < \varepsilon$ and hence $u_n = 0$ for almost all n. A detailed proof of the following theorem can be found in [6].

Theorem 2.2. Let the assumptions of Theorems 1.2 and 2.1 hold. Moreover, assume that

$$A(\lambda)Qu = -Qf(\lambda, Qu)$$

implies $Qu = 0$, where $Q = id - P(\lambda)$. Denote by C_p^k the component of nontrivial solutions of (0.1) in $\mathbb{R} \times Z_p^1$ containing $(\lambda_k, 0)$.

Then there exists a positive constant p_o - which may depend on k - such that for all $p \in (0,p_o)$, C_p^k meets the boundary of $(\lambda_o,\infty) \times Z_p^1$.

An example where both theorems apply is given by

$$y^{(4)} - 2\lambda y'' + (\lambda^2 - \lambda^4)y - g(y, y', y'', y''') = 0$$

where g satisfies

$$|g(x)| \leq \gamma_o(1 + |x_1|^\beta), \quad x \in \mathbb{R}^4 \text{ and } 0 < \beta < 1$$

$$g(x) = (x_1 + x_3)h(x), \quad h(-x_1, x_2, -x_3, x_4) = h(x)$$

$$h(x_1, 0, \delta x_1, 0) > 0 \text{ for all } \delta > 0.$$

The bifurcation of quasiperiodic solutions requires more external parameters. Assume, therefore, that $\lambda \in \mathbb{R}^k$ or $\lambda \in \mathbb{R}^{k-1}$ and assume that $A(\lambda)$ has exactly k pairs of simple eigenvalues $\pm i\omega_1(\lambda),\ldots,\pm i\omega_k(\lambda)$ which are imaginary and which for $\lambda = \lambda_o$ satisfy

$$\text{a)} \quad \det\left(\frac{\partial\omega_i}{\partial\lambda_k}\right)(\lambda_o) \neq 0 \text{ for } \lambda \in \mathbb{R}^k$$

$$\text{b)} \quad \det\left(\frac{\partial\omega_i}{\partial\lambda_j},\omega\right)(\lambda_o) \neq 0 \text{ for } \lambda \in \mathbb{R}^{k-1} \tag{2.1}$$

The rest of Σ_λ is supposed to be bounded away from the imaginary axis, then the following theorem holds.

<u>Theorem 2.3</u>. (J. Scheurle [11]) Let the system (0.1) be reversible and real analytic in λ and x where $\lambda \in \mathbb{R}^k$, or $\lambda \in \mathbb{R}^{k-1}$. Assume (2.1a) or (2.1b) to be fulfilled. Moreover, suppose that, for some $\tau > 1$ and $\alpha > 0$

$$\left|\sum_{j=1}^{k} j_i\omega_i(\lambda_o)\right| \geq \alpha|j|^{-\tau} \tag{2.2}$$

for all $j \in \mathbb{Z}^k$, $j \neq 0$, then there exists for every $c \in \mathbb{R}^k$, $|c| = 1$, a one-parameter family of invariant tori of the form

$$u(\varepsilon,z) = \varepsilon \sum_{j=1}^{k} c_j \mathrm{Re}(\phi_j e^{iz_j}) + \varepsilon\psi(\varepsilon,z)$$

of (0.1) for all ε with $|\varepsilon| < \varepsilon_o(c)$. Here, $\lambda = \lambda(\varepsilon) = \lambda_o + O(\varepsilon)$, ϕ_j is an eigenvector to $i\omega_j(\lambda_o)$ and the functions ψ and λ are analytic in ε. The function ψ is of order $O(\varepsilon)$ uniformly in $z = (z_1,\ldots,z_k)$ and 2π-periodic in each z_j.

For $c_j = 0$, u is independent of z_j, i.e., the dimension of the tori coincides with the number of nonvanishing c-components. The flow on the tori is quasiperiodic via

$$z_j = \omega_j(\lambda_o)t + d_j, \quad \text{for the case } \lambda \in \mathbb{R}^k$$

$$z_i = \alpha\omega_j(\lambda_o)t + d_j, \quad \text{for the case } \lambda \in \mathbb{R}^{k-1},$$

where $\alpha = \alpha(\varepsilon) = 1 + O(\varepsilon)$ is analytic in ε as well.

For the proof of this theorem we refer the reader to [11]. For k = 2 and (2.1b) the theorem yields a result for our original system, namely the bifurcation of quasi-periodic solutions of two independent frequencies. In general, we have bifurcation of invariant tori of every dimension m up to k inclusively, where for $m < k$ the diophantic condition (2.2) could be relaxed by replacing k by m. The ε-interval, where existence can be proved, generally depends on c, the "direction" of bifurcation. If ε_o would be independent of c one could classify all small solutions as being quasiperiodic by Theorem 1.1.

3. SINGULAR SOLUTIONS

Now we turn our interest to nonperiodic bounded solutions of (0.1). We shall call these solutions *singular*, and we include the constant solutions in this category. Moreover, let us assume

For $\lambda \in (\lambda_o,\lambda_1)$, $A(\lambda)$ has exactly two imaginary eigenvalues - denoted by $\pm i\omega(\lambda)$ - which are supposed to be simple and to satisfy $\omega'(\lambda) \neq 0$. For $\lambda < \lambda_o$ the spectrum Σ_λ should not intersect (3.1)

the imaginary axis.

We shall call the system (0.1) *noncritical* if $\lambda = \lambda_o$, $u = 0$ is an isolated solution of (0.1) in $\{\lambda_o\} \times Y^1$, i.e., $(\lambda_o,0)$ is no point of vertical bifurcation. If (0.1) is reversible then, according to Theorem 2.1, every $(\lambda,0)$, $\lambda \in (\lambda_o,\lambda_1)$, is a point of bifurcation of periodic solutions whereas $u = 0$ is seen to be E_1-isolated for $\lambda < \lambda_o$ by applying Theorem 1.1. The question what happens at $\lambda = \lambda_o$ will be answered in this section for reversible, non-critical systems.

To clarify the situation let us consider some simple examples in $\mathbb{R}^2$, namely the system (0.1) for

$$A(\lambda) = \begin{pmatrix} 0 & 1 \\ -\lambda & 0 \end{pmatrix}, \quad f(\lambda,u) = \begin{pmatrix} 0 \\ g(x,y) \end{pmatrix}$$

where $u = (x,y)$ and $g(x,y)$ is one of the following functions

$$g(x,y):\ x^2,\ -x^3,\ x^3,\ y^2,\ y^3.$$

These example correspond to the second-order equation

$$x_{tt} + \lambda x - g(x,x_t) = 0 \tag{3.2}$$

and we seek bounded solutions. Subsequently, we have drawn phase portraits for the cases x^2, $-x^3$, x^3 and x^2. For $\dot{x}^3$ one easily shows that, for $\lambda_o = 0$, the hyperbolic point (0,0) changes into a stable focus and no periodic solutions exist. The system is not reversible in this case.

For a) and c), (3.2) is reversible and noncritical and exhibits a nonconstant singular solution which bifurcates at $\lambda = 0$, $u = 0$. In case a) this is a pulse (connecting an equilibrium point to itself); in case c) it is a wave-front; both have interesting extremal properties, namely in case a): $x(0)$ is minimal (if restricted to even solutions); in case c), $y(0) = x_t(0)$ is extremal (if restricted to odd solutions). Example b) is reversible but critical, no singular solution exists, and example d) is reversible and critical, but the solutions bifur

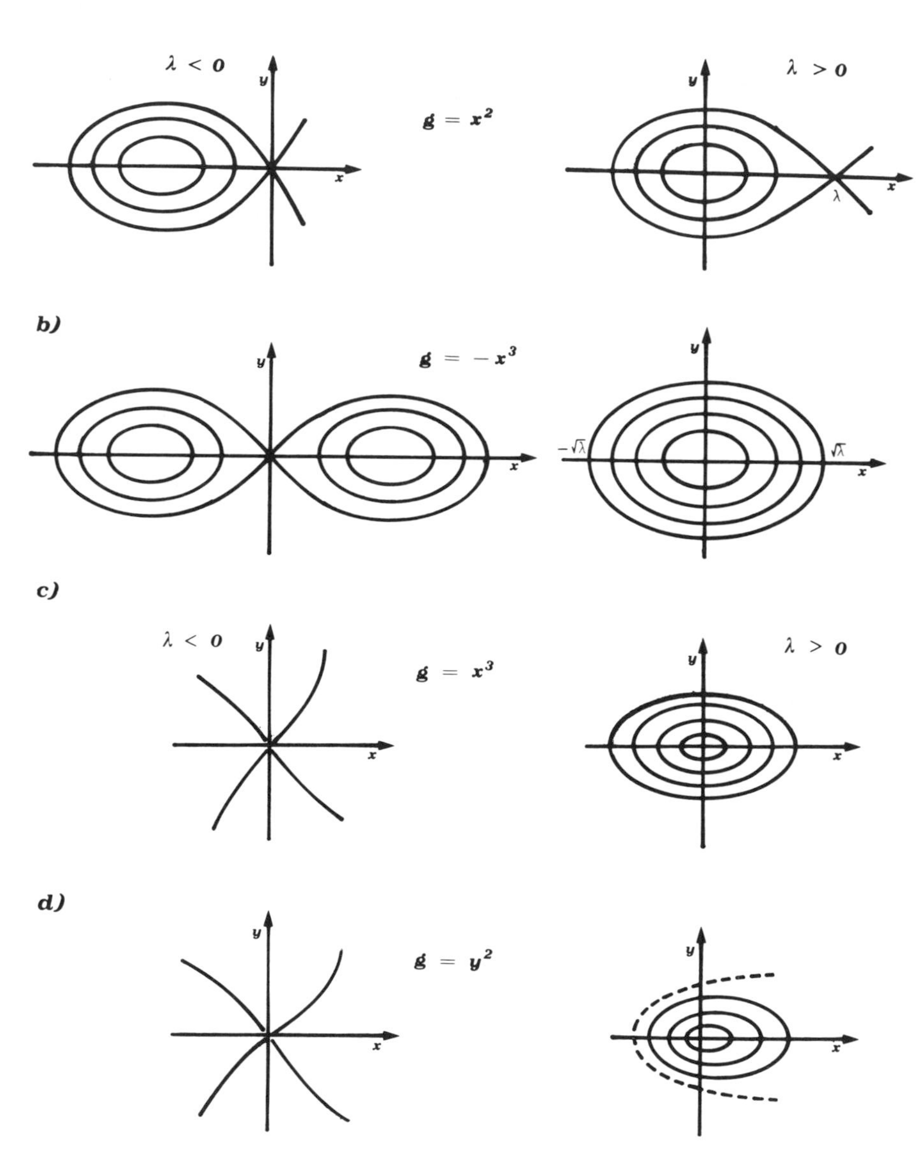

Fig. 1. Phase-space plots for the examples described at page 9.

cating from $\lambda = 0$, $u = 0$ are constants and a polynomially growing "singular" solution exists as an "envelope" of the periodic ones.

The subsequent theorem covers all cases except d). It is an interesting open problem to give conditions for the existence of polynomially growing "singular" solutions. Moreover, the above examples illustrate the results of Section 2 as well.

Theorem 3.1. Let the assumption (3.1) be valid. Moreover, suppose that the system (0.1) is reversible and noncritical. Then there exists a $\delta > 0$ such that, for every $\lambda \in (\lambda_o, \lambda_o + \delta)$, there exists a solution u_λ which is singular. Furthermore,

$$\lim_{\lambda \to \lambda_o + 0} (\lambda, u_\lambda) = (\lambda_o, 0).$$

We shall only sketch the proof. The reader who is interested in details should consult [4] or [5] for an infinite-dimensional version. Fix $\lambda \in (\lambda_o, \lambda_o + \delta)$ and set

$$v(s) = u(s/\rho)$$

then (0.1) is transformed into

$$\rho v' = A(\lambda)v + f(\lambda, v). \tag{3.3}$$

Now we treat (3.3) in $(\rho, v) \in \mathbb{R}^+ \times Z^1_{2\pi}$ where Z^k_p was defined in Section 2. The kernel of $\rho d/ds - A(\lambda)$ is nontrivial for $\rho_m = \omega(\lambda)/m$, $m \in \mathbb{N}$, and every $(\rho_m, 0)$ turns out to be a bifurcation point. Again the global results of Rabinowitz apply. Let C^m_λ denote the component of nontrivial solutions in $\mathbb{R}^+ \times Z^1_{2\pi}$ containing $(\rho_m, 0)$, then the following alternatives hold:

1. C^m_λ is unbounded in $[\alpha, \infty) \times Z^1_{2\pi}$ for some $\alpha > 0$;
2. C^m_λ contains another bifurcation point; (3.4)
3. qC^m_λ contains an interval $(0, a)$ for some $a > 0$.

Here q denotes the projection $(\lambda, u) \to \lambda$.

We henceforth restrict our attention to $C^1_\lambda = C_\lambda$ and define

$$\Gamma_\lambda = \{u \in Y^1 / \text{there exists } (\rho,v) \in C_\lambda \text{ such that } u(t) = v(\rho t)\}$$

and observe that from any bounded sequence $u_n \in \Gamma_\lambda$ we may select a subsequence converging in $\mathbb{H}^\sigma(K)$ for $\sigma \in (\frac{1}{2},1)$ and every compactum $K \subset \mathbb{R}$. Since f maps $\mathbb{H}^\sigma(K)$ into $\mathbb{H}^o(K)$, and since u_n solves (0.1), the sequence converges in $\mathbb{H}^1(K)$. Moreover, for every $\varepsilon > 0$ there exists a $\delta > 0$ such that

$$\sup_{u \in \Gamma_\lambda} E_1(u) < \varepsilon \text{ holds for } \lambda \in (\lambda_o, \lambda_o + \delta). \tag{3.5}$$

(3.5) follows by an easy argument from the assumption of non-criticality. Actually it provides an important a priori estimate which rules out (3.4,1). Observe that, for every compact interval $[\alpha,\beta] \subset \mathbb{R}^+$ there are positive constants γ_1, γ_2 such that

$$\gamma_1 \|v\|_{2\pi} \leq E_1(u) \leq \gamma_2 \|v\|_{2\pi} \tag{3.6}$$

holds for $(\rho,v) \in C_\lambda$, $\rho \in [\alpha,\beta]$, $u(t) = v(\rho t)$.

Lemma 3.2. There are positive numbers a, b and δ such that

$$(0,a) \subset qC_\lambda \subset (0,b].$$

Proof: First we show that qC_λ is bounded above. For this reason assume that, for some $\rho > 0$, there are sequences $\lambda_n \to \lambda_o + 0$ and $(v_n,\omega) \in C_{\lambda n}$. Then, by (3.5), $E_1(u_n) \to 0$. Setting

$$z_n = u_n/E_1(u_n)$$

we obtain a bounded sequence in Y^1 from which we may extract a subsequence converging in $\mathbb{H}^1_{\ell oc}(\mathbb{R})$. Since z_n has period $2\pi/\rho$ and since $\mathbb{H}^1_{\ell oc}(\mathbb{R}) \subset C^o_{\ell oc}(\mathbb{R})$, the (z_n) converge uniformly to some $2\pi/\rho$-periodic z. Since $f(E_1(u_n)z_u)/E_1(u_n) \to 0$ we obtain from (0.1): $\dot{z} = A(\lambda_o)z$, $E_1(z) = 1$ which leads to a contradiction to (3.1).

Now we show that qC_λ contains some interval (0,a). Since (3.6), together with (3.5) and the above considerations rule out (3.4,1.), it suffices to show that (3.4,2.) cannot occur. Let us match the ε in (3.5) with that of Theorem 1.1 for $\lambda = \lambda_o$ and define, for every $u \in \Gamma_\lambda$, the projection

$$\chi_t u = P(\lambda)u(t)$$

which, for every t, maps Y^1 into $\mathbb{R}^2$, thus defining a curve K_u if t ranges in $[0,2\pi/\rho]$, $u(t) = v(\rho t)$. The index of K_u with respect to 0 is given by

$$\mathrm{ind}(K_u,0) = \frac{1}{2\pi i}\int_{K_u} \frac{dz}{z} .$$

As u varies in $\Gamma_\lambda \setminus \{0\}$, $\mathrm{ind}(K_x,0)$ is defined and remains constant, since $P(\lambda)u(t_o) = (0,0)$ implies, via Theorem 1.1, that $u = 0$.

If C_λ contains, besides $(\omega(\lambda),0)$, another point of bifurcation, say $(\hat{\omega},0)$, where $\hat{\omega} \neq \omega(\lambda)$, then $\hat{\omega} = \rho_m = \omega(\lambda)/m$ for some $m \neq 1$. For $(\rho,v) \in C_\lambda$ near $(\omega(\lambda),0)$, respectively, $(\hat{\omega},0)$ and $u(t) = v(\rho t)$ we obtain

$$\mathrm{ind}(K_u,0) = 1 \text{ resp. } m$$

yielding a contradiction to the constancy of the index, and thus the proof of the lemma.

Proof of Theorem 3.1: Every ray through 0 meets K_u in at most one point, for every $u \in \Gamma_\lambda$, $\lambda \in (\lambda_o, \lambda_o + \delta)$ (see below). In view of $u_o = u(0) = Ru(0)$ and $RP(\lambda) = -P(\lambda)R$ we have $R\chi_o u = -\chi_o u$, and thus we may assume without loss of generality that

$$\chi_o(\Gamma_\lambda) \subset \mathbb{R} \times \{0\}.$$

$\chi_o(\Gamma_\lambda)$ is an open interval containing 0. To see this, let $a = \chi_o(u) > 0$. Take a connected closed subset C_λ^1 of C_λ containing $(\omega,0)$ as well ad (ρ,v), $u(t) = v(\rho t)$. The preceding lemma yields a point $(\hat{\rho},\hat{v}) \in C_\lambda$ not in C_λ^1. Connect $(\hat{\rho},\hat{v})$ and (ρ,v) by C_λ^2 in C_λ, set $C_\lambda^* = C_\lambda^1 \cup C_\lambda^2$ and denote by Γ_λ^*, respectively, Γ_λ^1 the corresponding sets in Y^1. Then $\chi_o(\Gamma_\lambda^1) \supset [0,a]$ and $\chi_o(\Gamma_\lambda^*) \supset [0, a+\eta]$ for some $\eta > 0$. Since K_u contains 0 in its interior, we may find $\alpha < 0$ and $\tau \in [0,2\pi/\rho]$ such that $\chi_\tau(u) = (\alpha,0)$. By a similar argument one concludes that $\chi_o(\Gamma_\lambda) \supset [\alpha,0]$ and hence $\chi_o(\Gamma_\lambda)$ is open.

If K_u would contain more than one point of $\mathbb{R}^+ \times \{0\}$, say $\chi_o(u)$ and $\chi_\tau(u)$, then, if $\chi_\tau(u) \in \chi_o(\Gamma_\lambda)$, there exists a $\tilde{u} \in \Gamma_\lambda$ such that $\chi_o(\tilde{u}) = \chi_\tau(u)$. Since orbits of different solutions do not intersect one obtains $u = \tilde{u}$ and $\tau = 0$. If $\chi_\tau \notin \chi_o(\Gamma_\lambda)$ we consider $\chi_\tau(\Gamma_\lambda)$ which is also an open interval containing $\chi_o(u)$ as well. For an arbitrary ray $re^{i\phi}$, $r > 0$, consider $e^{-i\phi}\chi_t(u)$ and argue as above.

Now define

$$\alpha = \inf \chi_o(\Gamma_\lambda), \quad \beta = \sup \chi_o(\Gamma_\lambda).$$

Since $\chi_o(\Gamma_\lambda)$ is bounded for $\lambda \in (\lambda_o, \lambda_o + \delta)$, α and β are finite. Choose sequences a_n with $a_n \to \beta - 0$, and $u_n \in \Gamma_\lambda$ with $\chi_o(u_n) = a_n$. Select a subsequence u_n, converging in $\mathbb{H}^1_{\ell oc}(\mathbb{R})$ to some u. Since $E_1(u_n) < \varepsilon$, u belongs to Y_1 and solves (2.1). Moreover, u is nonzero in view of $\chi_o(u) = \beta$. A similar argument shows the existence of a limiting u with $\chi_o(u) = \alpha$.

To prove that u is singular, assume that $\chi_t(u)$ is periodic but not constant. Per period, $\chi_t(u_n)$ has exactly two intersections with $\mathbb{R} \times \{0\}$ and $\chi_t(u)$ has at least two. But being the locally uniform limit of periodic solutions with indefinitely increasing periods this yields a contradiction. It is easily seen that $\chi_t(u) = $ const. implies $u = $ const., and thus the theorem is proved.

Remark 3.3. We conclude from the above proof that there are singular solutions u^+ an u^- satisfying

$$\chi_o(u^-) = \alpha < 0 < \beta = \chi_o(u^+).$$

Moreover, for all $u \in \Gamma_\lambda$

$$\chi_o(u^-) < \chi_o(u) < \chi_o(u^+) \text{ holds.}$$

Example. Consider the differential equation

$$y^{(4)} - 2y'' + (1 - \lambda^2)y - g(y, y', y'', y''') = 0 \tag{3.7}$$

where g is a smooth function and satisfies

$$g(x_1, x_2, x_3, x_4) = (x_3 - 2x_1)h(x_1,\dots,x_4)$$

$h(-x) = h(x)$, $h(0) = 0$, and $h(x) \geq 0$ for all $x \in \mathbb{R}^4$. Moreover, assume that $x_1 h(x_1, 0, 0, 0) = 0$ implies $x_1 = 0$. Transforming into a system we obtain

$$A(\lambda) = \begin{pmatrix} 0 & 1 & 0 & 0 \\ 0 & 0 & 1 & 0 \\ 0 & 0 & 0 & 1 \\ \lambda^2-1 & 0 & 2 & 0 \end{pmatrix}, \qquad f(u) = \begin{pmatrix} 0 \\ 0 \\ 0 \\ g(x) \end{pmatrix}.$$

(3.7) is reversible with respect to the reflexion $Rx = (-x_1, x_2, -x_3, x_4)$. The eigenvalues of $A(\lambda)$ are

$$\pm\sqrt{\lambda + 1} \text{ for } |\lambda| < 1, \text{ and } \pm i\sqrt{|\lambda| - 1},\ \pm\sqrt{|\lambda| + 1} \text{ for } |\lambda| > 1.$$

Hence we may set $\lambda_o = 1$. To show existence of singular solutions bifurcating from $\lambda_o = 1$, $x = 0$, we have to prove that (3.7) possesses no nonzero bounded solutions for $\lambda_o = 1$. Multiplication of (3.7) by $y'' - 2y$ yields for $\lambda_o = 1$

$$\frac{d^2}{dt^2}(y'' - 2y)^2 = 2\left(\frac{d}{dt}(y'' - 2y)\right)^2 + 2(y'' - 2y)^2 h \geq 0$$

hence $y'' - 2y = \text{const.}$, thus $y = \text{const.}$ and therefore $yh(y, 0, 0, 0) = 0$ implying $y = 0$. Since 0 is the only constant solution, Remark 3.3 shows that there exist for all $\lambda \in (1, 1 + \delta)$ at least two nonperiodic bounded solutions of (3.7).

4. SINGULAR SOLUTIONS FOR ELLIPTIC EQUATIONS

The results of Section 3 as well as the local results of Section 2 have been shown to be true even for certain second-order elliptic equations in an infinite strip [4]. To complete this exposition we discuss these results and show some applications to the existence of waves and pulses in two-dimensional layers.

Consider the boundary value problem

$$\begin{aligned} &\Delta u + \lambda u - f(u, u_x, u_y) = 0 \\ &u\big|_{\partial\Omega} = 0 \end{aligned} \tag{4.1}$$

in the domain $\Omega = (0,1) \times \mathbb{R}$. Here $\Delta = \frac{\partial^2}{\partial x^2} + \frac{\partial^2}{\partial y^2}$, $\lambda \in \mathbb{R}$, and f is a real valued C^2-function which, together with its gradient, vanishes at 0. The system

$$\frac{dw}{dy} = A(\lambda)w + F(w) \tag{4.2}$$

with

$$w = \begin{pmatrix} u \\ v \end{pmatrix}, \quad A(\lambda) = \begin{pmatrix} 0 & 1 \\ -\frac{\partial^2}{\partial x^2} - \lambda & 0 \end{pmatrix}, \quad F(w) = \begin{pmatrix} 0 \\ f(u, u_x, v) \end{pmatrix}$$

shows the connection to (0.1). If (4.2) is studied in the right space, taking into account the boundary conditions, the spectrum of $A(\lambda)$ consists of eigenvalues only. They are all real and nonzero for $\lambda < \pi^2$, and two simple eigenvalues move onto the imaginary axis as λ crosses π^2. Hence, we may consider (4.1) as an infinite-dimensional version of (0.1). We return to (4.1) and describe the results directly for this boundary value problem. The problem (4.1) is *reversible* if one of the following conditions is satisfied:

$$\begin{aligned} &\text{a)}\ f(u, p, -q) = f(u, p, q) \\ &\text{b)}\ f(-u, -p, q) = -f(u, p, q). \end{aligned} \tag{4.3}$$

The corresponding reflexions are $(u,v) \to (u, -v)$, respectively $(u,v) \to (-u,v)$. We use the spaces

$$Y^2 = \{u \in H^2_{\ell oc}(\Omega)/E_2(u) < \infty\}$$

where

$$E_2(u) = \sup_{j \in \mathbb{Z}^1} \|u\|_{H^2(\tilde{K}_j)}, \quad \tilde{K}_j = [0,1] \times K_j$$

K_j having been defined in Section 1. We say that (4.1) is *non-critical* if $\lambda_o = \pi^2$, $u = 0$ is an isolated solution of (4.1) in $\{\pi^2\} \times Y^2$. A solution is called *singular* if it is either independent of or nonperiodic in y. The role of the mapping χ_t of the previous section is now played by

$$\chi_y(u) = (u_1(y), u_1'(y))$$

$$u_1(y) = \sqrt{2}\int_0^1 u(x,y)\sin\pi x dx.$$

The function f defines a continuous map from $H^\sigma_{\ell oc}(\bar{\Omega})$ for some $\sigma < 2$ if we require the following growth condition:

$$|f(u, p, q)| \leq \gamma_1(u) + \gamma_2|p|^{r_1} + \gamma_3|q|^{r_2} \tag{4.4}$$

for some arbitrary exponents r_1, $r_2 \geq 0$; $\gamma_1(u)$ is any continuous function defined for all $u \in \mathbb{R}$.

Theorem 4.1. Suppose that (4.1) is reversible and noncritical and assume that (4.4) holds. Then there is a right-neighborhood U of π^2, positive numbers α, β and singular solutions u^+, u^- of (4.1) such that

(i) $u_1^+(0) = \beta$, $u_1^-(0) = -\alpha$ holds if (4.3a) is valid,

(ii) $u_1^{+\prime}(0) = \beta$, $u_1^{-\prime}(0) = -\alpha$ holds if (4.3b) is valid.

For the proof see [4], Section 6.

As applications let us consider semilinear wave equations for $(\xi,\eta) \in \Omega = (0,1) \times \mathbb{R}$, $t \in \mathbb{R}$:

$$\begin{aligned} &u_{tt} - u_{\xi\xi} - u_{\eta\eta} + g(u, u_\xi, u_\eta, u_t) = 0 \\ &u|_{\partial\Omega} = 0. \end{aligned} \tag{4.5}$$

We assume that g vanishes if all its arguments are zero, g is a C^2-function satisfying a growth condition similar to (3.2), and

$$-g_u^o \equiv -g_u(0, \ldots, 0) = a > 0.$$

This condition is inessential in as far as $g_u^o \neq 0$ holds. If $g_u^o > 0$ then the following analysis should be done near $a\lambda = -\pi^2$. Let us suppose that one of the following conditions is fulfilled for $f = g + au$:

$$\begin{aligned} &\text{a) } f(u, p, q, r) > 0 \text{ if } \delta > u^2 + p^2 + q^2 + r^2 > 0 \\ &\text{b) } uf(u, p, q, r) > 0 \text{ if } \delta > u^2 + p^2 + q^2 + r^2 > 0 \end{aligned} \tag{4.6}$$

and some $\delta > 0$. We seek solutions of the form

$$u(\xi, \eta, t) = \tilde{U}(\xi, k\eta - \omega t) \text{ for } |\omega| < k.$$

Setting $U(x, y) = \tilde{U}(x, y\sqrt{k^2 - \omega^2})$ we obtain

$$\begin{aligned} &\Delta U = a\lambda U - F(\lambda, U, U_x, U_y) = 0 \\ &U|_{\partial\Omega} = 0 \end{aligned} \tag{4.7}$$

where

$$\lambda = \frac{1}{k^2 - \omega^2}, \quad F(\lambda, U, U_x, U_y) = \lambda f(U, U_x, k\sqrt{\lambda}U_y, -\omega\sqrt{\lambda}U_y).$$

Since the singular solutions are constructed for fixed $a\lambda > \pi^2$ anyway, and since we treat k and ω as parameters, no additional difficulties arise. Let us show that no vertical bifurcation occurs for $a\lambda = \pi^2$. In case (4.6a) multiplication of (4.7) by $\sqrt{2}\sin\pi x$ yields

$$U_1'' = \sqrt{2}\int_0^1 F(\frac{\pi^2}{a}, U, U_x, U_y)\sin \pi x dx \geq 0.$$

Hence $U = 0$ if $U \in Y^2$. In case that (4.6b) holds, multiply by U and integrate

$$\frac{1}{2}\frac{d^2}{dy^2}\int_0^1 U^2 dx - \int_0^1 U_y^2 dx - \int_0^1 U_x^2 dx + \pi^2\int_0^1 U^2 dx =$$

$$\int_0^1 F(\frac{\pi^2}{a}, U, U_x, U_y) U dx \geq 0.$$

In view of the inequality

$$\pi^2\int_0^1 U^2 dx \leq \int_0^1 U_x^2 dx$$

we obtain

$$\frac{d^2}{dy^2}\int_0^1 U^2 dx \geq 2\int_0^1 UF(\frac{\pi^2}{a}, U, U_x, U_y) dx \geq 0.$$

Therefore, $U \in Y^2$ implies $U = 0$. Now, Theorem 4.1 shows the existence of singular solutions for $a\lambda > \pi^2$ if (4.7) is reversible. However, solutions which are independent of y are relatively uninteresting. To exclude these we consider a solution V(x) of (4.7):

$$V_{xx} + a\lambda V - F(\lambda, V, V_x, 0) = 0. \tag{4.8}$$

If (4.6a) holds, multiply by $\sqrt{2}\sin\pi x$ and integrate

$$(-\pi^2 + a\lambda)V_1 = \sqrt{2}\int_0^1 F(\lambda, V, V_x, 0)\sin\pi x dx \geq 0.$$

Since V_1 is constant we conclude $V_1 = u_1^+(0) = \beta > 0$. Therefore, Theorem 4.1 implies that U^- is a singular solution depending on y, if F satisfies (4.3a).

If (4.6b) holds and if the reversibility condition (4.3b) is satisfied for F, we may consider odd solutions and use (ii) in Theorem 2. In this case, both U^+ and U^- are singular solutions depending on y.

Theorem 4.2. Consider the semilinear wave equation (4.5). Assume that g satisfies (4.3), (4.4) and (4.6). Then there exist nontrivial wave solutions of the form $u(\xi, \eta, t) = \tilde{U}(\xi, k\eta - \omega t)$ for $|\omega| < k$ and

$$(k^2 - \omega^2)\pi^2 < -g_u^o < (k^2 - \omega^2)(\pi^2 + \delta)$$

for some $\delta > 0$.

Actually the solutions U of (4.7) are bifurcating from $(\pi^2, 0)$ and π^2 is the lowest point of the continuous spectrum of the linearization of (4.7). Finally let us consider some explicit examples. A two-dimensional Sine-Gordon equation in Ω is given by (c.f., [6])

$$\sigma_{\xi\xi} + \sigma_{\eta\eta} - \sigma_{tt} = A \sin \sigma, \quad A > 0.$$

Setting $\sigma = \pi + u$ and imposing $u|_{\partial\Omega} = 0$ as boundary condition we obtain an equation of the form (4.5) with $g = -A\cdot\sin(u)$, $g_u^o = -A$. Hence, case (ii) of Theorem 4.1 applies and there exist two odd singular solutions $u^\pm$. It can be shown that

$$\lim_{y\to\infty} u^+(x,y) = V(x), \quad \lim_{y\to-\infty} u^+(x,y) = -V(x)$$

holds for $x \in (0,1)$; sign reversed for u^-. Here, V satisfies (4.8), $V \geq 0$, with $a = A$, $F = A(\sin V - V)$. Therefore, our solutions are wavefronts moving to the right or to the left.

REFERENCES

1. M.G. Crandall: An Introduction to Constructive Aspect of Bifurcation and the Implicit Function Theorem, in: Applications of Bifurcation Theory, Proc. of Advanced Seminar, ed. by P.H. Rabinowitz, New York, Academic Press, 1-35, 1977.
2. M.G. Crandall, P.H. Rabinowitz: Nonlinear Sturm-Liouville Eigenvalue Problems and Topological Degree, J. Math. Mech. 19, 1083-1102, (1969/70).
3. P.H. Rabinowitz: Nonlinear Sturm-Liouville Problems for Second-order Ordinary Differential Equations, Comm. Pure Appl. Math. 23, 939-961, (1970).
4. K. Kirchgässner, J. Scheurle: On the Bounded Solutions of a Semilinear Elliptic Equation in a Strip, J. Diff. Eq. 32, 119-148, (1979).
5. K. Kirshgässner, J. Scheurle: Bifurcation from the Continuous Spectrum and Singular Solutions, to appear in Proc. ISIMM, Edinburgh, ed. by R. Knops, 1980.
6. K. Kirchgässner, J. Scheurle: Global Branches of Periodic Solutions of Reversible Systems, to appear in Recent Contributions to Nonlinear Partial Differential Equations, ed. by H. Brézis and H. Berestycki, London, Pitman Adv. Publ. Progr., 1981.
7. N. Kopell, L.N. Howard: Bifurcations and Trajectories Joining Critical Points, Adv. Math. 18, 306-358, (1975).
8. J. Moser: Quasiperiodic Motions, SIAM Rev. 8, 145-172, (1966).
9. P.H. Rabinowitz: Some Global Results for Nonlinear Problems, J. Functional Analysis 7, 487-513, (1971).
10. M. Renardy: Bifurcation of Singular and Transient Solutions, Spatially Nonperiodic Patterns for Chemichal Reaction Models in Infinitely Extended Domains, to appear in Recent Contributions to Nonlinear Partial Differential Equations, ed. by H. Brézis and H. Berestycki, London, Pitman Adv. Publ. Prog., 1981.

11. J. Scheurle: Verzweigung quasiperiodischer Lösungen bei reversiblen dynamischen Systemen, Habilitationschrift, Universität Stuttgart, 1980.

12. J.H. Wolkowisky: Branches of Periodic Solutions of the Nonlinear Hill's Equation, J. Diff. Eq. 11, 385-400, (1972).

SOME PROBLEMS OF REACTION-DIFFUSION EQUATIONS

V. Lakshmikantham[1]

Department of Mathematics
The University of Texas at Arlington

1. INTRODUCTION

Let T be the temperature and n the concentration of a combustible substance. A simple model governing the combustion of the material is given by

$$\left.\begin{aligned} \frac{\partial T}{\partial t} &= K_1 \Delta T + Qn \exp(-E/RT) \\ \frac{\partial n}{\partial t} &= K_2 \Delta n - n \exp(-E/RT) \end{aligned}\right\}, \tag{1.1}$$

where the constant Q is the heat of reaction; the constants K_1, K_2 are thermal, material diffusion coefficients; the term $\exp(-E/RT)$ is the Arrenhius rate factor; E is the activation energy; and R is the universal gas constant. Equations (1.1) are considered on a bounded domain Ω with the boundary conditions

$$T = T_o\ , \quad \frac{\partial n}{\partial \tau} = 0 \text{ on } \partial\Omega, \tag{1.2}$$

[1] Research partially supported by U.S. Army Research Grant DAAG29-80-C-0060.

ISBN 0-12-508780-2

together with initial conditions

$$T(x,0) = T_o(x),\ n(x,0) = n_o(x), \tag{1.3}$$

under the assumption $T_o(x) \geq T_o$. Here τ denotes an outward normal. A discussion of the derivation of the general equations of chemical kinetics may be found in the books by Gavalas [1] and Frank-Kamenetzky [2].

Let u, v denote certain measures of total population such as number of individuals, mass, area of shade cast by plants, etc. A simple model describing populations which are diffusing through Ω as well as interacting with each other is

$$\left.\begin{aligned} u_t &= \alpha_1 \Delta u + uM(u,v) \\ v_t &= \alpha_2 \Delta v + vN(u,v) \end{aligned}\right\}, \tag{1.4}$$

where $\alpha_1, \alpha_2 \geq 0$, Δ is the Laplace operator in x and M, N are suitable functions. Equations (1.4) are considered with the boundary conditions

$$\frac{\partial u}{\partial \tau} = 0,\ \frac{\partial v}{\partial \tau} = 0,\ \text{on } R_+ \times \partial\Omega \tag{1.5}$$

and the initial conditions

$$u(0,x) = u_o(x),\ v(0,x) = v_o(x) \text{ on } \Omega. \tag{1.6}$$

Models of the type (1.4) to (1.6) occurs in studies of population genetics [3,4], conduction of nerve impulses [3,4], chemical reactions [5,6] and several other biological questions [5,7].

If we confine our attention to equations which are independent of space variable x, the equations (1.4) become the Kolmogorov form of ordinary differential equations describing interactive growth of two populations. This approach has a long history dating from the pionnering work of Lotka and Voltera and still occupies a central portion in mathematical ecology [8,9].

It is therefore clear that many models in chemical, biological and ecological processes involve a system of parabolic differential equations of the form

$$u_t^i = a^i \Delta u^i + f_i(t,x,u_1,\ldots,u_N,u_{x_1}^i,\ldots,u_{x_n}^i) \tag{1.7}$$

subject to appropriate initial boundary conditions. Such systems are called weakly coupled systems.

No model of the form (1.7) can be relied on to be in quantitative agreement with real world systems. Voltera himself recognized this fact. He emphasized consistently that differential equations are, at best, only rough approximations of actual ecological systems. They would apply only to animals without age or memory, which eat all the food they encounter and immediately convert it into offspring. Nevertheless,system (1.7) is worth studying since it might offer better qualitative understanding of real-world problems and since such study might help in constructing and analyzing more complex models.

In Section 2, we discuss comparison results including the notion of quasi-solutions. Monotone iterative methods and Müller's type results form the content of Section 3. Flow invariance is considered in Section 4. Section 5 is devoted to syability considerations. The final section is concerned with traveling waves.

2. COMPARISON RESULTS

One of the important and effective methods in the qualitative analysis of dynamical systems is the comparison technique or the theory of differential inequalities [10]. This method is also a useful tool for the qualitative analysis of reaction-diffusion equations. Unfortunately, comparison theorems for systems impose a monotonicity requirement on the reaction terms which is physically unreasonable and thus not suited to the problems at hand. To circumvent this monotonic restriction, one has two alternatives:

(i) choose an appropriate cone other than the standard cone R^n_+ to work in a given situation,

(ii) extend suitably the important classical result of Müller [28] which leads to the new notion of quasi-solutions that have computational advantage.

Let us begin by stating one of the simplest results that can be deduced from the very general comparison results given in [29]. Consider the system of nonlinear reaction-diffusion equations of the form

$$u_t = A\Delta u + F(t,u), \tag{2.1}$$

in $R_+ \times \Omega$ with the initial condition

$$u(0,x) = u_o(x) \text{ on } \bar{\Omega} \tag{2.2}$$

and the Neumann boundary condition

$$\frac{\partial u}{\partial \tau}(t,x) = 0 \text{ on } (0,\infty) \times \partial\Omega. \tag{2.3}$$

Here Δ denotes the Laplace operator in $x \in R^n$, $u,F,u_o \in R^N$, A is a diagonal matrix and $\Omega \subset R^n$ is a bounded domain. Relative to the problem (2.1) to (2.3) we have

Theorem 2.1. Assume that

(i) $A \geq 0$ and $u(t,x)$ is any solution of (2.1) to (2.3) existing on $R_+ \times \bar{\Omega}$;

(ii) $F(t,u)$ satisfies a Lipschitz condition for a constant $L \geq 0$;

(iii) the boundary $\partial\Omega$ is regular, i.e., there exists a $h \in C$ such that $h(x) \geq 0$ on $\bar{\Omega}$, $\frac{\partial h(x)}{\partial \tau} \geq \gamma > 0$ on $\partial\Omega$ and h_x, h_{xx} are bounded. Then the following conclusions are valid:

(a) If $u_i = 0$, $u_j \geq 0$, $j \neq i$, $j = 1,2,\ldots,N$, implies $F_i(t,u) \geq 0$, then $u(t,x) \geq 0$ on $R_+ \times \bar{\Omega}$ provided $u_o(x) \geq 0$ on $\bar{\Omega}$;

(b) If $F(t,u)$ is quasimonotone nondecreasing in u relative to R^N_+, that is, for each i, $1 \leq i \leq N$, $F_i(t,u)$ is nondecreasing in u_j, $j \neq i$ and if the solutions $r(t)$, $\rho(t)$ of

$y' = F(t,y)$ with $r(0) = \bar{y}_0$, $\rho(0) = \underline{y}_0$ exist on R_+, then

$$\rho(t) \leq u(t,x) \leq r(t) \text{ on } R_+ \times \bar{\Omega}, \quad (2.4)$$

provided that $\underline{y}_0 \leq u_0(x) \leq \bar{y}_0$ on $\bar{\Omega}$;

(c) If $F(t,u)$ is quasimonotone nondecreasing in u, $F(t,0) \equiv 0$, then

$0 \leq u_0(x)$ on $\bar{\Omega}$ implies that $u(t,x) \geq 0$ on $R_+ \times \bar{\Omega}$,

$0 < u_0(x)$ on $\bar{\Omega}$ implies that $u(t,x) > 0$ on $R_+ \times \bar{\Omega}$,

$0 < u_0(x) \leq y_0$ on $\bar{\Omega}$ implies that $0 \leq u(t,x) \leq r(t)$ on $R_+ \times \bar{\Omega}$, where $r(t)$ is the same function assumed in (b);

(d) If $F(t,u)$ is not quasimonotone and if the closed set $\bar{w} = \{u \in R^N : a \leq u \leq b\}$ is flow invariant relative to (2.1) to (2.3), then the estimate (2.4) holds $r(t)$, $\rho(t)$ are now being the solutions of

$$r' = g_1(t,r),\ r(0) = \bar{y}_0,\ \rho' = g_2(t,\rho),\ \rho(0) = \underline{y}_0 ,$$

where $g_{1i}(t,u) = \max\{F_i(t,v) : a \leq v \leq u,\ v_i = u_i\}$ and $g_{2i}(t,u) = \min\{F_i(t,v) : u \leq v \leq b,\ v_i = u_i\}$, $1 \leq i \leq N$;

(e) If $F(t,u)$ is not quasimonotone and $\bar{w}$ is not known to be flow invariant, then (4.4) holds if $r(t)$, $\rho(t)$ satisfy the relations

$r_i' \geq F(t,\sigma)$ for all σ such that $\rho \leq \sigma \leq r$ and $\sigma_i = r_i$,

$\rho_i' \leq F_i(t,\sigma)$ for all σ such that $\rho \leq \sigma \leq r$ and $\sigma_i = \rho_i$, $1 \leq i \leq N$.

Let us next demonstrate the advantage of employing a suitable cone other than R_+^N in the study of reaction-diffusion equations. Consider the system

$$\left.\begin{aligned} \frac{\partial u_1}{\partial t} &= a_{11}\Delta u_1 + a_{12}\Delta u_2 + F_1(t,u_1,u_2) \\ \frac{\partial u_2}{\partial t} &= a_{21}\Delta u_1 + a_{22}\Delta u_2 + F_2(t,u_1,u_2) \end{aligned}\right\} \tag{2.5}$$

with (2.2) and (2.3) where $u = (u_1,u_2)$. Clearly this system is not weakly coupled in the sense of system (2.1). Consequently, if we choose to work relative to the cone R^2_+ , we can not draw any conclusions concerning (2.5) as in Theorem 2.1. However, suppose we notice that for some $\alpha,\beta > 0$ with $\beta > \alpha$, we have the relations

$$b_1 \equiv \alpha a_{22} - \alpha^2 a_{12} = \alpha a_{11} - a_{21} \geq 0$$

$$b_2 \equiv \beta a_{22} - \beta^2 a_{12} = \beta a_{11} - a_{21} \geq 0.$$

Then, considering the cone $K = \{u \in R^2 : u_2 \leq \beta u_1$, and $u_2 \geq \alpha u_1\}$ and noting that $S = \{\varphi : \varphi_1(u) = u_2 - \alpha u_1,\ \varphi_2(u) = \beta u_1 - u_2\}$ generates the cone K, we can write (2.5) as

$$\left.\begin{aligned} \frac{\partial}{\partial t}(u_2 - \alpha u_1) &= b_1\Delta(u_2 - \alpha u_2) + \tilde{F}_1(t,u_1,u_2) \\ \frac{\partial u}{\partial t}(\beta u_1 - u_2) &= b_2\Delta(\beta u_1 - u_2) + \tilde{F}_2(t,u_1,u_2) \end{aligned}\right\} \tag{2.6}$$

where $\tilde{F}_1 = F_2 - \alpha F_1$ and $\tilde{F}_2 = \beta F_1 - F_2$. It is easy to see that (2.6) is weakly coupled relative to K. We therefore have the following result observing that $K \subset R^2_+$.

<u>Theorem 2.2.</u> Assume that $\tilde{F} = (\tilde{F}_1,\tilde{F}_2)$ is quasimonotone nondecreasing relative to K and $\tilde{F}$ satisfies a uniqueness condition as in (ii) of Theorem 2.1. Suppose that (iii) of Theorem 2.1 holds. Then

$$\underline{y}_{20} - \alpha\underline{y}_{10} \leq u_{20}(x) - \alpha u_{10}(x) \leq \bar{y}_{20} - \alpha\bar{y}_{10} ,$$

$$\beta\underline{y}_{10} - \underline{y}_{20} \leq \beta u_{10}(x) - u_{20}(x) \leq \beta\bar{y}_{10} - \bar{y}_{20}, \text{ on } \bar{\Omega}$$

implies

$$\rho_1(t) \leq u_1(t,x) \leq r_1(t),$$

$$\rho_2(t) \leq u_2(t,x) \leq r_2(t) \text{ on } R_+ \times \bar{\Omega}.$$

For each i, $1 \leq i \leq N$, let p_i, q_i be two nonnegative integers such $p_i + q_i = N - 1$. We split $u \in R^N$ into $u = (u_i, [u]_{p_i}, [u]_{q_i})$. Then we can, following [11], define quasi-solutions of the problem (2.1) to (2.3). A typical result is as follows.

Theorem 2.3. Let v, w be coupled quasi-solutions of $y' = F(t,y)$, that is, $v_i' \leq F_i(t,v_i,[v]_{p_i},[w]_{q_i}$ and $w_i' \geq F_i(t,w_i,[w]_{p_i},[v]_{q_i})$ such that $v(0) \leq u_o(x) \leq w(0)$ on $\bar{\Omega}$ and u be any solution of (2.1) to (2.3). Suppose that there exists a $z > 0$ on $[0,T] \times \bar{\Omega}$, $\frac{\partial z}{\partial \tau} \geq \gamma > 0$ on $[0,T] \times \partial\Omega$ and for all $\varepsilon > 0$ sufficiently small

$$\varepsilon z_t^i > \varepsilon a^i \Delta z^i + F_i(t,w_i + \varepsilon z_i,[w+\varepsilon z]_{p_i},[v-\varepsilon z]_{q_i})$$
$$-F_i(t,w_i,[w]_{p_i},[v]_{q_i})$$

and

$$\varepsilon z_t^i > \varepsilon a^i \Delta z^i + F_i(t,v_i,[v]_{p_i},[w]_{q_i})$$
$$-F_i(t,v_i - \varepsilon z_i,[v-\varepsilon z]_{p_i},[w+\varepsilon z]_{q_i}).$$

Assume that $F_i(t,u_i,[u]_{p_i},[u]_{q_i})$ is nondecreasing in $[u]_{p_i}$ and nonincreasing in $[u]_{q_i}$ for each i. Then

$$v(t) \leq u(t,x) \leq w(t) \text{ on } [0,T] \times \bar{\Omega}.$$

3. MONOTONE ITERATIVE TECHNIQUES

Consider the system for $i = 1,2,\ldots,N$

$$u_t^i - L_i u_i = F_i(t,x,u),\ (t,x) \in (0,T) \times \Omega, \tag{3.1}$$

$$B_i[u_i] = \beta_i(t,x)\frac{\partial u_i}{\partial \tau} + \gamma_i(t,x)u_i = h_i(t,x),\quad (t,x) \in [0,T] \times \partial\Omega, \tag{3.2}$$

$$u_i(0,x) = u_{oi}(x),\ x \in \Omega, \tag{3.3}$$

where for each i, L_i is a uniformly elliptic operator. Assume that F_i is Hölder continuous on $[0,T] \times \bar{\Omega} \times R^N$, β_i, γ_i, h_i are continuous nonnegative functions on $[0,T] \times \partial\Omega$ with $\beta_i + \gamma_i \neq 0$. Suppose that v, w are lower and upper solutions for the problem (3.1) to (3.3) such that $v \leq w$ on $[0,T] \times \bar{\Omega}$. Let F_i be quasimonotone nondecreasing in u and $F_i(t,x,u_1,\ldots,u_i,\ldots,u_N) - F_i(t,x,u_1,\ldots,\bar{u}_i,\ldots,u_N) \geq -M_i(u_i - \bar{u}_i)$ for $v_i \leq \bar{u}_i \leq u_i \leq w_i$ consider the linear uncoupled problem

$$\frac{\partial u_i^{(k)}}{\partial t} - (L_i - M_i)u_i^{(k)} = F_i(t,x,u^{(k-1)}) + M_i u^{(k-1)},$$

$$B_i[u_i^{(k)}] = h_i(t,x) \text{ and } u_i^{(k)}(0,x) = u_{oi}(x),\ k = 1,2,3,\ldots.$$

Starting from $v_o = v$, $w_o = w$, we can construct monotone sequences $\{v^k\}$, $\{w^k\}$ such that $v^k \leq w^k$ on $[0,T] \times \bar{\Omega}$ and show that

$$v^* = \lim_{k\to\infty} v^k,\quad w^* = \lim_{k\to\infty} w^k$$

are minimal and maximal solutions of the problem (3.1) to (3.3) respectively. Furthermore, $v \leq v^* \leq w^* \leq w$ on $[0,T] \times \bar{\Omega}$. If F satisfies a Lipschitz conditions, then $v^* = w^*$ and the problem (3.1) to (3.3) has a unique solution.

Suppose that F is not quasi-monotone nondecreasing in u.

However F possesses a quasi-mixed monotone property (see [12]). Then assuming v, w are coupled quasi lower and upper solutions of the problem (3.1) to (3.3), it is still possible to construct monotone sequences $\{v^k\}$, $\{w^k\}$. But in this case, v^*, w^* will be coupled minimal, maximal solutions of (3.1) to (3.3).

One can construct upper and lower solutions based on the eigenfunction φ_i corresponding to the least eigenvalue λ_i of the linear eigenvalue problem

$$L_i\varphi_i + \lambda_i\varphi_i = 0, \; x \in \Omega,$$

$$B_i\varphi_i = 0, \; x \in \partial\Omega.$$

Since it is well known that the least eigenvalue λ_i is positive and the corresponding eigenfunction φ_i is also positive in Ω. In case $\beta_i > 0$, $\varphi_i > 0$ on $\bar{\Omega}$. Utilizing this observation explicit lower and upper solutions can be constructed so that steady state solution is globally asymptotically stable with respect to nonnegative initial perturbations. For details of the results discussed above, see [12,13,14].

If F does not posses any monotone property, it is not possible to construct monotone sequences. However one could still prove existence of a unique solution for the problem (3.1) to (3.3) by means of Schauder's fixed point theorem such that every solution u(t,x) of the parabolic system satisfies $\lim_{t\to\infty} u(t,x) = u(x)$ uniformly in x where u(x) is the unique solution of the corresponding elliptic problem. For a discussion of such results and further references see [15,16].

A monotone iterative approximation scheme for nonlinear elliptic boundary value problems can be shown to be an approximation to the evolution in time of the solution of a corresponding parabolic boundary value problem. This is done by using the backward time difference approximation scheme to the parabolic problem and exploiting convergence arguments for the difference scheme. Details are given in [17].

4. FLOW-INVARIANCE

Let S be a closed, convex set in R^N. The set S is said to be flow-invariant relative to (3.1) if for any solution u of (3.1) the property $u(\Gamma) \subset S$ implies $u(H) \subset S$ where $\Gamma = [0,\infty) \times \partial\Omega$ and $H = (0,\infty) \times \Omega$. In (3.1), we can allow F to depend on u_x in this case. We need the following assumptions:

$$\text{for any } z \in \partial S \text{ and any outer normal } n(z)\text{, we have } n(z)F(t,x,z,p) \leq 0 \text{ for all } (t,x) \in H \text{ and all } p \in R^{Nn}, \text{ with } p_i \in R^N \text{ such that } n(z)p_i = 0, \; i = 1,2,\ldots,n, \tag{4.1}$$

$$g \in C[R_+ \times R_+,R], \; (z-z_0)[F(t,x,z,p) - F(t,x,z_0,p)] \leq |z - z_0|g(t,|z - z_0|) \text{ whenever } z_0 \in \partial S, \; n(z_0) = \frac{(z-z_0)}{|z-z_0|} \text{ is an outer normal, and } n(z_0)p_i = 0, \; i = 1,2,\ldots,n, \tag{4.2}$$

$$\text{if } v \in C[R_+, R_+] \text{ such that } v(0) \leq 0 \text{ and } D_-v(t) \leq g(t,v(t)) \text{ whenever } v(t) > 0, \text{ then } v(t) \leq 0 \text{ for } 0 \leq t < \infty. \tag{4.3}$$

Theorem 4.1. If u is any solution of (3.1) and the assumptions (4.1) to (4.3) hold, then S is flow-invariant for (3.1).

The extensions if this result to mixed boundary value problems and unbounded domains Ω are also possible. The proofs of these invariance results are essenrially elementary and are in the spirit of differential inequalities. Hence the proofs require no existence theory as was employed in [18]. Also the convexity of the set S is essential for parabolic systems. For a thorough discussion see [19].

5. STABILITY AND ASYMPTOTIC BEHAVIOUR

We specialize (2.1) for convenience, so that $F(t,u) = F(u) = Bu + g(u)$ with $g(0) = 0$ and $g_u(0) = 0$. Define the linear system

$$u_t = A\Delta u + Bu. \tag{5.1}$$

Assume that $A_i > 0$, $i = 1,2,\ldots,N$. Let $0 = \lambda_o \leq \lambda_1 \leq \ldots \leq \lambda_k \ldots$ denote the eigenvalues and $\varphi_o, \varphi_1, \ldots, \varphi_k, \ldots,$ the corresponding normalized eigenfunctions of Laplace equation in Ω with $\frac{\partial \varphi_k}{\partial \tau} = 0$ on $\partial\Omega$ and $\int_\Omega \varphi_k^2(x)dx = 1$. For each nonnegative integer n, let z_{on} be the k-vector $z_{on} = \int_\Omega \alpha(x)\varphi_n(x)dx$ and let the $k \times k$ matrix $e^{B_n t}$ be the matrix solution to the differential equation

$$\frac{d}{dt} e^{B_n t} = (B - \lambda_n A)e^{B_n t}, \tag{5.2}$$

where $B_n = B - \lambda_n A$ with the initial condirion $e^{A_n 0} = I$. Then the solution of (5.1) can be given in the form $u(t,x) = \sum_{n=0}^{\infty} \varphi_n(x)e^{B_n t} z_{no}$ and this defines a semigroup T_t through the definition $u(t,x) = (T_t\alpha)(x)$. One can now prove the following result concerning the stability of (5.1).

<u>Theorem 5.1.</u> (i) The zero solution is globally asymptotically stable if for each nonnegative integer n the eigenvalues of $B = \lambda_n A$ have negative real parts. Further there exist positive constants K, ω such that for any $t > 0$,

$$\|z(t,x)\| \leq Ke^{-\omega t}\|\alpha(x)\|.$$

(ii) The zero solution is stable if for each nonnegative integer n the eigenvalues of $B - \lambda_n A$ have nonpositive real parts and those with zero real parts have simple elementary divisors.

(iii) The zero solution is unstable if for some n there exists an eigenvakue of $B - \lambda_n A$ with either positive real part or zero real part with a nonsimple elementary divisor.

It is easy to construct examples with unequal diffusion coefficients to show that diffusion could turn a stable solution into an unstable one. Concerning the problem (2.1) we have

Theorem 5.2. The zero solution of (2.1) is asymptocically stable if the zero solution of the linearized problem (5.1) is asymptotically stable.

Let us note that in the case of equal diffusion coefficients, the hypothesis of Theorem 5.2 reduces to the requirement that B have only eigenvalues with negative real parts. In this case it is possible to use a Lyapunov function approach to estimate the extent of asymptoric stability. For further details see [20]. For application of Lyapunov methods to reaction diffusion equations see also [21,22,23].

6. TRAVELING WAVES

Let us consider the scalar reaction-diffusion equation

$$u_t = u_{xx} + f(u), \quad x \in R, \ t > 0 \tag{6.1}$$

with the initial condition

$$u(0,x) = \varphi(x), \quad x \in R \tag{6.2}$$

A traveling front solution of (6.1) is of the form $u(t,x) = u(x-ct)$ for some c, the velocity, with the limits $u(\pm\infty)$ exist and unequal. An interesting question is to find conditions such that the solution $u(t,x)$ of (6.1) tends to traveling front solution as $t \to \infty$. A typical result in this direction is as follows.

Theorem 6.1. Let $f \in C^1[0,1]$ satisfy, for some $\alpha \in (0,1)$, $f(0) = f(1) = 0$, $f(u) \leq 0$ for $u \in (0,\alpha)$, $f(u) > 0$ for

$u \in (\alpha,1)$, $\int_0^1 f(u)du > 0$. Then there exists a travelling front solution $u(x-ct)$ of (6.1), unique modulo traslation and necessarily monotonic, and if $\varphi \in C^1(-\infty, \infty)$ with $\varphi(-\infty) = 0$, $\varphi(+\infty) = 1$, $\varphi'(x) > 0$ for all x, then there exists a function $\gamma \in C^1[0,\infty)$, with $\gamma'(t) \to 0$ as $t \to \infty$, such that, uniformly in x,

$$|u(x,t) - U(x-ct-\gamma(t))| = 0(1)$$

as $t \to \infty$, where u is the solution of the initial value problem (6.1), (6.2) corresponding to the initial function φ.

The principal tools that are employed in the proof are comparison theorems for parabolic equations, although the parabolic equation is now one for $p = u_x$ in terms of t,u. The problem is complicated in that this parabolic equation is degenerate since the coefficient of p_{uu} is p^2, and the boundary conditions demand that p vanishes at the end points (0,1) of the range of u. For necessary analysis see [24,25,26].

In conclusion, we are glad to mention the recent book of Paul Fife [27] on this subject which deals with several important topics and includes a large number of references.

REFERENCES

1. G.R. Gavalas: Nonlinear Differential Equations of Chemically Reaction Systems. New York, Springer, 1968.
2. D.A. Frank-Kamenetzky: Diffusion and Heat Exchange in Chemical Kinetics. Translated by N. Thon. Princeton, N.J., Princeton Univ. Press, 1955.
3. D.G. Aronson and H.F. Weinberger: Nonlinear Diffusion in Population Genetics, Combustion and Nerve Propagation. New York, Springer Verlag, Lecture Notes, Vol. 446, 1975.
4. P.E. Waltman: The equations of growth, Bull. Math. Biophs. 26 (1964), 39-43.

5. R. Aris: The Mathematical Theory of Diffusion and Reaction in Permeable Catalysts, Oxford, England, Clarendon Press, 1975.
6. P.C. Fife: Pattern Formation in Reacting and Diffusing Systems. J. Chem. Phys. 64 (1976), 554-564.
7. A.M. Turing: On The Chemical Basis of Morphogenesis, Phil. Trans. Roy. Soc. London Ser. B, 237 (1952), 37-52.
8. J. Maynard-Smith: Models in Ecology, Cambridge, Cambridge Univ. Press, 1974.
9. R. May: Stability and Complexity in Model Ecosystems, Princeton: Princeton Univ. Press, 1973.
10. V. Lakshmikantham, S. Leela: Differential and Integral Inequalities, Vols. I and II. New York, Academic Press, 1979.
11. K. Deimling, V. Lakshmikantham: Quasi Solutions and Their Role in Qualitative Theory of Differential Equations, J. Nonlinear Analysis, 4, 655-663 (1980).
12. G.S. Ladde, V. Lakshmikantham, A.S. Vatsala: Existence and Asymptotic Behaviour of Reaction-Diffusion Systems via Coupled Quasi-Solutions, Proc. Int. Conf. on Nonlinear Phenomena in Mathematical Sciences, held in Arlington, Texas, June 1980. New York: Academic Press (to appear).
13. J. Chandra, Norman, F. Dressel: A Monotone Method for a System of Nonlinear Parabolic Differential Equations, (to appear).
14. C.V. Pao: Stability Analysis of Some Nonlinear Reaction-Diffusion Systems in Chemical and Nuclear Kinetics, (to appear).
15. J. Hernàndez: Some Existence and Stability Results for Solutions of Reaction-Diffusion Systems with Nonlinear Boundary Conditions. Proc. Int. Conf. on Nonlinear Phenomena in Mathematical Sciences, held in Arlington, Texas, June 1980. New York, Academic Press, (to appear).
16. J. Hernàndez: Existence and Global Stability Results for Reaction-Diffusion Systems with Nonlinear Boundary Conditions, (to appear).

17. J. Chandra, P.W. Davis: On the Relation between Monotone Method and Temporal Evolution in Fully Nonlinear Boundary Value Problems, J. Inst. Maths. Applics. 25, 231-240, (1980).
18. H. Weinberger: Invariant Sets for Weakly Coupled Parabolic and Elliptic Systems, Rendiconti di Matematica 8, Ser VI, 295-310, (1975).
19. R. Redheffer, W. Walter: Invariant Sets for Systems of Partial Differential Equations. I. Parabolic Equations. Arch. Rat. Mech., Anal. 67, 41-52 (1978), II. First Order and Elliptic Equations, Ibid., 73, 19-29, (1980).
20. R.G. Casten, C.J. Holland: Stability Properties of solutions to Systems of Reaction-Diffusion Equations, SIAM J. Appl. Math. 33 , 353-364 (Sept. 1977).
21. N. Chafee: A Stability Analysis for a Semilinear Parabolic Partial Differential Equations, J. Diff. Eqs. 15, 522-540, (1974).
22. P. de Mottoni,A. Tesei: Asymptotic Stability Results for a System of Quasilinear Parabolic Equations, Applicable Analysis 9, 7-21, (1979).
23. F. Rothe, P. de Mottoni: A Simple System of Reaction-Diffusion Equations Describing Morphogenesis. Asymptotic Behaviour. Ann. Pat. Pura Appl. (IV), 122, 141-157 (1979).
24. P.C. Fife, J.B. McLeod: A Phase Plane Discussion of Convergence to Travelling Fronts for Nonlinear Diffusion. MRC Technical Report 1986, University of Wisconsin.
25. P.C. Fife, J.B. McLeod: The Approach of Solutions of Nonlinear Diffusion Equations to Travelling Front Solutions. Arch. Rat. Mech. Anal. 65, 335-361 (1977), also Bull. Amer. Math. Soc., 81, 1075-1078 (1975).
26. K. Uchiyama: The Behaviour of Solutions of Some Nonlinear Diffusion Equations for Large Time, J. Mat. Kyoto Univ. (J.M.K. Yaz) 18-3, 453-508, (1978).
27. P.C. Fife: Mathematical Aspects of Reacting and Diffusion Systems. Lecture Notes in Biomathematics, Springer-Verlag, 1979.
28. M. Müller: Über das Fundamentaltheorem in der Therie der gewöhnlinhen Differentialgleichungen, Math. Z. 26, 619-645, (1926).

29. V. Lakshmikantham: Comparison Results for Reaction-Diffusion Equations in a Banach Space. Proc. of Conf. "A Survey on the Theoretical and Numerical Trends in Nonlinear Analysis", held in Bari, Italy, 1979.

THE ROLE OF QUASI-SOLUTIONS IN THE STUDY OF DIFFERENTIAL EQUATIONS

S. Leela[1]

State University of New York at Geneseo

1. INTRODUCTION

It is well known that quasimonotone property plays a crucial role in the study of comparison theorems, existence of extremal solutions and the monotone iterative techniques for initial and boundary value problems [1,3,5,10]. Also, in the method of vector Lyapunov functions to investigate the stability of large scale systems, an unpleasant fact is that quasimonotone property is required for comparison systems since comparison systems with a desired stability property exist without being quasimonotone [9,13,14]. However, in many physical situations, we often find that quasimonotone property is not satisfied. For example, consider a simple model governing the combustion of gas [2,7] represented by the system

$$\left.\begin{aligned} T_t &= k_1 \Delta T + QC \exp(-E/RT) \\ C_t &= k_2 \Delta C - C \exp(-E/RT) \end{aligned}\right\} \qquad (1.1)$$

together with appropriate initial and boundary conditions.

[1] Research partially supported by U.S. Army Grant DAAG29 - 80 - C0060.

ISBN 0-12-508780-2

Here T denotes the temperature and C the concentration of the combustible substance and Q, E, R, k_1, k_2 are appropriate positive constants. We note that the system (1.1) is not quasimonotone but satisfies a kind of mixed quasimonotone property. The idea of quasi-solutions introduced in [8] and developed further in [6] can be gainfully employed to study differential systems which possess a certain mixed quasi monotonicity. The aim of this article is to give a brief account of the role of quasi-solutions in the study of comparison theorems, monotone methods for initial ans boundary value problems and the stability of large scale systems [4,6,11,12].

2. QUASI-SOLUTIONS OF INITIAL VALUE PROBLEMS FOR ODE

Consider the initial value problem

$$x' = f(t,x), \quad x(0) = x_0 \tag{2.1}$$

where $f \in C[J\times R^n, R^n]$ and $J = [0,T]$. To define quasisolutions of (2.1), we fix for each $i = 1,2,\ldots,n$, two nonnegative integers p_i, q_i such that $p_i + q_i = n-1$ and split $x \in R^n$ into $x = (x_i, [x]_{p_i}, [x]_{q_i})$. With this (2.1) can be written as

$$x_i' = f_i(t, x_i, [x]_{p_i}, [x]_{q_i}), \quad 1 \le i \le n$$
$$x_i(0) = x_{oi}.$$

Let us note that all relations between vectors are assumed to hold componentwise. We can now define various types of quasi-solutions of (2.1) [6].

<u>Definition 1</u>. Let $a \in C[J,R^n]$. Then $y \in C^1[J,R^n]$ is a quasisolution of (2.1) with respect to a, if for each $i = 1,2,\ldots,n$,

$$y_i' = f_i(t, y_i, [y]_{p_i}, [a]_{q_i}), \quad y_i(0) = x_{oi}. \tag{2.2}$$

The function y is said to be a lower quasisolution of (2.1) relative to a if for each $i = 1,2,\ldots,n$,

$$y_i' \leq f_i(t,y_i,[y]_{p_i},[a]_{q_i}), \quad y_i(0) \leq x_{oi} \tag{2.3}$$

and y is a upper quasisolution of (2.1) if both inequalities in (2.3) are reversed.

In some considerations, the concempts of coupled quasisolutions and coupled lower and upper quasisolutions are useful.

Definition 2. The functions u, $v \in C^1[J,R^n]$ with $v(t) \leq u(t)$ on J are coupled upper and lower quasisolutions of (2.1), if for each $i = 1,2,\ldots,n$,

$$\left.\begin{array}{l} u_i' \geq f_i(t,u_i,[u]_{p_i},[v]_{q_i}), u_i(0) \geq x_{oi} \\ v_i' \leq f_i(t,v_i,[v]_{p_i},[u]_{q_i}), v_i(0) \leq x_{oi} \end{array}\right\} \tag{2.4}$$

The functions u, v are coupled quasi solutions of (2.1) if all the inequalities in (2.4) are replaced by equalities.

We can define maximal and minimal quasisolutions of (2.1) (with respect to a) to be the maximal and minimal solutions of the system (2.2) whenever they exist. The flexibility of the idea of quasi-solutions lies in the fact that when $q_i = 0$ for all i, quasi-solutions of (2.1) are just solutions and in case $p_i = 0$ for all i, the quasi solutions are most easily determined. The appropriate choice of a(t) makes all the difference in any situation. For the error estimates between (i) quasi-solutions and the choice function a; (ii) quasi-solutions and solutions, see [11].

Definition 3. The function f is said to be mixed quasimonotone if for each $i = 1,2,\ldots,n$, $f_i(t,x_i,[x]_{p_i},[x]_{q_i})$ is monotone nondecreasing in $[x]_{p_i}$ and monotone nonincreasing in $[x]_{q_i}$.

It is clear that if $q_i = 0$ for all i, then f is quasimonotone nondecreasing. The following is a known comparison result in this context [6].

Theorem 2.1. Let $a \in C[J,R^n]$ and $Q_a = \{(t,x): t \in J$ and $x \geq a(t)\}$. Let $f \in C[Q_a,R^n]$ be mixed quasimonotone. Let

$m \in C[J,R^n]$ be such that $m(t) \in Q_a(t) = \{x: (t,x) \in Q_a\}$ for all t and $D_-m(t) \leq f(t,m(t))$, $t \in J$. Then $m(0) \leq x_0$ implies $m(t) \leq r(t)$, $t \in J$, where $r(t)$ is the maximal quasi-solutions (with respect to a) of (2.1).

For another result concerning the existence of coupled quasi–solutions and quasi–solutions relative to a, see [11]. The comparison theorem (Theorem 2.1) can be gainfully employed in place of the usual comparison theorems [10] whenever the system is mixed quasimonotone.

3. MONOTONE METHOD FOR INITIAL VALUE PROBLEMS

In order to develop monotone iterative technique for the initial value problem (2.1), let us make the following assumptions:

(A_1) f is mixed quasimonotone;

(A_2) $\underline{u}$, $\bar{u} \in C^1[J,R^n]$ such that $\underline{u}(t) \leq \bar{u}(t)$, $t \in J$ are coupled lower and upper quasi-solutions of (2.1);

(A_3) for each $i = 1,2,\ldots,n$, and $\underline{u}_i(t) \leq y_i \leq x_i \leq \bar{u}_i(t)$ on J, we have $f_i(t,x_i,[x]_{p_i},[x]_{q_i}) - f_i(t,y_i,[x]_{p_i},[x]_{q_i}) \geq -M_i(x_i - y_i)$, where $M_i \geq 0$.

For any $\alpha, \beta \in C^1[J,R^n]$ with $\underline{u} \leq \alpha$, $\beta \leq \bar{u}$ on J, we define

$$G_i(t,x) = f_i(t,\alpha_i,[\alpha]_{p_i},[\beta]_{q_i}) - M_i(x_i - \alpha_i)$$

and consider the uncoupled linear system

$$x_i' = G_i(t,x), \quad x_i(0) = x_{0i}. \tag{3.1}$$

Clearly, for any given α, β the system (3.1) possesses a unique solution $x(t)$ defined on J. For each $\alpha, \beta \in C^1[J,R^n]$ such that $\underline{u} \leq \alpha$, $\beta \leq \bar{u}$ on J, define the mapping A by $A[\alpha,\beta] = x$, where x is the unique solution of (3.1). In view of the assumptions (A_1), (A_2), (A_3), it can be shown that $\underline{u}(t) \leq x(t) \leq \bar{u}(t)$. Moreover, the mapping A is such that

(i) $\underline{u} \le A[\underline{u},\bar{u}]$ and $\bar{u} \ge A[\bar{u},\underline{u}]$;
(ii) for any α, β with $\underline{u} \le \alpha$, $\beta \le \bar{u}$, $A[\alpha,\beta] \le A[\beta,\alpha]$.
This enables one to define the monotonic sequences $\{\alpha_n\}$, $\{\beta_n\}$ given by

$$\alpha_n = A[\alpha_{n-1},\beta_{n-1}], \quad \beta_n = A[\beta_{n-1},\alpha_{n-1}] \tag{3.2}$$

which converge uniformly and monotonically to coupled minimal and maximal quasi-solutions of (2.1).

The following is an important result proved in [11] and demonstrates the role of coupled lower and upper quasi-solutions in the monotone method.

Theorem 3.1. Let the assumptions (A_1), (A_2) and (A_3) hold. Then the sequences $\{\alpha_n\}$, $\{\beta_n\}$ defined by (3.2) converge uniformly and monotonically to coupled minimal and maximal quasi-solutions of (2.1), that is, if $\eta = \lim_{n\to\infty} \alpha_n$, $\mu = \lim_{n\to\infty} \beta_n$ and x, y are any coupled quasi-solutions of (2.1) such that $\underline{u} \le x$, $y \le \bar{u}$, we have

$$\underline{u} \le \alpha_1 \le \dots \le \alpha_n \le \eta \le x, \; y \le \mu \le \dots \le \beta_n \le \dots \le \beta_1 \le \bar{u}. \tag{3.3}$$

Moreover, any solution x of (2.1) such that $\underline{u} \le x \le \bar{u}$ also satisfies (3.3).

Note that when $q_i = 0$ for each i, f is quasimonotone nondecreasing and consequently η, μ reduce to the minimal and maximal solutions of (2.1).

4. MONOTONE METHOD FOR THE SECOND ORDER BOUNDARY VALUE PROBLEMS

Consider the boundary value problem (BVP for short)

$$\left.\begin{aligned} &-x_i'' = f_i(t,x,x_i'), \; t \in [0,1], \; i = 1,2,\dots,n \\ &B_\mu x(\mu) = \alpha_\mu x(\mu) + (-1)^{\mu+1}\beta_\mu x'(\mu) = b_\mu, \; \mu = 0,1 \end{aligned}\right\} \tag{4.1}$$

where $f \in C[I\times R^n \times R, R^n]$, $I = [0,1]$ and satisfies mixed quasi-monotone property, α_μ, $\beta_\mu \in R^n$ such that $\alpha_0, \alpha_1 \geq 0$ and β_0, $\beta_1 > 0$. As in Section 2, we shall rewrite (4.1) as

$$\left.\begin{aligned} &-x_i'' = f_i(t,x_i,[x]_{p_i},[x]_{q_i},x_i'),\ i = 1,2,\ldots,n, \\ &B_\mu x(\mu) = b_\mu,\ \mu = 0,1. \end{aligned}\right\}$$

we can now define various types of quasi-solutions for the BVP (4.1) analogous to the definitions in Section 2. For example, the functions v, $w \in C^2[I,R^n]$ with $v(t) \leq w(t)$ on I are said to be coupled lower and upper quasi-solutions of (4.1) if for each $i = 1,2,\ldots,n$,

$$\left.\begin{aligned} &-v_i'' \leq f_i(t,v_i,[v]_{p_i},[w]_{q_i},v_i'),\ B_\mu v(\mu) \leq b_\mu \\ &-w_i'' \geq f_i(t,w_i,[w]_{p_i},[v]_{q_i},w_i'),\ B_\mu w(\mu) \geq b_\mu. \end{aligned}\right\}$$

The following is an existence result proved in [12].

<u>Theorem 4.1</u>. Assume that

(A_1') f is mixed quasimonotone, that is, for each $i = 1,2,\ldots,n$, $f_i(t,x_i,[x]_{p_i},[x]_{q_i},x_i')$ is monotone nondecreasing in $[x]_{p_i}$ and monotone nonincreasing in $[x]_{q_i}$;

(A_2') v, w are coupled lower and upper quasisolutions of (4.1);

(A_3') for $t \in I$, $v(t) \leq x \leq w(t)$ and $x_i' \in R$, $|f_i(t,x,x_i')| \leq h_i(|x_i'|)$, where $h_i \in C[R_+,(0,\infty)]$;

(A_4') there exists an $N \geq 0$ depending only on v, w and h such that $\displaystyle\int_{\lambda_i}^{N_i} \frac{s}{h_i(s)}\,ds > \max_I w_i(t) - \min_I v_i(t)$, where

$\lambda_i = \max(|v_i(0) - w_i(1)|,\ |v_i(1) - w_i(0)|)$ and $N = (N_1,\ldots,N_n)$.

Then, the BVP (4.1) has a solution $x \in C^2[I,R^n]$ such that $v(t) \leq x(t) \leq w(t)$ on I. Moreover, there exists an $N_0 > 0$ such that $|x'(t)| \leq N_0$ on I.

The proof depends upon the construction of a modified function and using Scorza-Dragoni's theorem. By a well known result in [1] one has a $N \in R_+^n$ depending only on v, w and h such that $|x'(t)| \leq N$ on I, where $x(t)$ is any solution of (4.1). Choosing $N_0 > \max(N, \max_I |v'(t)|, \max_I |w'(t)|)$ and $d > N_0$, we can define the modified functions as follows:

$$F(t,x,x') \equiv f(t,p(t,x),q(x')) - Q(x) \tag{4.2}$$

where

$$q_i(x') = \max(-d_i,\min(x_i',d_i)),$$

$$p_i(t,x) = \max(v_i(t),\min(x_i,w_i(t))),$$

and

$$Q_i(x) = \begin{cases} \dfrac{x_i - w_i(t)}{1 + x_i^2} & \text{if } x_i > w_i(t) \\ 0 & \text{if } v_i(t) \le x_i \le w_i(t) \\ \dfrac{x_i - v_i(t)}{1 + x_i^2} & \text{if } x_i < v_i(t) \end{cases}$$

For details of the proof, see [12].

A further assumption is needed for f in order to develop the monotone method for (4.1). Suppose that (A_5') for each i, $f_i(t,x,x_i')$ is continuonsly differentiable in x_i and x_i' for $t \in I$, $v(t) \le x \le w(t)$ and $|x_i'| \le d$ and $M_i = \max_I \left|\dfrac{\partial f_i(t,x,x_i')}{\partial x_i}\right|$. For any $\eta_1, \eta_2 \in C[I,R^n]$ such that $v(t) \le \eta_1, \eta_2 \le w(t)$ on I, we shall define

$$G_i(t,x,x_i') = F_i(t,\eta_{1i},[\eta_1]_{p_i},[\eta_2]_{q_i},x_i') - M_i(x_i - \eta_{1i})$$

where F is the modified function defined by (4.2) and consider the BVP

$$-x_i'' = G_i(t,x,x_i'), \quad B_\mu x(\mu) = b_\mu. \tag{4.3}$$

Under the assumptions $(A_1') - (A_5')$, it can be shown that there exists a unique solution $x \in C^2[I,R^n]$ to the BVP (4.3) such that $v(t) \le x(t) \le w(t)$ on I and there exists an $N_0 > 0$ such that $|x'(t)| \le N_0$ on I. For details, see [12]. As in the case of the IVP in Section 3, we define the mapping A by $A[\eta_1,\eta_2] = x$

for any $\eta_1, \eta_2 \in C^2[I,R^n]$ with $v(t) \leq \eta_1$, $\eta_2 \leq w(t)$ on I and x is the unique solution of (4.3). This mapping has the property that $v \leq A[v,w]$ and $w \geq A[w,v]$. Once again, the monotone sequences $\{v_n\}$, $\{w_n\}$ can be defined by

$$v_n = A[v_{n-1}, w_{n-1}], \quad w_n = A[w_{n-1}, v_{n-1}]$$

with $v_o = v$ and $w_o = w$. Using standard arguments, it can be shown that $\{v_n(t)\}$, $\{w_n(t)\}$ converge uniformly and monotonically to coupled minimal and maximal quasi-solutions α, β of the BVP (4.1), that is, if x, y are coupled quasi-solutions of (4.1) such that $v \leq x$, $y \leq w$ on I and $|x'|$, $|y'| \leq N_o$, then $v \leq v_1 \leq \ldots \leq v_n \leq \ldots \leq \alpha \leq x, y \leq \beta \leq \ldots \leq w_n \leq \ldots \leq w_1 \leq w$ on I. For detailed proof, see [12]. Also, if x is any solution of (4.1) such that $v(t) \leq x(t) \leq w(t)$, the monotonic iterative scheme holds.

5. LARGE SCALE SYSTEMS AND QUASI SOLUTIONS

Consider the large scale system

$$x' = F(t,x), \quad x(0) = x_o \tag{5.1}$$

where $x \in R^N$, N being a large natural number. Suppose that the system (5.1) can be decomposed into isolated subsystems

$$y_k' = g_k(t,y_k) + R_k(t,x), \quad y_k(0) = x_{ok}$$

where $y_k \in R^{N_k}$ such that $\sum_{k=1}^{n} N_k = N$ and the interaction terms given by $R_k(t,x)$ are of the form $\sum_{j=1}^{n} R_{kj}(t,y_j)$. The method of Lyapunov can be effectively employed to study the stability of the zero solution of the system (4.1) if there exist functions V_k, d_k, w_k, $k = 1,\ldots,n$ such that

$$D^+V_k(t,y_k) \leq -d_k(t,y_k) + w_k(t,y_k,x)$$

and the right hand side can be further majorized to yield

$$D^+V(t,y) \leq g(t,V(t,y))$$

with the comparison function g being quasimonotone and satisfying other appropriate properties. In many applications, one gets g(t,v) of the form - AV where the matrix A has positive diagonal and nonpositive off diagonal elements. It is a well known fact [14] that the system u' = - Au is stable if and only if A is a Metzler matrix. The classical method of vector Lyapunov functions fails if this matrix A has positive off diagonal elements. However, we can employ Theorem 2.1 in this case. Since all $V_k \geq 0$, we can choose the function $a \equiv 0$ to get the estimate

$$V(t,y(t)) \geq r(t), \quad t \geq 0$$

where r is the maximal quasi-solution of the system

$$r_i' = g_i(t,r_i,[r]_{p_i},[0]_{q_i}), \quad r(0) = V(0,y_0)$$

and obtain the stability information for the system (4.1) when the stability behavior of the quasi solution is known.

In the study of large scale systems, the idea of decomposition of a large system into a suitable system of simpler, isolated subsystems is crucial. It is shown in [11] that quasi-solutions always make such a decomposition possible.

REFERENCES

1. S. Bernfeld, V. Lakshmikantham: An Introduction to Nonlinear Boundary Value Problems, New York, Academic Press, (1974).
2. J. Chandra: Some Comparison Theorem for Reaction Diffusion Equations. To appear in Proc. Intern. Conf. on Recent Trends in Differential Equations, Trieste (1978); Academic Press.

3. J. Chandra, P.W. Davis: A Monotone Method for Quasi-linear Boundary Value Problems, Arch. Rat. Mech. Anal., 54 257-266, (1974).
4. J. Chandra, F. Dressel, P.D. Norman: A Monotone Method for a System of Nonlinear Parabolic Differential Equations. To appear in Proc. Roy. Soc. Edinburg.
5. J. Chandra, V. Lakshmikantham, S. Leela: A Monotone Method for Infinite Systems of Nonlinear Boundary Value Problems. Arch. Rat. Mech. Anal., 68 179-190, (1978).
6. K. Deimling, V. Lakshmikantham: Quasi Solutions and their Role in Qualitative Theory of Differential Equations. Non linear Analysis, TMA, 4 657-663, (1980).
7. D.A. Frank-Kamantzky: Diffusion and Heat Exchange in Chemical Kinetics. Translated by N. Thon, Princeton, N.J. Princeton Univ. Press, (1955).
8. V. Lakshmikantham: Comparison Results for Reaction Diffusion Equations in a Banach Space. UTA Tech. Report n. 94 (1978).
9. V. Lakshmikantham: Vector Lyapunov Functions. Proc. of 12th Annual Allerton Conf. on Circuit and System Theory, 71-77, (1974).
10. V. Lakshmikantham, S. Leela: Differential and Integral Inequalities, Vol I. New York, Academic Press, (1969).
11. V. Lakshmikantham, S. Leela, M.N. Oguztöreli: Quasi Solutions, Vector Lyapunov Functions and Monotone Method, Proc. IEEE Trans. Auto control (to appear).
12. V. Lakshmikantham, A.S. Vatsala: Quasi Solutions and Monotone Method for Systems of Nonlinear Boundary Value Problems, Jour. Math. Anal. Appl. (to appear).
13. A.N. Michel, R.K. Miller: Qualitative Analysis of Large Scale Dynamical Systems, New York, Academic Press, (1977).
14. D.D. Siljak: Large Scale Dynamical Systems, New York, North Holland, (1978).

SEMILINEAR EQUATIONS OF GRADIENT TYPE IN HILBERT SPACES AND APPLICATIONS TO DIFFERENTIAL EQUATIONS

Jean Mawhin

Institut Mathématique
Université de Louvain
Belgium

1. INTRODUCTION

An existence and uniqueness theorem for equations of the form

$$Lu - N(u) = f$$

in a Hilbert space H, with L: dom L $\subset$ H $\to$ H linear self-adjoint and N: H $\to$ H a gradient mapping of class C^1 has been formulated in [6] and proved by a simple application of the Banach fixed point theorem. It was motivated by and gave a very simple proof of earlier results of Lazer and Sanchez [5] on the existence of 2π-periodic solutions of second order systems of ordinary differential equations. Moreover, it dit not require any compactness assumption on the resolvant of L and could therefore be applied to the more difficult problem of the existence of periodic solutions of semi-linear wave equations [7,8]. An extension of the result to the case of continuous gradient mapping N was then given by Brézis and Nirenberg [2] using the alternative method, a perturbation argument and maximal mono-

NONLINEAR DIFFERENTIAL EQUATIONS:
INVARIANCE, STABILITY, AND BIFURCATION

ISBN 0-12-508780-2

tone operators and more recently by Amann [1] as a consequence of his saddle point reduction method which combines the alternative method with minimax theorems of Rockafellar type and degree arguments.

The aim of this paper is to show how the basic existence results of Amann's paper [1] can be proved in a simple and straigthforward way by combining the simple approach of [6] with the alternative method and degree arguments. This considerably simplifies the treatment and furnishes, when the corresponding bifurcation equation is trivial, an iterative process converging to the solution. Refereing the reader to the above quoted papers for applications of the abstract theory to the existence of periodic solutions of Hamiltonian systems of ordinary differential equations or of semilinear wave equations with one space-variable, we end the paper by shortly describing an application to the existence and uniqueness of periodic solutions of a semi-linear wave equation with more than one space-variable.

2. A REDUCTION THEOREM FOR SEMILINEAR GRADIENT EQUATIONS IN HILBERT SPACES

Let H be a real Hilbert space, with inner product (,) and corresponding norm $|.|$. The following result is a slight generalization of a result due to Brézis and Nirenberg [2] and the proof, given here for the reader's convenience, is modeled on their original approach.

Lemma 1. Let $F: H \to H$ be the gradient of a C^1 function $f: H \to \mathbb{R}$ and be such that, for some $a \leq b$ and all u and v in H, one has

$$a|u - v|^2 \leq (F(u) - F(v), u - v) \leq b|u - v|^2. \quad (1)$$

Then one has, for all u and v in H,

$$|F(u) - F(v)| \leq \max(|a|,|b|)|u - v|. \quad (2)$$

Proof. First step. The result is true if F is Gâteaux differentiable.

Using (1) with $v = u - tw$ where the real $t \neq 0$ and $w \in H$, dividing by t^2 and letting t going to zero, we obtain

$$a|w|^2 \leq (F'(u)w,w) \leq b|w|^2,$$

where $F'(u)$ is the Gâteaux differential of F at u, and hence $F'(u)$ being self-adjoint, we have, for all $u \in H$,

$$|F'(u)| \leq \max(|a|,|b|).$$

By the mean value theorem, we then obtain

$$|F(u) - F(v)| \leq \sup_{z \in H}|F'(z)||u - v| \leq \max(|a|,|b|)|u - v|.$$

Second step. The result is true if H has finite dimension. We identify H in this case with $\mathbb{R}^n$. Let (ρ_k) be a sequence of nonnegative mollifiers on H which converge to the delta function, and let $f_k = \rho_k * f$ be the corresponding convolution product. Then, for all $k \in \mathbb{N}^*$ and $u \in H$,

$$f_k'(u) = (\rho_k * F)(u) = F_k(u),$$

so that F_k is of class C^∞ and pointwise converges to F if $k \to \infty$. Moreover, for all u and $v \in H$, we have

$$(F_k(u) - F_k(v), u - v) = \int_H \rho_k(w)(F(u - w), u - v)dw,$$

which implies, using (1), the nonnegativity of ρ_k and the fact that

$$\int_H \rho_k(w)dw = 1,$$

the relation

$$a|u - v|^2 \leq (F_k(u) - F_k(v), u - v) \leq b|u - v|^2,$$

for all $k \in \mathbb{N}^*$, u and $v \in H$. Then, using step one, we have for all $k \in \mathbb{N}^*$, u and v in H,

$$|F_k(u) - F_k(v)| \leq \max(|a|,|b|)|u - v|,$$

and then the result follows by letting $k \to \infty$.

Third step. The result is true if H is infinite dimensional. Let u and v be fixed in H, and let X be the finite-dimensional subspace of H spanned by u, v, F(u) and F(v), with P the corresponding orthogonal projector on X. Denote by f_X the restriction of f to X. Then, we find easily that, for $x \in X$, f_X has a Gâteaux differential given by

$$f'_X(x) = PF(x).$$

Therefore we have, for all x and y in X,

$$(f'_X(x) - f'_X(y), x - y) = (PF(x) - PF(y), x - y) =$$
$$(F(x) - F(y), x - y),$$

which implies, by assumption, that

$$a|x - y|^2 \leq (f'_X(x) - f'_X(y), x - y) \leq b|x - y|^2,$$

and then, by step two, we have, for all x and y in X,

$$|PF(x) - PF(y)| = |f'_X(x) - f'_X(y)| \leq \max(|a|,|b|)|x - y|,$$

and, in particular,

$$|F(u) - F(v)| = |PF(u) - PF(v)| \leq \max(|a|,|b|)|u - v|,$$

which completes the proof.

Now let L: dom $L \subset H \to H$ be a linear self-adjoint operator, with spectrum $\sigma(L)$, and let N: $H \to H$ be a continuous gradient operator. We shall make the following

Assumption (A): There exist real numbers α and β in $\mathbb{R} \setminus \sigma(L)$, with $\alpha \leq \beta$, such that, for all u and v in H, one has,

$$\alpha|u - v|^2 \leq (N(u) - N(v), u - v) \leq \beta|u - v|^2.$$

If $\{E_\lambda : \lambda \in \mathbb{R}\}$ denotes the spectral resolution of L, let us denote by P the orthogonal projector defined on H by

$$P = \int_\alpha^\beta dE_\lambda,$$

and let X = Im P, Y = Im Q where

$$Q = I - P = \int_{]-\infty,\alpha]} dE_\lambda + \int_{[\beta,\infty[} dE_\lambda,$$

so that H is the orthogonal direct sum of X and Y. We can now state and prove the following reduction theorem first given in [1].

Theorem 1. If assumption (A) holds for L and N above then, for each $x \in X$ and each $f \in H$, the equation

$$Ly - QN(x + y) = Qf \tag{3}$$

has a unique solution

$$y = R(x) \in \text{dom } L \cap Y,$$

and the corresponding mapping $R: X \to Y$ is Lipschitzian.

Proof. As $R \setminus \sigma(L)$ is open, we can find λ_- and λ_+ such that $[\lambda_-,\alpha] \cap \sigma(L)$ and $[\beta,\lambda_+] \cap \sigma(L)$ are empty, and hence

$$P = \int_{\lambda_-}^{\lambda_+} dE_\lambda, \qquad Q = \int_{-\infty}^{\lambda_-} dE_\lambda + \int_{\lambda_+}^{+\infty} dE_\lambda.$$

Now, for every $\tau \in]\lambda_-,\lambda_+[$, equation (3) is equivalent to the fixed point problem

$$y = (L - \tau I)^{-1} Q[N(x + y) - \tau y + f] = T_x(y),$$

where moreover

$$(L - \tau I)^{-1}Q = \int_{-\infty}^{\lambda_-} (\lambda - \tau)^{-1} dE_\lambda + \int_{\lambda_+}^{+\infty} (\lambda - \tau)^{-1} dE_\lambda,$$

so that

$$|(L - \tau I)^{-1}Q| \leq [\min(\tau - \lambda_-, \lambda_+ - \tau)]^{-1}. \tag{4}$$

If we now define, for each $x \in X$, $F_x: Y \to Y$ by

$$F_x(y) = Q[N(x + y) - \tau y + f],$$

and $f_x: Y \to \mathbb{R}$ by

$$f_x(y) = n(x + y) - (\tau/2)|y|^2 + (f,y)$$

where $n: H \to \mathbb{R}$ is such that $n'(u) = N(u)$ for all $u \in H$, then it is easy to check that $F_x(y) = f_x'(y)$ for all $x \in X$ and $y \in Y$ and that

$$(\alpha - \tau)|y - z|^2 \leq (F_x(y) - F_x(z), y - z) \leq (\beta - \tau)|y - z|^2 \tag{5}$$

for all $x \in X$, $y \in Y$ and $z \in Y$. By (5) and Lemma 1, we have

$$|F_x(y) - F_x(z)| \leq \max(|\alpha - \tau|, |\beta - \tau|)|y - z|$$

for all $x \in X$, $y \in Y$ and $z \in Y$, hence,

$$|T_x(y) - T_x(z)| \leq [\min(\tau - \lambda_-, \lambda_+ - \tau)]^{-1} \max(|\alpha - \tau|, |\beta - \tau|)|y - z|, \tag{6}$$

and it is elementary to check that the above Lipschitz constant for T_x is strictly smaller than one if and only if we choose τ such that

$$(1/2)(\lambda_- + \beta) < \tau < (1/2)(\lambda_+ + \alpha),$$

which is always possible. Now the Banach fixed point theorem implies that, for each $x \in X$, T_x has a unique fixed point $R(x) \in Y$. Now, if x and x' are in X, we have

$$|R(x) - R(x')| = |T_x(R(x)) - T_{x'}(R(x'))| \leq$$
$$\leq |T_x(R(x)) - T_x(R(x'))| + |T_x(R(x')) - T_{x'}(R(x'))| \leq$$
$$\leq \chi(\tau)|R(x) - R(x')| + |(L - \tau I)^{-1}Q[N(x + R(x')) +$$
$$- N(x' + R(x'))]|,$$

where $\chi(\tau) < 1$ is the Lipschitz constant in (6). Consequently,

$$|R(x) - R(x')| \leq (1 - \chi(\tau))^{-1}[\min(\tau - \lambda_-, \lambda_+ - \tau)]^{-1}$$
$$\max(|\alpha|,|\beta|)|x - x'|,$$

and the proof is complete.

Remark 1. It follows also from the Banach fixed point theorem that, under the conditions of Theorem 1, $R(x)$ is the limit of the sequence of successive approximations $(R_k(x))$ where $R_0(x) \in \text{dom } L \cap Y$ is arbitrary and, for each $k \in \mathbb{N}$, $R_{k+1}(x) \in \text{dom } L \cap Y$ is the unique solution of the equation

$$(L - \tau I)R_{k+1}(x) = Q[N(x + R_k(x)) - \tau R_k(x) - f].$$

Remark 2. With the notations introduced above, it is clear that equation

$$Lu - N(u) = f \tag{7}$$

is, by projection, equivalent to the system of equations

$$Ly - QN(x + y) = Qf, \tag{8}$$

$$Lx - PN(x + y) = Pf, \tag{9}$$

where $x = Pu$, $y = Qu$. Theorem 1 solves equation (8) in y for each fixed $x \in X$ and hence the existence of a solution for (7) is reduced to the existence of $x \in X$ which solves the equation in X

$$Lx - PN(x + R(x)) = Pf. \tag{10}$$

We therefore have an example of application of the generalized Lyapunov-Schmidt method or alternative method introduced in a general setting by Cesari (see e.g. [3,4]), in that, in contrast with the classical Lyapunov-Schmidt method, the projection is not made along the kernel of L and its orthogonal complement.

3. EXISTENCE THEOREMS FOR SEMILINEAR GRADIENT EQUATIONS IN HILBERT SPACES

Let H be a real Hilbert space with inner product (,) and corresponding norm $|.|$. Let $L: \text{dom } L \subset H \to H$ be a linear self-adjoint operator, with spectrum $\sigma(L)$, and $N: H \to H$ be a continuous gradient operator. We first have the following trivial consequence of Theorem 1 and Remark 2 which generalizes or completes results of [1], [2] and [6].

Theorem 2. Assume that there exist real numbers p and q with $p \leq q$ such that

$$[p,q] \cap \sigma(L) = \emptyset$$

and such that, for all u and v in H, one has

$$p|u - v|^2 \leq (N(u) - N(v), u - v) \leq q|u - v|^2.$$

Then, for each $f \in H$, equation

$$Lu - N(u) = f$$

has a unique solution $u \in \text{dom } L$ which is Lipschitzian in f and can be obtained as the limit of the sequence of successive approximations (u_k) where $u_0 \in \text{dom } L$ is arbitrary and, for $k \in \mathbb{N}$, $u_{k+1} \in \text{dom } L$ is the unique solution of the equation

$$(L - \tau I)(u_{k+1}) = (N - \tau I)(u_k) + f,$$

with

$$(1/2)(\lambda_- + q) < \tau < (1/2)(\lambda_+ + p),$$

$]\lambda_-, \lambda_+[$ being the largest interval in $\mathbb{R} \setminus \sigma(L)$ containing $[p,q]$.

Proof. Choosing $\alpha = p$, $\beta = q$ in Theorem 1, and using its notations, we obtain $P = 0$, $Q = I$, and the existence result follows; the convergence of successive approximations comes from Remark 1. Finally, if f and f' are in H and if u and u' are the respective solutions of

$$Lu - N(u) = f, \quad Lu' - N(u') = f',$$

then

$$|u - u'| = |(L - \tau I)^{-1}[(N - \tau I)(u) - (N - \tau I)(u') - (f - f')]| \leq \chi(\tau)|u - u'| + |(L - \tau I)^{-1}||f - f'|,$$

where $\chi(\tau) < 1$, and hence u is Lipschitzian in f.

We shall now consider a more general situation and introduce the following assumption, due to Amman [1].

Assumption (B): There exist real numbers α and β with $\alpha \leq \beta$ such that $[\alpha,\beta] \cap \sigma(L)$ is made of a positive finite number of eigenvalues of L having finite multiplicity and such that, for al u and v in H, one has

$$\alpha|u - v|^2 \leq (N(u) - N(v), u - v) \leq \beta|u - v|^2.$$

Of course Assumption (B) implies Assumption (A) and we can use the notations and the results of Section 2, the conditions of Assumption (B) implying moreover that the space $X = \mathrm{Im}\, P$ is finite-dimensional. Therefore, by Theorem 1, the mapping $S: X \to X$ defined by

$$S(x) = Lx - PN(x + R(x)) - Pf \tag{11}$$

is continuous. We then have the following

Theorem 3. If assumption (B) holds for L and N above and if there exists an open bounded set $\Omega \subset X$ such that $S(x) \neq 0$ for $x \in \partial\Omega$ and the Brouwer degree

$$d(S,\Omega,0)$$

is non-zero, then equation (7) has at least one solution.

Proof. By Theorem 1 and Remark 2, the existence of a solution for (7) is reduced to the existence of a solution $x \in X$ of equation (10) i.e. of a zero for S, which follows from our assumptions using degree theory [9].

As an application we can give a simplified proof of a result of Amann [1].

Corollary 1. If assumption (B) holds for L and N above and if there exist real numbers ν, γ and δ such that

$$\nu \in [\alpha,\beta] \setminus \sigma(L), \quad 0 \leq \gamma < \mathrm{dist}(\nu, \sigma(L)), \quad \delta \geq 0, \tag{12}$$

and such that, for all $u \in H$, one has

$$|N(u) - \nu u| \leq \gamma|u| + \delta,$$

then equation (7) has at least one solution for each $f \in H$.

Proof. We shall show that the conditions of Theorem 2 are satisfied if we take for Ω an open ball $B(r)$ of center 0 and radius r sufficiently large. It is clear that equation $S(x) = 0$ with S given by (11) is equivalent to the equation $T(x) = 0$ where $T: X \to X$ is given by

$$T(x) = x - (L - \nu I)^{-1}P[N(x + R(x)) - \nu x + f] =$$
$$= (L - \nu I)^{-1}PS(x),$$

and, by this relation and the properties of degree, their degrees with respect to a given bounded open set in X and zero are simultaneously defined and have the same absolute value. We then show that $d(T,B(r),0)$ is defined and different from zero for sufficiently large r. Using (12), (13) and Theorem 1, we obtain, for every $x \in X$,

$$(T(x),x) = (x - (L - \nu I)^{-1}P[N(x + R(x)) - \nu x + f],x)$$
$$+ (R(x) - (L - \nu I)^{-1}Q[N(x + R(x)) - \nu R(x) + f], R(x))$$
$$= (x + R(x) - (L - \nu I)^{-1}[N(x + R(x)) - \nu(x + R(x)) + f],$$
$$x + R(x)) \geq |x + R(x)|^2 - |(L - \nu I)^{-1}||(N - \nu I)$$
$$(x + R(x)) + f||x + R(x)| \geq |x + R(x)|^2 - [\mathrm{dist}(\nu,$$
$$\sigma(L))]^{-1}[\gamma|x + R(x)| + \delta + |f|]|x + R(x)| = (1 - \gamma$$
$$[\mathrm{dist}(\nu,\sigma(L))]^{-1})|x + R(x)|^2 - (\delta + |f|)[\mathrm{dist}$$
$$(\nu,\sigma(L))]^{-1}|x + R(x)| \geq (1 - \gamma[\mathrm{dist}(\nu,\sigma(L))]^{-1}|x|^2$$
$$- (\delta + |f|)[\mathrm{dist}(\nu,\sigma(L))]^{-1}[(1 + K)|x| + K'],$$

where K is the Lipschitz constant of R and $K' = R(0)$. Consequently, there exists $r > 0$ such that, for all $x \in X$ with $|x| \geq r$, one has

$$(T(x),x) > 0,$$

and this implies, by Poincaré-Bohl theorem [9], that

$$d(T,B(r),0) = 1,$$

which completes the proof.

Remark 3. Theorem 2 is false without the assumption that N is a gradient operator as shown by the following example where $H = \mathbb{R}^2$, $L(u_1,u_2) = (-u_1,u_2)$ and $N(u_1,u_2) = (u_2,-u_1)$; the conditions of Theorem 2 hold with $p = q = 0$ as immediately checked. On the other hand,

$$\det(L - N) = 0$$

so that $L - N$ is neither one-to-one nor onto.

4. APPLICATION TO A SEMILINEAR MULTIDIMENSIONAL WAVE EQUATION

Let $J = (]0,2\pi[)^3$, $f \in L^2(J) = H$, with the usual inner product $(,)$ and corresponding norm $|.|$, and $g: J \times \mathbb{R} \to \mathbb{R}$ a Caratheodory function such that $g(.,.,.,0) \in H$. By generalized solution for the periodic problem on J of the equation

$$u_{tt} - u_{xx} - u_{yy} = g(t,x,y,u) + f(t,x,y) \qquad (14)$$

(higher dimensions for the space variables could be considered as well) we mean a function $u \in H$ such that one has

$$(v_{tt} - v_{xx} - v_{yy}, u) = (g(.,.,.,u(.,.,.)),v) + (f,v)$$

for all $v \in C^2(J)$ which are 2π-periodic in each variable. If

$$v_{mnp}(t,x,y) = \exp i(mt + nx + py) \quad (m,n,p \in \mathbb{Z})$$

and if $\sum_{m,n,p\in\mathbb{Z}} u_{mnp}v_{mnp}$ denotes the corresponding Fourier series of $u \in H$, we define $L: \text{dom } L \subset H \to H$ as follows:

$$\text{dom } L = \{u \in H: \sum_{m,n,p\in\mathbb{Z}} (n^2 + p^2 - m^2)^2 |u_{mnp}|^2 < \infty\},$$

$$Lu = \sum_{m,n,p\in\mathbb{Z}} (n^2 + p^2 - m^2) u_{mnp} v_{mnp}.$$

One can show then that $u \in H$ is a generalized solution of the periodic problem on J for the equation

$$u_{tt} - u_{xx} - u_{yy} = f(t,x,y)$$

if and only if $u \in \text{dom } L$ and $Lu = f$. Moreover, L is self-adjoint and

$$\sigma(L) = \{n^2 + p^2 - m^2 : m,n,p \in \mathbb{Z}\},$$

so that elementary number theoretical considerations show that $\sigma(L) = \mathbb{Z}$ and that each eigenvalue has infinite multiplicity. Assume now that there exist real numbers $p \leq q$ such that

$[p,q] \cap \mathbb{Z} = \emptyset$ and such that

$$p(u - v)^2 \leq [g(t,x,y,u) - g(t,x,y,v)](u - v) \leq q(u - v)^2$$

for a.e. $(t,x,y) \in J$ and all u and v in R. Then, the mapping N defined on H by the relation

$$(N(u))(t,x,y) = g(t,x,y,u(t,x,y))$$

for a.e. $(t,x,y) \in J$ maps continuously H into itself, satisfies the inequalities

$$p|u - v|^2 \leq (N(u) - N(v), u - v) \leq q|u - v|^2$$

for all u and v in H, and is such that N is the gradient of $n: H \to \mathbb{R}$ given by

$$n(u) = \int_J \int_0^{u(t,x,y)} g(t,x,y,s)\,ds\,dt\,dx\,dy.$$

As the problem of the existence of a generalized solution of the periodic problem on J for equation (14) is clearly equivalent to finding a solution of equation $Lu - N(u) = f$ with L and N defined above, it immediately follows from Theorem 2 that, under the conditions we have listed, equation (14) has, for each $f \in H$, a unique generalized solution satisfying, in the considered generalized sense, the periodic conditions on J.

REFERENCES

1. H. Amann: Saddle Points and Multiple Solutions of Differential Equations, Math. Z. 109, 127-166 (1979).
2. H. Brezis, L. Nirenberg: Characterizations of the Ranges of Some Nonlinear Operators, and Applications to Boundary Value Problems, Ann. Scuola Norm. Sup. Pisa (4) 5, 225-326 (1978).

3. L. Cesari: Functional Analysis and Galerkin's Method, Michigan Math. J. 11, 385-414 (1964).
4. L. Cesari: Functional Analysis, Nonlinear Differential Equations and the Alternative Method, in Nonlinear Functional Analysis and Differential Equations, ed. by L. Cesari, R. Kannan and D. Schuur, New York, Dekker, 1-197 (1976).
5. A.C. Lazer, D.A. Sanchez: On Periodically Perturbed Conservative Systems, Michigan Math. J. 16, 193-200 (1969).
6. J. Mawhin: Contractive Mappings and Periodically Perturbed Conservative Systems, Arch. Math. (Brno) 12, 67-74 (1976).
7. J. Mawhin: Recent Trends in Nonlinear Boundary Value Problems, in VII. Internationale Konferenz Über nichtlineare Schwingugen, Berlin, 1975, Band I.2, Berlin, Akademie-Verlag, 52-70 (1977).
8. J. Mawhin: Solutions périodiques d'équations aux dérivées partielles hyperboliques non linéaires, in Mélanges Th. Vogel, ed. by Rybak, Janssens et Jessel, Bruxelles, Presses Univ. Bruxelles, 301-319, (1978).
9. J. Mawhin: Tolopogical Degree Methods in Nonlinear Boundary Value Problems, Regional Confer. in Math. n. 40, Providence, American Math. Soc., (1979).

SUR LA DECOMPOSITIONS ASYMPTOTIQUE DES SYSTEMES DIFFERENTIELS FONDEE SUR DES TRANSFORMATIONS DE LIE

Ju.A. Mitropolsky

Académie des Sciences
Kiev, U.R.S.S.

1.

Il est bien connu que les méthodes asymptotiques de mécanique non linéaire, comme les méthodes de la théorie des perturbations qui ont des applications diverses, supposent qu'un certain "système-étalon" soit donné et l'on compare ensuite la résolution formelle de ce dernier avec la résolution concrète du système exact de départ.

Habituellement on peut obtenir un système-étalon d'un système exact si l'on égale un paramètre ε à 0. Ce paramètre ε (en général, petit) caractérise la perturbation du système traité. Ainsi, en ce cas, le mouvement non perturbé est décrit par le système-étalon des équations différentielles et le mouvement perturbé ($\varepsilon \neq 0$) est décrit par le système exact de départ.

Dans bien des cas cette méthode requiert la construction d'un système-étalon. Par exemple, en appliquant la méthode de centrage à un système d'équations sous forme standard nous obtenons un système centré, c'est-à-dire un système-étalon.

ISBN 0-12-508780-2

Il est connu aussi que dans la méthode de centrage développée par N.N. Bogolioubov on suppose l'introduction d'un changement de variables particulier. Ayant l'équation différentielle

$$\frac{dy}{dt} = \varepsilon F(t,y), \tag{1.1}$$

où y est un vecteur de dimension n, ε est un petit paramètre positif, il faut effectuer le changement de variables

$$y = x + \varepsilon F_1(t,x) + \varepsilon^2 F_2(t,x) + \dots \tag{1.2}$$

qui amène le système (1.1) à la forme

$$\frac{dx}{dt} = \varepsilon P_1(x) + \varepsilon^2 P_2(x) + \dots \tag{1.3}$$

Ce faisant, le premier terme du deuxième membre de l'équation (1.3) est obtenu à l'aide de l'opération du centrage

$$P_1(x) = \lim_{T\to\infty} \frac{1}{T}\int_0^T F(t,x)dt \tag{1.4}$$

Les deuxièmes membres des équations de ce système ne contiennent évidemment pas la variable t. Et c'est précisément pour cela que le système simplifié, dit système-étalon, est facile à traiter.

L'idée de N.N. Bogolioubov servant de base à la méthode de centrage peut être généralisée de la manière suivante.
Soit

$$\Phi_0 y + \varepsilon\Phi_1 y = 0, \tag{1.5}$$

un systéme perturbé où Φ_0, Φ_1 sont des opérateurs différentiels ou aux dérivées partielles. A l'aide d'une substitution particulière de variables

$$y = x + \varepsilon F_1(x) + \varepsilon^2 F_2(x) + \dots \quad (1.6)$$

il est nécessaire de projeter le systéme perturbé (1.5) sur le système

$$\Phi_o x + \varepsilon \Phi_{10} x + \varepsilon^2 \Phi_{20} x + \dots = 0, \quad (1.7)$$

qui possède certaines qualités spéciales.

Le système (1.7) peut, par exemple, posséder une variété intégrale dont la dimension est moins grande que celle de l'espace de phase, tout entier, du système (1.7). C'est dans ce cas que l'on peut réduire le processus de grande dimension à une suite de quelques processus de moindre dimension. Ce problème est d'une grande importance et c'est précisément pour cela que dans la pratique il y a des systèmes de grandes dimensions.

Ces dernier temps plusieurs travaux intéressants sont accomplis dans la direction de cette généralisation. C'est tout d'abord la méthode de centrage lieé aux utilisations des séries et des transformations de Lie [13]. En ce cas, on réduit le système

$$\frac{dy}{dt} = F(y, \varepsilon) \quad (1.8)$$

où

$$F(y, \varepsilon) = \sum_m \frac{\varepsilon^m}{m!} F_m(y), \qquad F_m(y) = \left. \frac{\partial^m F}{\partial \varepsilon^m} \right|_{\varepsilon = 0},$$

y, F sont des n-vecteurs, à l'aide de la transformations presque identique

$$y = x + \varepsilon Q_1(x) + \varepsilon^2 Q_2(x) + \ldots \tag{1.9}$$

au système

$$\frac{dx}{dt} = \sum_m \frac{\varepsilon^m}{m!} G_m(x) \tag{1.10}$$

où $G_m(x)$ sont des fonctions vectorielles ne contenant que des termes variant lentement.

A titre de deuxième exemple du centrage servant à obtenir des équations centrées sous forme particulière, étudions un problème sur le centrage pour un système d'équations sous la forme canonique [12].

Supposons les équations exactes de départ sous forme standard et écrites sous la forme d'Hamilton

$$\frac{dq_k}{dt} = \varepsilon \frac{\partial H}{\partial P_k}, \quad \frac{dP_k}{dt} = -\varepsilon \frac{\partial H}{\partial q_k} \quad (k = 1,2,\ldots,2n) \tag{1.11}$$

le hamiltonien $H = H(p_i, q_i, t)$ étant une fonction périodique de t de période 2π.

Si nous centrons les deuxièmes membres du système (1.11) d'après des règles habituelles nous obtenons des équations centrées qui, dans le cas général, ne sont pas sous la forme de équations canoniques d'Hamiltons. Par conséquent, nous ne pouvons pas trouver tout de suite l'intégrale première.

Cependant, on peut centrer les systèmes du type (1.11) d'une manière immédiate, partant d'un hamiltonien

$$\varepsilon H(p_i, q_i, t) \tag{1.12}$$

qui leur correspond, au lieu de partir des équations (1.11) sous la forme standard.

Par un procédé élaboré, on trouve une fonction génératrice $S(P_i, q_i t)$ qui tansforme les variables canoniques p_i, q_i en les nouvelles variables P_i, Q_i selon les formules bien connues

$$p_i = \frac{\partial S}{\partial q_i}, \quad Q_i = \frac{\partial S}{\partial P_i} \tag{1.13}$$

$$\frac{\partial S}{\partial t} + \varepsilon H(p_i, q_i, t) = \varepsilon H'(P_i, Q_i) \tag{1.14}$$

de façon que le hamiltonien transformé soit déjà indépendant du temps d'un manière évidente.

On trouve la fonction génératrice S ci dessus et la fonction d'Hamilton transformée $H'(P_i,Q_i)$, sous la forme de séries de puissances de ε

$$S = S_o + \varepsilon S_1(q_i,P_i,t) + \varepsilon^2 S_2(q_i,P_i,t) + \ldots \qquad (1.15)$$

$$H'(P_i,Q_i) = H_1(P_i,Q_i) + \varepsilon^2 H_2(P_i,Q_i) + \ldots \qquad (1.16)$$

où $S_o = \sum q_i P_i$ est une transformation identique.

Maintenant, on peut fixer à un dégré arbitraire près le système d'équations centrées

$$\frac{dQ_i}{dt} = \varepsilon\frac{\partial H'}{\partial P_i}, \qquad \frac{dP_i}{dt} = -\varepsilon\frac{\partial H'}{\partial Q_i} \qquad (1.17)$$

qui correspond au système des équations exactes (c'est-à-dire, au système (1.11)) dans lequel on avait remplacé p_i, q_i par P_i, Q_i d'après le formules (1.13).

Pour centrer le système (1.17) nous pouvons obtenir tout de suite l'intégrale première, et le faire aussi pour le système (1.11).

Développant des méthodes dues à N.M. Krylov et N.N. Bogolioubov, A. Povzner a proposé en [8] une méthode d'étude des systèmes à petit paramètre de la forme

$$\varepsilon^p \frac{dy_k}{dt} = f_k(y) + \varepsilon\tilde{f}_k(y), \qquad (k = \overline{1,n}) \qquad (1.18)$$

qui est analogue à la théorie de perturbations de Schrödinger pour des opérateurs linéaires. En [8], au lieu de l'équation (1.18), on ne traite qu'une équation différentielle aux dérivées partielles d'ordre un

$$\varepsilon^p \frac{\partial f}{\partial t} + \sum_{k=1}^{n} [f_k(y) + \varepsilon\tilde{f}_k(y)]\frac{\partial f}{\partial y_k} = 0, \qquad (1.19)$$

laquelle, comme on sait, est équivalente au système des équations (1.18) en ce qui concerne la découverte des solutions générales.

Donc, le problème consiste à réduire l'équation (1.19) en une équation dont la structure est plus simple. Nous l'accomplissons à l'aide du changement asymptotique

$$y = x + \varepsilon Q_1(x) + \varepsilon^2 Q_2(x) + \ldots \qquad (1.20)$$

Comme la transformation (1.20) ne dépend pas de t, on peut considérer au lieu d'un opérateur générant l'équation (1.19) l'opérateur différentiel ci-dessous

$$Y = Y_o + \varepsilon \tilde{Y}_1, \qquad (1.21)$$

où

$$Y_o = \sum_k f_k(y)\frac{\partial}{\partial y_k}, \qquad \tilde{Y}_1 = \sum_k \tilde{f}_k(y)\frac{\partial}{\partial y_k} \qquad (1.22)$$

Comme transformation (1.20) il faut prendre le groupe de transformations à un paramètre

$$y = [\exp \varepsilon S]x, \qquad S = S_1 + \varepsilon S_2 + \ldots \qquad (1.23)$$

Ici les opérateurs S_1, S_2, ... sont des opérateurs différentiels d'ordre un aux coefficients non déterminés.

Ayant effectué la transformation (1.23), nous représentons l'opérateur (1.21) sous la forme

$$Y = Y_o + \varepsilon H_1 + \varepsilon^2 H_2 + \ldots , \qquad (1.24)$$

où les opérateurs différentiels sont définis par la condition de leur commutativité avec l'opérateur Y_o. Le problème de la détermination des opérateurs H_1, H_2, H_3, ... et des coefficients du changement de variables (1.23) S_1, S_2, ... , est soluble si l'on applique la théorie spectrale des opérateurs différentiels d'ordre un.

Il est clair que l'essence et l'algorithme de la méthode de centrage due à N.N. Bogolioubov (il s'agit précisément de la projection de système perturbé (1.5) sur le système non perturbé (1.17)) sortent d'un cadre de recherche des processus oscilatoires propres. Avec cela on peut s'intéresser non seulement aux propriétés analytiques d'une variété intégrale du système non perturbé (1.7) et aux propriétés liées au nombre de paramètres décrivant cette variété mais à la structure de la variété intégrale (c'est-à-dire, aux propriétés indépendantes du choix du système des coordonnées).

Maintenant nous nous fixons sur un développement de l'algorithme du centrage, dans le sens ci-dessous. Posons le problème de la projection d'un algèbre β' des opérateurs différentiels d'ordre un sur l'algèbre β découlant de cette dernière à conditions que $\varepsilon = 0$.

Nous allons aussi montrer que cette conception équivaut au problème de la perturbation pour un système des équations aux dérivées partielles d'un type assez général, pour lequel la multiplicité en approximation nulle possède certaines qualités.

C'est un cas particulier et très important du problème ci-dessus qui est un problème de décomposition dont l'essence consiste à substituer une résolution d'un système d'équations différentielles, d'ordre élevé, à la résolution successive de systèmes de moindre dimension. Si les systèmes obtenus sont solubles, la décomposition, dans ce cas, présente un grand intérêt.

Il y a beaucoup de systèmes concrets, dont les composantes, formant un système compliqué, ne travaillent pas toujours indépendamment, mais exercent essentiellement une influence réciproque. L'indépendance réciproque absolue des composantes est une exigence indispensable pour la réductibilité du système de départ. Et s'il y a une influence réciproque, même petite, il est possible que des profonds changements apparaissent avec le temps dans le comportement de la solution de tout le système. A cause de cela le petit paramètre y surgit et nous avons besoins d'analyser l'influence des perturbations petites, de petites influences reciproques etc.

Ces problèmes apparaissent en thèorie du réglage automatique, en économie, en problèmes sur la planification et le contrôle de procédés industriels.

2.

Nous nous fixons maintenant sur le problème principal. Soit un système de Pfaff

$$\begin{aligned} \omega_1(d) &= \omega_{11}dy_1 + \ldots + \omega_{1k}dy_k + \ldots + \omega_{1n}dy_n = 0, \\ \omega_2(d) &= \omega_{21}dy_1 + \ldots + \omega_{2k}dy_k + \ldots + \omega_{2n}dy_n = 0, \\ &\ldots\ldots\ldots\ldots\ldots\ldots\ldots\ldots \\ \omega_k(d) &= \omega_{k1}dy_1 + \ldots + \omega_{kk}dy_k + \ldots + \omega_{kn}dy_n = 0, \end{aligned} \tag{2.1}$$

défini dans l'espace $R^{(n)}$, des variables $y_1,\ldots,y_n$. Considérons un domaine Ψ de l'espace $R^{(n)}$ et faisons la supposition que les coefficients ω_{ij} $(i = 1,2,\ldots k;\ j = 1,2,\ldots,n)$ du système (2.1) soient des fonctions holomorphes de y au voisinage de chaque point du domain Ψ.

Faisons aussi la supposition qu'il y ait dans le domaine Ψ, pour le système de Pfaff (2.1), une variété M_m de dimension m dans laquelle tous les formes du système (2.1) s'annulent identiquement. La dimension de cette variété peut varier de 1 à n-1. On peut caractériser la variété non seulement par la dimension, mais aussi par bien d'autres propriétés qui seront l'objet de nos recherches, par exemple, par l'invariance relativement à certaines trasformations de groupe ou par la décomposition (par des variables), par la propriété de compacité ou par d'autres propriétés.

Etudions le système (2.1) en le comparant avec un système perturbé de Pfaff

$$\begin{aligned} \tilde{\omega}_1(d) &= (\omega_{11} + \varepsilon\tilde{\omega}_{11})dy_1 + \ldots + (\omega_{1n} + \varepsilon\tilde{\omega}_{1n})dy_n = 0, \\ &\ldots\ldots\ldots\ldots\ldots\ldots\ldots\ldots \\ \tilde{\omega}_k(d) &= (\omega_{k1} + \varepsilon\tilde{\omega}_{k1})dy_1 + \ldots + (\omega_{kn} + \varepsilon\tilde{\omega}_{kn})dy_n = 0. \end{aligned} \tag{2.2}$$

La question suivante se pose tout de suite: est-ce que les propriétés de la variété intégrale du système de départ

(2.1) M_m et les propriétés de celle du système perturbé (2.2) M_m^ε sont semblables pour les petites valeurs de ε? Ce problème est typique pour la théorie de la perturbation.

Chacun des systèmes (2.1), (2.2) équivaut aux équations correspondantes aux dérivées partielles. Trouver une variété intégrale est un problème identique à celui de valeurs initiales choisies particulièrement. D'ici découle importance de la considération des système de Pfaff (2.1), (2.2).

La deuxième particularité essentielle des système de Pfaff est qu'ils permettent d'étudier de manière unifiée les systèmes mixtes, composés d'équations ordinaires et d'équations aux dérivées partielles, ce qui est bien utile dans la solution de problèmes pratiques.

Les variables $y_1, \ldots, y_n$ qui entrent dans les systèmes des Pfaff ci-dessus sont de même nature. Il est plus commode d'introduire des variables dépendantes et indépendantes qui sont naturelle au point de vue des applications.

Soit k le rang du système (2.1) (ou (2.2)) dans le domaine Ψ, c'est-à-dire, par exemple,

$$\det \begin{pmatrix} \omega_{11}, & \ldots\ldots, & \omega_{1k} \\ \\ \omega_{k1}, & \ldots\ldots, & \omega_{kk} \end{pmatrix} \neq 0$$

en chaque point du domaine Ψ. Alors on prend $y_{k+1}, y_{k+2}, \ldots$ $\ldots, y_n$ pour variables indépendantes et $y_1, y_2, \ldots, y_k$ pour variables dépendantes. Ayant résolu les systèmes (2.1) et (2.2) par rapport aux différentielles des variables dépendantes nous obtenons le système particulier de Pfaff

$$\begin{aligned} dy_1 &= q_{11}dy_{k+1} + \ldots + q_{1n-k}dy_n, \\ &\ldots\ldots\ldots\ldots\ldots\ldots\ldots\ldots, \\ dy_k &= q_{k1}dy_{k+1} + \ldots + q_{kn-k}dy_n \end{aligned} \tag{2.3}$$

et le système perturbé correspondant

$$dy_1 = (q_{11} + \varepsilon\tilde{q}_{11})dy_{k+1} + \ldots + (q_{1n-k} + \varepsilon\tilde{q}_{1n-k})dy_n,$$
$$\ldots\ldots\ldots\ldots\ldots\ldots\ldots\ldots\ldots, \quad (2.4)$$
$$dy_k = (q_{k+1} + \varepsilon\tilde{q}_{k1})dy_{k+1} + \ldots + (q_{kn-k} + \varepsilon\tilde{q}_{kn-k})dy_n$$

Nous allons rechercher une fonction définie sur la variété du système (2.3) M_{n-k}, c'est-à-dire, par exemple une fonction f des variables $y_1, \ldots, y_n$ satisfaisant le système (2.3) et calculer sa différentielle

$$df = \frac{\partial f}{\partial y_1}dy_1 + \frac{\partial f}{\partial y_k}dy_k + \frac{\partial f}{\partial y_{k+1}}dy_{k+1} + \ldots + \frac{\partial f}{\partial y_n}dy_n. \quad (2.5)$$

En faisant la substitution dans l'égalité (2.5) des formes explicites des différentielles, déterminées par (2.3) nous pouvons écrire

$$df = (\frac{\partial f}{\partial y_1} + q_{11}\frac{\partial f}{\partial y_1} + \ldots + q_{k1}\frac{\partial f}{\partial y_k})dy_{k+1} + \ldots$$
$$\ldots + (\frac{\partial f}{\partial y_n} + q_{1n-k}\frac{\partial f}{\partial y_1} + \ldots + q_{kn-k}\frac{\partial f}{\partial y_k})dy_n. \quad (2.6)$$

Désignons les opérateurs différentiels d'ordre un dans les deuxièmes membres (2.6) par $\hat{Y}_1, \hat{Y}_2, \ldots, \hat{Y}_{n-k}$ et faisons l'hypothèse que l'ensemble de ces opérateurs soit linéairement indépendant. Il est alors clair que l'ensemble des opérateurs différentiels

$$\frac{\partial}{\partial y_1}, \ldots, \frac{\partial}{\partial y_k}, \quad \hat{Y}_1, \ldots, \hat{Y}_{n-k} \quad (2.7)$$

est complet. Autrement dit, chacune des parenthèses de Poisson de ces opérateurs est exprimée par des opérateurs différentiels de cet ensemble.

Posons

$$Y_{oi} = \frac{\partial}{\partial y_i} + q_{1i}\frac{\partial}{\partial y_1} + \ldots + q_{ki}\frac{\partial}{\partial y_k} = \frac{\partial}{\partial y_i} + \hat{Y}_i \quad (i = \overline{1,n-k}) \quad (2.8)$$

ce qui permet d'écrire l'expression (2.6) sous la forme

$$df = Y_{o1}fdy_{k+1} + \ldots + Y_{on-k}fdy_n. \quad (2.9)$$

Nous allons supposer que f dans (2.9) prenne dans l'ordre les valeurs $y_1, \ldots\ldots, y_k$. Il est clair qu'en ce cas les équations de Pfaff (2.3), à l'aide des opérateurs $Y_{o1}, \ldots, Y_{om}$, s'exprime sous la forme

$$dy_i = Y_{o1}y_i dy_{k+1} + \ldots + Y_{on-k}y_i dy_n \quad (i = \overline{1,k}) \qquad (2.10)$$

Pareillement, en introduisant des opérateurs

$$Y_i = \frac{\partial}{\partial y_i} + (q_{1i} + \varepsilon\tilde{q}_{1i})\frac{\partial}{\partial y_1} + \ldots + (q_{ki} + \varepsilon\tilde{q}_{ki})\frac{\partial}{\partial y_k} \quad (i = \overline{1,k}) \qquad (2.11)$$

nous allons obtenir des équations de Pfaff

$$dy_i = Y_1 y_i dy_{k+1} + \ldots + Y_{n-k}y_i dy_n \quad (i = \overline{1,k}) \qquad (2.12)$$

Il se trouve que c'est par la structure des opérateurs différentiels (2.8) que des propriétés de la variété intégrale du système de Pfaff (2.10), qui nous intéressent sont déterminées. Pour être précis, par l'algèbre β des opérateurs différentiels générés par les opérateurs (2.7). L'algèbre β peut être compacte ou décomposable en somme directe des sous-algèbres (c'est-à-dire, réductible), ou admet d'autres propriétés liées à la structure de la variété intégrale M_m.

Ainsi nous allons supposer que les opérateurs Y_{oi} en approximation nulle du système perturbé (2.12) (autrement dit des opérateurs du système non perturbé (2.10)) soient doués d'une propriété particulière. Cette propriété peut être formulée comme l'appartenance de l'opérateur Y_{oi} à l'algèbre β. Ci-dessous si l'on écrit qu'un opérateur différentiel X appartient à l'alqèqre β ($X \in \beta$) il s'agit justement du fait qu'il possède une propriété intéressante, qui est caracteristique pour tous les éléments de β.

Accomplissons maintenant le changement de variables dans l'opérateur différentiel perturbé qui est inversible dans le domaine Ψ et représentable sous la forme de certaines suites asymptotiques formelles par rapport au paramètre ε :

$$y = x + \varepsilon\psi_1(x) + \varepsilon^2\psi_2(x) + \ldots , \tag{2.13}$$

où $x = (x_1, x_2, \ldots, x_n)$ est un vecteur de nouvelles variables. Ayant fait cette substitution nous avons des opérateurs différentiels (2.11) sous la forme

$$Y_i = \sum_{s=1}^{n} \widetilde{(Y_i(x_s))} \frac{\partial}{\partial x_s} . \tag{2.14}$$

Le signe "∿" au-dessus de celui de fonction $Y_i(x_s)$ marque le résultat de la substitution de variables y selon les expressions (2.13) dans la même fonction.
Ainsi le problème principal consiste dans ce qui suit: profitant de l'arbitraire de choix du changement (2.13) il faut assortir des fonctions $\psi_i(x)$ $(i = 1,2,\ldots)$ de telle manière que des opérateurs perturbés transformés Y_i (2.14) appartiennent l'algèbre β. Autrement dit il faut projeter les opérateurs Y_i (2.11) sur l'algèbre β.

A l'aide des opérateurs (2.13) on détermine un système de Pfaff conformément aux nouvelles variables x:

$$dx_i = (Y_i x_i)dx_{k+1} + \ldots + (Y_{n-k} x_i)dx_n \quad (i = \overline{1,k}). \tag{2.15}$$

Les opérateurs Y_i $(i = \overline{1,n-k})$ appartenant à l'algèbre β, la variété intégrale de ce système

$$x_i = \varphi_i(x_{k+1}, \ldots, x_n) \qquad (i = \overline{1,k}) \tag{2.16}$$

est douée des qualités typiques pour la variété intégrale du système non perturbé (2.10) (c'est-à-dire, pour le système en nulle approximation).
Si l'on introduit les valeurs x_i $(i = \overline{1,n})$ découlant des formules (2.13) dans la variété intégrale (2.16) on obtient

$$x_i(\varepsilon,y) = \varphi_i(\varepsilon,x_{k+1}(\varepsilon,y), \ldots, x_n(\varepsilon,y)) \quad (i = \overline{1,k}). \tag{2.17}$$

Ayant résolu (2.17) rélativement à $y_1, y_2, \ldots, y_k$ nous avons la représentation asymptotique pour la variété du système perturbé (2.12)

$$y_i = \psi_i(\varepsilon, y_{k+1}, \ldots, y_n) \quad (i = \overline{1,k}). \tag{2.18}$$

3.

Exposons en détail l'algorithme de la décomposition asymptotique. Etudions l'algèbre β des opérateurs différentiels qui caractérisent un système non perturbé ($\varepsilon = 0$), engendrée par des éléments de base

$$Y_{oi} = a_{i1}(y)\frac{\partial}{\partial y_1} + \ldots + a_{in}(y)\frac{\partial}{\partial y_n} \quad (i = \overline{1,n}) \tag{3.1}$$

où $a_{ij}(y)$ $(i,j = \overline{1,n})$ sont des fonctions holomorphes des variables $y_1, \ldots, y_n$ dans le domaine Ψ.
En ce cas il est plus convenable de ne pas distinguer les variables indépendantes et les variables dépendantes les unes des autres. L'ensemble d'opérateurs Y_{oi} est considéré comme linéairement indépendant dans le domaine Ψ.

On va écrire pour les opérateurs perturbés

$$Y_i = Y_{oi} + \varepsilon\tilde{Y}_i \quad (i = \overline{1,n}), \tag{3.2}$$

où

$$\tilde{Y}_i = \tilde{a}_{i1}(y)\frac{\partial}{\partial y_1} + \ldots + \tilde{a}_{in}(y)\frac{\partial}{\partial y_n},$$

$\tilde{a}_{ij}(y)$ $(i,j = \overline{1,n})$ sont des fonctions holomorphes de y dans le domaine Ψ, ε est un paramètre petit positif.

Indiquons par β' l'algèbre engendrée par les opérateurs perturbés (3.2).

On pose d'abord le problème de trouver un changement inversible sur le domaine

$$y = x + \varepsilon F_1(x) + \varepsilon^2 F_2(x) + \ldots, \tag{3.3}$$

où x, y sont des vecteurs à n dimensions, F_1, F_2, ... sont des fonctions vectorielles. Ce changement réduit les opérateurs différentiels (3.2) à la forme

$$Y_i = (Y_{oi} + \varepsilon Y_{1i} + \varepsilon^2 Y_{2i} + \dots) \quad (i = \overline{1,n}), \tag{3.4}$$

où Y_{1i}, Y_{2i}, ... appartiennet déjà à l'algèbre β et, conséquemment, le deuxième membre de (3.4) appartient à l'algèbre β.

Introduisons les opérateurs différentiels de premier ordre

$$S_i = \sum_j s_{ij}(x)\frac{\partial}{\partial x_j} \qquad (i,j = \overline{1,n})$$

Nous supposons que les coéfficients de ces opérateurs soient des fonctions pour le moment non spécifiées, holomorphes par rapport aux variables x dans le domaine Ψ.

Nous allons chercher le changement de variables (3.3) sous la forme d'une exponentielle dépendant d'une suite asymptotique en le paramètre ε:

$$S = S_1 + \varepsilon S_2 + \varepsilon^2 S_3 + \dots \tag{3.5}$$

c'est-à-dire sous la forme

$$y = \exp(-\varepsilon S)x = (1 - \frac{1}{1!}\varepsilon S + \frac{1}{2!}\varepsilon^2 S^2 + \dots)x. \tag{3.6}$$

Il faut assortir les coefficients indéfinis des opérateurs S_i (i = 1,2,...) pour résoudre ce problème-là.

Le hasard n'intervient pas dans le choix de la forme du changement (3.6). Celui-ci est déterminé par les propriétés importantes pour nos recherches, qui sont propres à cette forme de changement de variables. Il est bien connu que l'opérateur différentiel du premier ordre (en ce cas Y_i) est influencé par le changement (3.6) de la manière suivante

$$Y_i = Y_{oi} - \varepsilon S]Y_i + \varepsilon^2 \frac{1}{2!} S]^2 Y_i + \dots \quad (i = \overline{1,n}) \tag{3.7}$$

Nous abrégons la notation en écrivant les parenthèses de Poisson

$$Y]X = YX - XY = [Y,X], \quad Y]^2X = [Y,[Y,X]],\ldots,$$

La particularité essentielle de la transformation (3.6) consiste en ce que l'opérateur transformable en approximation nulle ($\varepsilon = 0$) est invariant par le changement. C'est-à-dire, on remplace seulement des symboles de variables y par ceux de x sans changer la forme des fonctions. Cela est marqué par l'index x tout près du terme d'ordre zéro Y_{ix} du développement en le paramètre ε de l'opérateur Y_i.

Maintenant, dans la relation (3.7), nous allons substituer la valeur de Y_i selon (3.2) et celle de S selon la représentation asymptotique (3.5). Après des calculs usuels, nous obtenons ci-dessous les valeurs des coefficients en le paramètre ε à la puissance correspondante:

$$\varepsilon^o\colon\ Y_{oi} = a_{i1}(x)\frac{\partial}{\partial x_1} + \ldots + a_{in}(x)\frac{\partial}{\partial x_n} \quad (i = \overline{1,n}), \tag{3.8}$$

$$\varepsilon^1\colon\ Y_{oi}]S_1 + \tilde{Y}_i, \tag{3.9}$$

$$\varepsilon^2\colon\ Y_{oi}]S_2 - S_1]\tilde{Y}_i + \tfrac{1}{2}S_1]^2Y_{oi} \tag{3.10}$$

. . . .

$$\varepsilon^p\colon\ Y_{oi}]S_p + F_p(\tilde{Y}_i,Y_{oi},S_1,\ldots,S_{p-1}), \tag{3.11}$$

où la fonction F est connue après que l'on ait trouvé S_1, S_2, ..., S_{p-1}.

Désignons par P le projecteur sur l'algèbre β, autrement dit, $PY \in \beta$ si $Y \notin \beta$.

Exigons aussi que l'opérateur perturbé transformé (3.2) engendre l'algèbre β.

En ce cas il faut déterminer les opérateurs S_1, S_2, ... d'après le système d'équations différentielles

$$Y_{oi}]S_1 = -\tilde{Y}_i + P\tilde{Y}_i, \tag{3.12}$$

$$Y_{oi}]S_2 = -S_1]\tilde{Y}_i + \tfrac{1}{2}S_1]^2Y_{\ i} + PS_1]\tilde{Y}_i, \tag{3.13}$$

$$Y_{oi}]S_p = -F_p(\tilde{Y}_i,Y_{oi},S_1,\ldots,S_{p-1}) - PF_p(\tilde{Y}_i,Y_{oi},S_1,\ldots,S_{p-1}) \tag{3.14}$$

Ayant obtenu les opérateurs S_1, S_2, ... selon le système (3.12) - (3.14) nous avons le développement asymptotique de l'opérateur différentiel Y_i (3.7) sous la forme

$$Y_i = Y_{oi} + \varepsilon P\tilde{Y}_i + \varepsilon^2 PS_1]\tilde{Y}_i + \ldots +$$
$$\varepsilon^p PF_p(\tilde{Y}_i, Y_{oi}, S_1, \ldots, S_{p-1}) + \ldots \qquad (3.15)$$

où $Y_i \in \beta$ et, conséquemment, nous venons de résoudre le problème de la projection de l'algèbre d'opérateurs β' sur β.

Ensuite, en substituant les valeurs des opérateurs $S_1, S_2, \ldots, S_p, \ldots$ trouvées en vertu du système (3.12) - (3.14), dans l'expression (3.6) nous obtenons la transformation réduisante en forme explicite.

4.

A titre de premier exemple on peut appliquer la théorie exposée à l'étude du système décrit par l'équation de Van der Pol avec un terme supplémentaire corrèspondant à la perturbation périodique:

$$\frac{d^2u}{dt^2} - \varepsilon(1 - u^2)\frac{du}{dt} + u = B\cos\nu t + B_o \qquad (4.1)$$

où ε est une valeur petite positive, $B\cos\nu t + B_o$ est une force extérieure contenant une composante constante.

Pour les systèmes d'auto-oscillation décrits par de telles équations on peut selon le rapport de fréquence propre et de celle d'excitation exterieure s'attendre à ce que surgisse, par exemple, une oscillation presque périodique ou des battements ou le phénomène de capture de la fréquence (les captures ultra-harmoniques ou sous-harmoniques) et si ces fréquences-là ne se distinguent pas sensiblement l'une de l'autre on peut observer le phénomène de synchronisation (la capture harmonique)[1].

---

[1]On peut trouver l'analyse en détail de toutes ces oscillations en [11].

En introduisant dans l'équation (4.1) au lieu de u une nouvelle variable $v = u - B_0$ nous la réduisons à la forme

$$\frac{d^2v}{dt^2} - \mu(1 - \beta v - \gamma v^2)\frac{dv}{dt} + v = B\cos\nu t, \tag{4.2}$$

$$\mu = (1 - B_0^2)\varepsilon, \quad \beta = \frac{2B_0}{1 - B_0^2}, \quad \gamma = \frac{1}{1 - B_0^2} .$$

Comme le système décrit par l'équation (4.2) est auto-oscillatoire, il faut supposer que $\mu > 0$ et, conséquemment, $B_0^2 < 1$.

Dans le cas de la capture harmonique (la fréquence forçante ν est proche de 1) on cherche habituellement à trouver la solution de l'équation (4.2) sous la forme

$$v(t) = b_1(t)\sin\nu t + b_2(t)\cos\nu t, \tag{4.3}$$

et quand on s'attend l'apparition de la capture ultra-harmonique ou de la capture sous-harmonique (la fréquence de l'oscillation capturée est plus grande ou moins grande que la fréquence forçante ν pour un facteur entier), on cherche à la trouver sous la forme

$$v(t) = b_1(t)\sin\nu t + b_2(t)\cos n\nu t + \frac{B}{1 - \nu^2}\cos\nu t \tag{4.4}$$

où $n = 1,2,3,\ldots$ pour les oscillation ultra-harmoniques et $n = \frac{1}{2},\frac{1}{3},\ldots$ pour les oscillations sous-harmoniques.

Etudions d'abord le cas $B = B_0 = 0$, c'est-à-dire l'équation homogène de Van der Pol

$$\frac{d^2u}{dt^2} - \varepsilon(1 - u^2)\frac{du}{dt} + u = 0. \tag{4.5}$$

En introduisant b_1 et b_2 au lieu de u et $\frac{du}{dt}$ d'après les formules

$$\begin{aligned} u &= b_1\cos t + b_2\sin t, \\ \frac{du}{dt} &= -b_1\sin t + b_2\cos t, \end{aligned} \tag{4.6}$$

et en centrant les équations obtenues exprimées en les nouvelles variables b_1, b_2 sous la forme standard, nous avons le système d'équations à l'ordre ε^2 près:

$$\begin{aligned} \frac{db_1}{dt} &= \frac{\varepsilon}{2}(1 - \frac{r^2}{4})b_1 + \frac{\varepsilon^2}{8}(1 - \frac{3}{2}r^2 + \frac{11}{32}r^4)b_2, \\ \frac{db_2}{dt} &= \frac{\varepsilon}{2}(1 - \frac{r^2}{4})b_2 - \frac{\varepsilon^2}{8}(1 - \frac{3}{2}r^2 + \frac{11}{32}r^4)b_1, \end{aligned} \tag{4.7}$$

où

$$r^2 = b_1^2 + b_2^2.$$

Le système d'équations (4.7) se met d'après le changement naturel $\tau = \varepsilon t$, $r^2 = x_1$, $b_2^2/b_1^2 = x_2$, sous la forme

$$\begin{aligned} \frac{dx_1}{d\tau} &= (1 - \frac{x_1}{4})x_1, \\ \frac{dx_2}{d\tau} &= -\frac{\varepsilon}{8}(1 - \frac{3}{2}x_1 + \frac{11}{32}x_1^2)(1 + x_2^2). \end{aligned} \tag{4.8}$$

Quand $\varepsilon = 0$ les variables du système (4.8) se séparent.

Conformément à la théorie exposée, effectuons cette séparation a l'ordre ε près.

Au lieu du système (4.7) on peut écrire l'équation différentielle aux dérivées partielles équivalente

$$\frac{\partial f}{\partial \tau} + (1 - \frac{x_1}{4})x_1\frac{\partial f}{\partial x_1} - \frac{\varepsilon}{8}(1 - \frac{3}{2}x_1 + \frac{11}{32}x_1^2)(1 + x_2^2)\frac{\partial f}{\partial x_2} = 0. \tag{4.8}$$

Alors les opérateurs (3.1) et (3.2) ont la forme

$$Y = Y_o + \varepsilon\tilde{Y}, \tag{4.9}$$

où

$$Y_o = \frac{\partial}{\partial \tau} + (1 - \frac{x_1}{4})x_1\frac{\partial}{\partial x_1}, \tag{4.10}$$

$$\tilde{Y} = -\frac{1}{8}(1 - \frac{3}{2}x_1 + \frac{11}{32}x_1^2)(1 + x_2^2)\frac{\partial}{\partial x_2}. \tag{4.11}$$

Quand $\varepsilon = 0$ l'opérateur (4.9) $Y = Y_o$ appartient à l'algèbre β et engendre un système aux variables separées. Nous cherchons à trouver un changement de variables transformant l'opérateur (4.9) $(\varepsilon \neq 0)$ en une forme qui engendre ce système-là avec une approximation au plus d'ordre ε.
Pour cela il faut prendre le changement sous la forme

$$x = \exp(-\varepsilon S)z \tag{4.12}$$

où

$$x = (x_1, x_2),\ z = (z_1, z_2),\ S = S_1 + \varepsilon S_2 + \varepsilon^2 S_3 + \ldots \tag{4.13}$$

S_1, S_2, S_3, ... sont des opérateurs de premier ordre aux coefficients indéterminés.

En faisant les substitutions (4.12) et (4.13) dans (4.9) nous obtenons pour Y l'expression

$$Y_z = Y_{oz} + \varepsilon\{Y_{oz}]S_1 + \tilde{Y}_z\} + \varepsilon^2\{S_2]Y_{oz} - S_1]\tilde{Y}_z + \frac{1}{2}S_1]^2 Y_{oz}\} + \ldots \tag{4.14}$$

où l'index z indique que x est exprimé partout en la nouvelle variable z. Avec cela on a évidemment

$$Y_{oz} = \frac{\partial}{\partial \tau} + (1 - \frac{z_1}{4}) z_1 \frac{\partial}{\partial z_1}, \tag{4.15}$$

$$\tilde{Y}_z = -\frac{1}{8}(1 - \frac{3}{2}z_1 + \frac{11}{32}z_1^2)(1 + z_2^2)\frac{\partial}{\partial z_2} \tag{4.16}$$

à l'ordre ε près.

Soit P un projecteur sur l'algèbre β (l'algèbre dont la qualité est une propriété de décomposition). Alors, en posant

$$PY_z = -\frac{1}{8}(1 + z_2^2)\frac{\partial}{\partial z_2}, \tag{4.17}$$

l'équation (3.12)

$$Y_{oz}]S_1 = -\tilde{Y}_z + P\tilde{Y}_z \tag{4.18}$$

se met sous la forme

$$Y_{oz}]S_1 = -\frac{1}{8}(\frac{3}{2}z_1 - \frac{11}{32}z_1^2)(1 + z_2^2)\frac{\partial}{\partial z_2}. \tag{4.19}$$

Prenons maintenant comme opérateurs de base

$$Z_1 = (1 - \frac{z_1}{4})z_1\frac{\partial}{\partial z_1}, \qquad Z_2 = \frac{\partial}{\partial z_2}, \tag{4.20}$$

(ces opérateurs doivent commuter). Cherchons l'opérateur S_1 sous la forme

$$S_1 = \gamma_1 Z_1 + \gamma_2 Z_2, \tag{4.21}$$

où γ_1 et γ_2 sont des fonctions scalaires inconnues des variables z_1, z_2 (γ_i ne doit pas dépendre de τ parce que les deuxièmes membres du système (4.8) ne dépendent pas de τ).

En substituant dans (4.19) S_1 par sa valeur donnèe en (4.12) et décomposant le deuxième membre de (4.19) par rapport aux opérateurs de base (4.20) nous obtenons

$$Y_{oz}(\gamma_1 Z_1 + \gamma_2 Z_2) = -\frac{1}{8}(\frac{3}{2}z_1 - \frac{11}{32}z_1^2)(1 + z_2^2)Z_2. \tag{4.22}$$

En tenant compte de ce que les coefficients des mêmes opérateurs de base sont égaux nous avons le système d'équations différentielles pour déterminer γ_1 et γ_2:

$$\begin{aligned} Y_{oz}\gamma_1 &= 0, \\ Y_{oz}\gamma_2 &= -\frac{1}{8}(\frac{3}{2}z_1 - \frac{11}{32}z_1^2)(1 + z_2^2). \end{aligned} \tag{4.23}$$

Chacune des équations du système (4.23) se résout indépendamment. On trouve par la première

$$\gamma_1 \equiv 0 \tag{4.24}$$

Pour déterminer γ_2 nous avons

$$(1 - \frac{z_1}{4})z_1\frac{\partial\gamma_2}{\partial z_1} = -\frac{1}{8}(\frac{3}{2}z_1 - \frac{11}{32}z_1^2)(1 + z_2^2) \tag{4.25}$$

d'où nous obtenons

$$\gamma_2 = (1 + z_2^2)\int^{z_1} \frac{-\frac{1}{8}(\frac{3}{2}z_1 - \frac{11}{32}z_1^2)}{(1 - \frac{z_1}{4})z_1} dz_1 + C_{\gamma_2}, \tag{4.26}$$

où C_{γ_2} est une constante arbitraire.

Si nous substituons γ_1 et γ_2 à leurs valeurs selon les formules (4.24) et (4.26) dans le deuxième membre (4.21) nous obtenons pour l'opérateur S_1 l'expression

$$S_1 = \{(1 + z_2^2)\int^{z_1} \frac{-(\frac{3}{2}z_1 - \frac{11}{32}z_1^2)}{2(4 - z_1)z_1} dz_1 + C_{\gamma_2}\}\frac{\partial}{\partial z_2}. \tag{4.27}$$

La substitution (4.12) sous la forme

$$\begin{aligned} x_1 &= (1 - \varepsilon S_1 + \varepsilon^2 \ldots)z_1, \\ x_2 &= (1 - \varepsilon S_1 + \varepsilon^2 \ldots)z_2, \end{aligned} \tag{4.28}$$

en vertu de (4.27) se change en

$$\begin{aligned} x_1 &= z_1 + \varepsilon^2 \ldots \\ x_2 &= z_2 + \varepsilon(1 + z_2^2)\int^{z_1} \frac{(\frac{3}{2}z_1 - \frac{11}{32}z_1^2)}{2(4 - z_1)z_1} dz_1 + \varepsilon C_{\gamma_2} + \varepsilon^2 \ldots \end{aligned} \tag{4.29}$$

Alors, l'opérateur (4.14), après ce changement, est de la forme suivante

$$Y_z = \frac{\partial}{\partial \tau} + (1 - \frac{z_1}{4})z_1\frac{\partial}{\partial z_1} - \frac{\varepsilon}{8}(1 + z_2^2)\frac{\partial}{\partial z_2} + \varepsilon^2 \ldots \tag{4.30}$$

et l'équation (4.8) prend la forme

$$\frac{\partial f}{\partial \tau} + (1 - \frac{z_1}{4})z_1\frac{\partial f}{\partial z_1} - \frac{\varepsilon}{8}(1 + z_2^2)\frac{\partial f}{\partial z_2} = 0 \tag{4.31}$$

Par conséquent, le système d'équations différentielles qui correspond à cette dernière, a la forme

$$\frac{dz_1}{d\tau} = (1 - \frac{z_1}{4})z_1,$$

$$\frac{dz_2}{d\tau} = -\frac{\varepsilon}{8}(1 + z_2^2). \tag{4.32}$$

Les nouvelles variables z_1, z_2 y sont sèparées à l'ordre ε près. Il est clair que le changement de variables dans (4.8) par les formules (4.29) mène au système (4.32) avec une erreur au plus d'ordre ε.

Etudions maintenant le cas d'une perturbation périodique du système décrit par l'équation de Van der Pol.

Posons $B = \varepsilon E$, $B_o = 0$ dans l'équation (4.1): nous obtenons

$$\frac{d^2u}{dt^2} - \varepsilon(1 - u^2)\frac{du}{dt} + u = \varepsilon E\cos\nu t. \tag{4.33}$$

Passons aux nouvelles variables b_1 et b_2 à l'aide de la transformation

$$x = b_1\cos\nu t + b_2\sin\nu t,$$

$$\frac{dx}{dt} = -b_1\nu\sin\nu t + b_2\nu\cos\nu t. \tag{4.34}$$

Ayant centré les équations obtenues sous la forme standard nous trouvons le système

$$\frac{db_1}{dt} = \Delta b_2 + \frac{\varepsilon}{2}\{b_1(1 - \frac{r^2}{4}) + \frac{E}{\nu}(1 + \frac{\Delta}{2\nu}) - \frac{\Delta}{4\nu}b_1r^2\} +$$
$$\frac{\varepsilon^2}{8\nu}\{b_2(1 + 4\nu\Delta^2) - \frac{3E}{2\nu}b_1b_2 + \frac{11}{32}b_2r^4 - \frac{3}{2}b_2r^2\},$$

$$\frac{db_2}{dt} = -\Delta b_1 + \frac{\varepsilon}{2}\{b_2(1 - \frac{r^2}{4}) - \frac{\Delta}{4\nu}b_2r^2\} -$$
$$\frac{\varepsilon^2}{8\nu}\{b_1(1 + 4\nu\Delta^2) + \frac{E}{\nu}(1 - \frac{5b_1^2 - b_2^2}{4}) + \frac{11}{32}b_1r^4 - \frac{3}{2}b_1r^2\}, \tag{4.35}$$

où $\frac{\nu^2 - 1}{2\nu} = \Delta$, $b_1^2 + b_2^2 = r^2$.

Introduisons de nouveau les nouvelles variables x_1, x_2, selon les formules

$$b_1^2 + b_2^2 = r^2 = x_1, \quad \frac{b_2}{b_1} = x_2 \tag{4.36}$$

et par conséquent,

$$b_1 = \sqrt{\frac{x_1}{1 + x_2^2}}, \quad b_2 = x_2\sqrt{\frac{x_1}{1 + x_2^2}} . \tag{4.37}$$

En substituant b_1 et b_2 par leurs valeurs selon (4.37) dans (4.35) et en faisant des calculs usuels nous obtenons

$$\begin{aligned}\frac{dx_1}{dt} &= \varepsilon\{[1 - \tfrac{1}{4}(1 + \tfrac{\Delta}{\nu})x_1]x_1 + \tfrac{E}{\nu}(1 + \tfrac{\Delta}{2\nu})\sqrt{\frac{x_1}{1 + x_2^2}}\} + \\ &\quad \frac{\varepsilon^2 E}{16\nu^2}(x_1 + 4)x_2\sqrt{\frac{x_1}{1 + x_2^2}}, \\ \frac{dx_2}{dt} &= -\Delta(1 + x^2) - \frac{\varepsilon}{2\nu}E(1 + \frac{\Delta}{2\nu})x_2\sqrt{\frac{1 + x_2^2}{x_1}} - \frac{\varepsilon^2}{8\nu}\{(1 + 4\nu\Delta^2) \cdot \\ &\quad (1 + x_2^2) + \tfrac{11}{32}x_1^2(1 + x_2^2) - \tfrac{3}{2}x_1(1 + x_2^2) + \frac{E}{4\nu}[\frac{4 - 5x_1}{\sqrt{x_1} \cdot \sqrt{1 + x_2^2}} + \\ &\quad \frac{4x_2^2 - 7x_2^2x_1}{\sqrt{x_1} \cdot \sqrt{x_2^2 + 1}}]\}.\end{aligned} \tag{4.38}$$

Si $E = 0$, $\nu = 1$ le système (4.38) se transforme en (4.8).

Il est évident que l'opérateur perturbé (3.2), correspondant au système d'équations (4.38), est de la forme

$$Y = Y_o + \varepsilon\tilde{Y}_1 + \varepsilon^2\tilde{Y}_2, \tag{4.39}$$

où

$$Y_o = \frac{\partial}{\partial t} - \Delta(1 + x_2^2)\frac{\partial}{\partial x_2}, \tag{4.40}$$

$$\begin{aligned}\tilde{Y}_1 &= \{[1 - \tfrac{1}{4}(1 + \tfrac{\Delta}{\nu})x_1]x_1 + \tfrac{E}{\nu}(1 + \tfrac{\Delta}{2\nu})\sqrt{\frac{x_1}{1 + x_2^2}}\}\frac{\partial}{\partial x_1} - \\ &\quad - \frac{E}{2\nu}(1 + \frac{\Delta}{2\nu})x_2\sqrt{\frac{1 + x_2^2}{x_1}}\frac{\partial}{\partial x_2},\end{aligned} \tag{4.41}$$

$$\tilde{Y}_2 = c_1(x_1,x_2)\frac{\partial}{\partial x_1} + c_2(x_1,x_2)\frac{\partial}{\partial x_2}, \tag{4.42}$$

et $c_1(x_1,x_2)$, $c_2(x_1,x_2)$ sont des coefficients à ε^2 près dans les deuxièmes membres des équations (4.38).

L'opérateur (4.40) appartient à l'algèbre β et engendre le système aux variables separées. Nous allons déterminer le changement des variables transformant l'opérateur Y. Pour cela cherchons comme ci-dessus le changement de variables sous la forme

$$\begin{gathered} x = \exp(-\varepsilon S)\cdot z \\ x = (x_1,x_2), \quad z = (z_1,z_2), \quad S = S_1 + \varepsilon S_2 + \ldots \end{gathered} \tag{4.43}$$

et définissons S_1 par l'équation (3.12)

$$Y_{oz}]S_1 = Y_{1z} + P\tilde{Y}_{1z}. \tag{4.44}$$

Ecrivons l'opérateur

$$\begin{aligned} \tilde{Y}_{1z} = \{ [1 - \tfrac{1}{4}(1 + \tfrac{\Delta}{\nu})z_1]z_1 + \tfrac{E}{\nu}(1 + \tfrac{\Delta}{2\nu})\sqrt{z_1}(1 - \tfrac{z_2^2}{2} + \tfrac{3}{8}z_2^4 \ldots)\} \\ \cdot\frac{\partial}{\partial z_1} + \frac{E}{2\nu}(1 + \frac{\Delta}{2\nu})z_2\sqrt{\frac{1 + z_2^2}{z_1}}\frac{\partial}{\partial z_2} + \varepsilon \ldots . \end{aligned} \tag{4.45}$$

Choisissons $P\tilde{Y}_{1z}$ sous la forme

$$\begin{aligned} P\tilde{Y}_{1z} = \{ [1 - \tfrac{1}{4}(1 + \tfrac{\Delta}{\nu})z_1]z_1 + \tfrac{E}{\nu}(1 + \tfrac{\Delta}{2\nu})\sqrt{z_1}\}\frac{\partial}{\partial z_1} - \\ \frac{E}{2\nu}(1 + \frac{\Delta}{2\nu})z_2\sqrt{\frac{1 + z_2^2}{z_1}} \cdot \frac{\partial}{\partial z_2}, \end{aligned} \tag{4.46}$$

et accomplissons la décomposition de l'opérateur $-\tilde{Y}_{1z} + P\tilde{Y}_{1z}$ par les opérateurs de base Z_{1z} et Z_{2z}

$$Z_{1z} = \frac{\partial}{\partial z_1}, \quad Z_{2z} = -\Delta(1 + z_2^2)\frac{\partial}{\partial z_2}. \tag{4.47}$$

Il en résulte

$$-\tilde{Y}_{1z} + P\tilde{Y}_{1z} = -\frac{E}{\nu}(1 + \frac{\Delta}{2\nu})\sqrt{z_1}\,(-\frac{z_2^2}{2} + \frac{3}{8}z_2^4 \ldots)\frac{\partial}{\partial z_1}. \tag{4.48}$$

Nous cherchons l'opérateur S_1 sous la forme

$$S_1 = \gamma_1(z_1, z_2)Z_{1z} + \gamma_2(z_1, z_2)Z_{2z}. \tag{4.49}$$

L'équation (4.44) se transforme d'après (4.48) et (4.49) en

$$Y_{oz}\gamma_1 Z_{1z} + Y_{oz}\gamma_2 Z_{2z} = c(z_1, z_2)Z_{1z} \tag{4.50}$$

où

$$c(z_1, z_2) = -\frac{E}{\nu}\left(1 + \frac{\Delta}{2\nu}\right)\sqrt{z_1}\left(-\frac{z_2^2}{2} + \frac{3}{8}z_2^4 \ldots\right). \tag{4.51}$$

Les coefficients correspondants des mêmes opérateurs de base doivent être égaux. Alors

$$\begin{aligned} Y_{oz}\gamma_1 &= c(z_1, z_2), \\ Y_{oz}\gamma_2 &= 0. \end{aligned} \tag{4.52}$$

Nous en dèduisons

$$\begin{aligned} \gamma_2(z_1, z_2) &= 0, \\ \gamma_1(z_1, z_2) &= \frac{E}{\Delta\nu}\left(1 + \frac{\Delta}{2\nu}\right)\sqrt{z_1}\int^{z_2} \frac{-\frac{z_2^2}{2} + \frac{3}{8}z_2^4 \ldots}{(1 + z_2^2)} dz_2 + C\gamma_1, \end{aligned} \tag{4.53}$$

e par conséquent,

$$S_1 = \left\{\frac{2}{\Delta\nu}\left(1 + \frac{\Delta}{2\nu}\right)\sqrt{z_1}\int^{z_2} \frac{-\frac{z_2^2}{2} + \frac{3}{8}z_2^4 \ldots}{(1 + z_2^2)} dz_2 + C_{\gamma_1}\right\}\frac{\partial}{\partial z_1} \tag{4.54}$$

Enfin nous trouvons pour les changements de variables (4.28) les expressions

$$\begin{aligned} x_1 &= z_1 - \varepsilon\frac{E}{\Delta\nu}\left(1 + \frac{\Delta}{2\nu}\right)\sqrt{z_1}\int^{z_2} \frac{-\frac{z_2^2}{2} + \frac{3}{8}z_2^4 \ldots}{1 + z_2^2} dz_2 + \varepsilon C_{\gamma_1}, \\ x_2 &= z_2 \end{aligned} \tag{4.55}$$

qui transforment le système (4.38) à celui ci-dessous

$$\begin{aligned}\frac{dz_1}{dt} &= \varepsilon\{[1 - \frac{1}{4}(1 + \frac{\Delta}{\nu})z_1]z_1 + \frac{E}{\nu}(1 + \frac{\Delta}{2\nu})\sqrt{z_1}\},\\ \frac{dz_2}{dt} &= -\Delta(1 + z_2^2) - \frac{\varepsilon}{2}\cdot\frac{E}{\nu}(1 + \frac{\Delta}{2\nu})z_2\sqrt{\frac{1 + z_2^2}{z_1}}\end{aligned} \quad (4.56)$$

à l'ordre ε près.

Donc, en ce cas, la décomposition n'est pas complète. Mais la première équation du système transformé (4.56) ne dépend que de z_1 et peut être intégrée indépendamment.
Ce procédé peut se répéter, c'est-à-dire, on peut définir S_2 et accomplir la division des variables à l'ordre ε^2 près etc.

5.

A titre de deuxième exemple concret nous allons étudier le problème de la décomposition d'un mouvement perturbé d'un appareil volant. Nous supposons que l'appareil volant, du type avion, ait un plan vertical de symmétrie $x_1 0 y_1$ (v. dessin).

Soit le système de coordonnées immobile terrestre $Ax_3y_3z_3$. Dèsignons par $0x_1y_1z_1$ le système de coordonnées lié avec l'appareil volant, et par $0x^*y^*z^*$ le système de coordonnées semi-rapide.

Les équations du mouvement par rapport au système de coordonnées semi-rapide contiennent 13 variables.

Les variables longitudinales sont indiquées par

V - la vitesse du centre de gravité par rapport à l'air

θ - l'angle d'inclinaison d'une trajectoire avec l'horizon

ω_{z_1} - la projection du vecteur de la vitesse angulaire sur l'axe $0z_1$

ϑ - l'angle de tanguage

x - la coordonnée du centre de gravité le long del'axe $0x_1$

H - l'altitude de vol

m - la masse de l'appareil volant.

Les variables transversales consistent en

Ψ - l'angle de virage de la trajectoire

ω_x, ω_y - la projection du vecteur de la vitesse angulaire sur l'axe $0x_1$ ou sur l'axe $0y_1$, respectivement

ψ - l'angle de lacet

γ - l'angle d'inclinaison transversale de l'appareil volant

z - la coordonnée de long de l'axe $0z_1$

Le mouvement longitudinal de l'appareil volant est formé du mouvement de centre de gravité le long des axes $0x_1$, $0y_1$ (c'est-à-dire, en plan de symétrie $0x_1$, $0y_1$) et du mouvement rotatif par rapport à l'axe $0z_1$.

Le mouvement transversal consiste en un mouvement du centre de gravité de l'appareil volant le long d'axe $0z_1$ et des mouvements rotatifs par rapport aux axes $0x_1$ et $0y_1$.

Le mouvement résultant de l'appareil volant est composé des deux mouvements indiqués et dans le cas général, ceux-ci peuvent exercer une action réciproque.

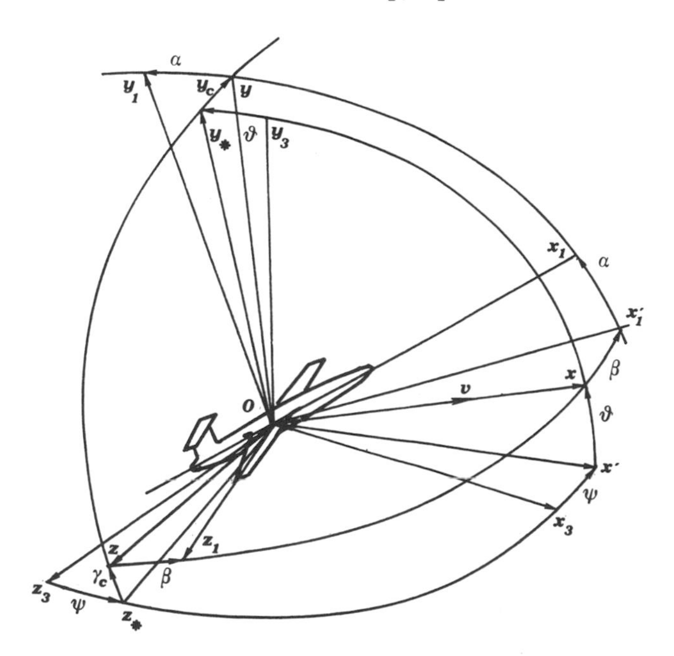

Fig. 1.

En tenant compte des notations citées les équations du mouvement de l'appareil volant sont écrites sous la forme ci-dessous. Précisément nous avons 7 équations contenant des paramètres du mouvement longitudinal

$$\begin{aligned}
\frac{dV}{dt} &= F_V(V,\theta,\vartheta,H,m,\Psi,\psi,\gamma),\\
\frac{d\theta}{dt} &= F_\theta(V,\theta,\vartheta,H,m,\Psi,\psi,\gamma),\\
\frac{d\omega_z}{dt} &= F_{\omega_z}(V,\theta,\omega_z,\vartheta,H,\Psi,\omega_x,\omega_y,\psi,\gamma),\\
\frac{d\vartheta}{dt} &= F_\vartheta(\omega_z,\omega_y,\gamma),\\
\frac{dx}{dt} &= F_x(V,\theta,\Psi),\\
\frac{dH}{dt} &= F_H(V,\theta),\\
\frac{dm}{dt} &= F_m(V,H,t).
\end{aligned} \tag{5.1}$$

et 6 équations contenant des paramètres du mouvement transversal

$$\begin{aligned}
\frac{d\Psi}{dt} &= F_\Psi(V,\theta,\vartheta,H,m,\Psi,\psi,\gamma),\\
\frac{d\omega_x}{dt} &= F_{\omega_x}(V,\theta,\omega_z,\vartheta,H,\Psi,\omega_x,\omega_y,\psi,\gamma),\\
\frac{d\omega_y}{dt} &= F_{\omega_y}(V,\theta,\omega_z,\vartheta,H,\Psi,\omega_x,\omega_y,\psi,\gamma),\\
\frac{d\psi}{dt} &= F_\psi(\omega_z,\vartheta,\omega_y,\gamma),\\
\frac{d\gamma}{dt} &= F_\gamma(\omega_z,\vartheta,\omega_x,\omega_y,\gamma),\\
\frac{dz}{dt} &= F_z(V,\theta,\Psi).
\end{aligned} \tag{5.2}$$

(v. par exemple, [2]).

Nous avons à introduire maintenant dans le système (5.1)-(5.2) les notations

$$F_V = \frac{P}{m}\cos\cdot\cos\beta - \frac{X}{m} - \frac{G}{m}\sin\theta,$$

$$F_\theta = \frac{P}{mV}(\sin\alpha\cdot\cos\gamma + \cos\alpha\cdot\sin\beta\cdot\sin\gamma_c) - \frac{Y}{mV}\cos\gamma_c - \frac{z}{mV}\sin\gamma_c - \frac{G}{mV}\cos\theta, \qquad (5.3)$$

$$F_{\omega_z} = \frac{M_z}{I_z} - \frac{I_y - I_x}{I_z}\omega_x\omega_y,$$

$$F_\vartheta = \omega_y\sin\gamma + \omega_z\cos\gamma,$$

$$F_x = V\cos\theta\cos\Psi,$$

$$F_H = V\sin\theta,$$

$$F_m = -m_c(V,H,t),$$

$$F_\Psi = -\frac{P}{mV\cos\theta}(\sin\alpha\cdot\sin\gamma_c - \cos\alpha\cdot\sin\beta\cdot\cos\gamma_c) - \frac{Y}{mV\cos\theta}\sin\gamma_c - \frac{Z}{mV\cos\theta}\cos\gamma_c, \qquad (5.4)$$

$$F_{\omega_x} = \frac{M_x}{I_x} - \frac{I_z - I_y}{I_x}\omega_y\omega_z,$$

$$F_{\omega_y} = \frac{M_y}{I_y} - \frac{I_x - I_z}{I_y}\omega_z\cdot\omega_x,$$

$$F_\psi = \frac{1}{\cos\vartheta}(\omega_y\cos\gamma - \omega_z\sin\gamma),$$

$$F_\gamma = \omega_x - \mathrm{tg}\vartheta(\omega_y\cos\gamma - \omega_z\sin\gamma),$$

$$F_z = -V\cos\theta\sin\Psi,$$

où nous désignons par

P - la force de propulsion du réacteur

X,Y,Z - la projection de la force aérodynamique résultante sur les axes de vitesse

α,β - l'angle d'attaque et celui de glissement

γ_c - l'angle d'inclinaison transversale par rapport aux axes de vitesse

Y_x, Y_y, Y_z - les moments d'inertie par rapport aux axes liés

M_x, M_y, M_z - des projections des moments des forces extérieures (celui de roulis, celui de lacet et celui de tangua ge)

G - le poids d'appareil volant

Nous allons écrire le système des équations (5.1) - (5.2) sous la forme vectorielle

$$\frac{d\eta}{dt} = F(t,\eta) \tag{5.5}$$

où on a désigné par $\eta(\eta_1,\eta_2,\ldots,\eta_{13})$, $F(F_{\eta_1},F_{\eta_2},\ldots,F_{\eta_{13}})$ les variables et les fonctions qui se trouvent aux deuxièmes membres des équations de mouvement, rétablissant l'ordre d'introduction des notions ci-dessus.

Soit maintenant $\eta = \eta_*$ un certain mouvement commandé

$$\frac{d\eta_*}{dt} \equiv F(t,\eta_*). \tag{5.6}$$

Etudions le mouvement au voisinage du mouvement commandé

$$\eta = \eta_* + \varepsilon\Delta\eta \tag{5.7}$$

où le petit paramètre $\varepsilon > 0$ caractérise l'ordre de la petitesse du mouvement perturbé.

En substituant η par sa valeur selon (5.7) dans l'équation (5.5) et en prenant en considération les identités (5.6) nous avons le système des équations du mouvement

$$\frac{d\Delta\eta_i}{dt} = \sum_{j=1}^{13} \Delta\eta_j \frac{\partial F_{\eta_i}(t,\eta_*)}{\partial\eta_j} + \frac{\varepsilon}{2}\Big(\sum_{j=1}^{13} \Delta\eta_j \frac{\partial}{\partial\eta_j}\Big)^2 F_{\eta_j}(t,\eta_*) + \ldots$$

$$(i = \overline{1,13}) \tag{5.8}$$

En tenant compte de la symétrie de l'appareil volant par rapport au plan $x_1 0 y_1$ on peut considérer les fonctions comme fonctions paires des paramètres $\Psi, \omega_x, \omega_y, \psi, \gamma$. Par conséquent, les développements en séries de Taylor de ces fonctions ne doivent pas contenir les puissances non paires de ces variables. De plus $\beta_* = 0$ si le mouvement essentiel commandé ne contient pas de glissement.

Introduisons maintenant l'écriture abrégée pour les opérateurs ci-dessous

$$\sum \Delta\eta\frac{\partial}{\partial\eta} = \Delta V\frac{\partial}{\partial V} + \Delta\theta\frac{\partial}{\partial\theta} + \Delta\omega_z\frac{\partial}{\partial\omega_z} + \Delta\vartheta\frac{\partial}{\partial\vartheta} + \Delta H\frac{\partial}{\partial H} + \Delta m\frac{\partial}{\partial m} + \Delta\Psi\frac{\partial}{\partial\Psi} + \Delta\omega_x\frac{\omega}{\partial\omega_x} + \Delta\omega_y\frac{\partial}{\partial\omega_y} + \Delta\psi\frac{\partial}{\partial\psi} + \Delta\gamma\frac{\partial}{\partial\gamma}, \tag{5.9}$$

$$\sum_n \Delta\eta\frac{\partial}{\partial\eta} = \Delta V\frac{\partial}{\partial V} + \Delta\theta\frac{\partial}{\partial\theta} + \Delta\omega_z\frac{\partial}{\partial\omega_z} + \Delta\vartheta\frac{\partial}{\partial\vartheta} + \Delta H\frac{\partial}{\partial H} + \Delta m\frac{\partial}{\partial m} \tag{5.10}$$

$$\sum_\delta \Delta\eta\frac{\partial}{\partial\eta} = \Delta\Psi\frac{\partial}{\partial\Psi} + \Delta\omega_x\frac{\partial}{\partial\omega_x} + \Delta\omega_y\frac{\partial}{\partial\omega_y} + \Delta\psi\frac{\partial}{\partial\psi} + \Delta\gamma\frac{\partial}{\partial\gamma}. \tag{5.11}$$

L'opérateur (5.9) comprend toutes les variables dont dépendent les deuxièmes membres des équations (5.1) - (5.2). Les opérateurs (5.10) et (5.11) contiennent par conséquent les variables du mouvement longitudinal et du mouvement transversal. On désigne par la notation

$$(\sum \Delta\eta\frac{\partial}{\partial\eta})F \tag{5.12}$$

le résultat de ce procédé: au début on accomplit la dérivation de F et ensuite la substitution des paramètres du mouvement non perturbé η_* au lieu des variables η.

Après avoir introduit ces notations, compte tenu des considérations précédentes, on peut écrire le système d'équations (5.8) sous la forme abrégée

$$\begin{aligned} \frac{d\Delta\eta_i}{dt} &= (\sum_n \Delta\eta\frac{\partial}{\partial\eta})F_{\eta_i} + \frac{\varepsilon}{2}(\sum \Delta\eta\frac{\partial}{\partial\eta})^2 F_{\eta_i} + \varepsilon^2\ldots, \\ \frac{d\Delta\eta_j}{dt} &= (\sum_\delta \Delta\eta\frac{\partial}{\partial\eta})F_{\eta_j} + \frac{\varepsilon}{2}(\sum \Delta\eta\frac{\partial}{\partial\eta})^2 F_{\eta_j} + \varepsilon^2\ldots \end{aligned} \tag{5.13}$$

$(i = \overline{1,7};\ j = \overline{8,13})$.

En approximation linéaire (c'est-à-dire, dans le cas $\varepsilon = 0$) le système (5.13) se décompose en deux sous-systèmes qui décrivent le mouvement longitudinal et le mouvement transversal indépendamment l'un de l'autre.

En vertu du procédé exposé faisons une décomposition asymptotique.

L'équation aux dérivées partielles étant conforme au système (5.13) a la forme

$$\frac{\partial f}{\partial t} + \sum_{i=1}^{7} (\sum_n \Delta\eta\frac{\partial}{\partial\eta}) F_{\eta_i}\frac{\partial f}{\partial\eta_i} + \sum_{j=8}^{13} (\sum_\delta \Delta\eta\frac{\partial}{\partial\eta}) F_{\eta_j}\frac{\partial f}{\partial\eta_j} +$$

$$\frac{\varepsilon}{2}\{ \sum_{i=1}^{7} (\sum \Delta\eta\frac{\partial}{\partial\eta})^2 F_{\eta_i}\frac{\partial f}{\partial\eta_i} + \sum_{j=8}^{13} (\sum \Delta\eta\frac{\partial}{\partial\eta})^2 F_{\eta_j}\frac{\partial f}{\partial\eta_j}\} = 0 \qquad (5.14)$$

et l'opérateur satisfaisant (5.14) a la forme

$$Y = Y_o + \varepsilon\tilde{Y}, \qquad (5.15)$$

où

$$Y_o = \frac{\partial}{\partial t} + \sum_{i=1}^{7} (\sum_n \Delta\eta\frac{\partial}{\partial\eta}) F_{\eta_i}\frac{\partial}{\partial\eta_i} + \sum_{j=8}^{13} (\sum_\delta \Delta\eta\frac{\partial}{\partial\eta}) F_{\eta_j}\frac{\partial}{\partial\eta_j}, \qquad (5.16)$$

$$\tilde{Y} = \frac{1}{2}\sum_{i=1}^{7} (\sum \Delta\eta\frac{\partial}{\partial\eta})^2 F_{\eta_i}\frac{\partial}{\partial\eta_i} + \frac{1}{2}\sum_{j=8}^{13} (\sum \Delta\eta\frac{\partial}{\partial\eta})^2 F_{\eta_j}\frac{\partial}{\partial\eta_j}. \qquad (5.17)$$

Conformément à la théorie générale nous introduisons le nouvelles variables $\bar{\eta}$ au lieu de η d'après les formules

$$\eta = \exp(-\varepsilon S)\cdot\bar{\eta} \qquad (5.18)$$

où

$$S = S_1 + \varepsilon S_2 + \varepsilon^2 S_3 + \ldots \qquad (5.19)$$

Après le changement (5.18) l'opérateur (5.15) est sous la forme

$$Y_{\bar{\eta}} = Y_{o\bar{\eta}} + \varepsilon\{Y_{o\bar{\eta}}]S_1 + \tilde{Y}_{\bar{\eta}}\} + \varepsilon^2 \ldots, \qquad (5.20)$$

où l'index $\bar{\eta}$ signifie que l'on fait partout la substitution par les nouvelles variables $\bar{\eta}$.

L'opératuer S_1 découle de l'équation (3.12)

$$Y_{o\bar{\eta}}]S_1 = -\tilde{Y}_{\bar{\eta}} + P\tilde{Y}_{\bar{\eta}}, \tag{5.21}$$

où $P\tilde{Y}_{\bar{\eta}}$ est une projection de l'opérateur $\tilde{Y}_{\bar{\eta}}$ sur l'algèbre β:

$$P\tilde{Y}_{\bar{\eta}} = \frac{1}{2}\sum_{i=1}^{7}(\textstyle\sum_n \Delta\bar{\eta}\frac{\partial}{\partial\bar{\eta}})^2 F_{\bar{\eta}_i}\frac{\partial}{\partial\bar{\eta}_i} + \frac{1}{2}\sum_{j=8}^{13}(\textstyle\sum_\delta \Delta\eta\frac{\partial}{\partial\bar{\eta}})^2 F_{\bar{\eta}_j}\frac{\partial}{\partial\eta_j}. \tag{5.22}$$

Pour le deuxième membre de l'équation (5.21) nous obtenons l'expression

$$-\tilde{Y}_{\bar{\eta}} + P\tilde{Y}_{\bar{\eta}} = \frac{1}{2}\sum_{i=1}^{7}[(\textstyle\sum_n \Delta\bar{\eta}\frac{\partial}{\partial\bar{\eta}})^2 F_{\bar{\eta}_i} - (\sum \Delta\bar{\eta}\frac{\partial}{\partial\bar{\eta}})^2 F_{\bar{\eta}_i}]\frac{\partial}{\partial\bar{\eta}_i} +$$
$$\frac{1}{2}\sum_{j=8}^{13}[(\textstyle\sum_\delta \Delta\bar{\eta}\frac{\partial}{\partial\bar{\eta}})^2 F_{\bar{\eta}_j} - (\sum \Delta\bar{\eta}\frac{\partial}{\partial\bar{\eta}})^2 F_{\bar{\eta}_j}]\frac{\partial}{\partial\bar{\eta}_j}. \tag{5.23}$$

Nous cherchons à trouver l'opérateur S_1 sous la forme

$$S_1 = \sum_{x=1}^{13}\gamma_x Z_x, \tag{5.24}$$

où Z_x $(x = \overline{1,13})$ sont des opérateurs de base

$$Z_x = \frac{\partial}{\partial\eta_x} \qquad (x = \overline{1,13}). \tag{5.25}$$

Ayant accompli les substitutions (5.25), (5.24) et (5.23) dans l'équation (5.21), après bien des transformations, nous obtenons le système d'équations pour déterminer des fonctions $\gamma_1, \ldots, \gamma_{13}$:

$$Y_{\gamma_1} = \sum_{x=1}^{7}\gamma_x Z_x(\textstyle\sum_n \Delta\bar{\eta}\frac{\partial}{\partial\bar{\eta}})F_{\bar{\eta}_1} + (\sum \Delta\bar{\eta}\frac{\partial}{\partial\bar{\eta}})^2 F_{\bar{\eta}_1},$$

$$\cdots\cdots\cdots\cdots\cdots\cdots \tag{5.26}$$

$$Y_{\gamma_7} = \sum_{x=1}^{7}\gamma_x Z_x(\textstyle\sum_n \Delta\bar{\eta}\frac{\partial}{\partial\bar{\eta}})F_{\bar{\eta}_7} + (\sum \Delta\eta\frac{\partial}{\partial\bar{\eta}})^2 F_{\bar{\eta}_7};$$

$$Y_{\gamma_8} = \sum_{x=8}^{13} \gamma_x Z_x (\sum_\delta \Delta\bar{\eta}\frac{\partial}{\partial\bar{\eta}}) F_{\bar{\eta}_8} + (\sum \Delta\bar{\eta}\frac{\partial}{\partial\bar{\eta}})^2 F_{\bar{\eta}_8},$$

$$\cdots\cdots\cdots\cdots \qquad (5.27)$$

$$Y_{\gamma_{13}} = \sum_{x=8}^{13} \gamma_x Z_x (\sum_\delta \Delta\bar{\eta}\frac{\partial}{\partial\bar{\eta}}) F_{\bar{\eta}_{13}} + (\sum \Delta\bar{\eta}\frac{\partial}{\partial\bar{\eta}})^2 F_{\bar{\eta}_{13}}.$$

Le système d'équations (5.26) ne contient que des variables inconnues $\gamma_1,\ldots,\gamma_7$ et le système (5.27) comprend seulement les variables $\gamma_8,\ldots,\gamma_{13}$.

L'intégration du système (5.26) équivaut à la résolution de l'équation

$$Yf + \{\sum_{x=1}^{7} \gamma_x Z_x (\sum_n \Delta\bar{\eta}\frac{\partial}{\partial\bar{\eta}}) F_{\bar{\eta}_1} + (\sum \Delta\bar{\eta}\frac{\partial}{\partial\bar{\eta}})^2 F_{\bar{\eta}_1}\}\frac{\partial f}{\partial\gamma_1} + \ldots$$
$$\{\sum_{x=1}^{7} \gamma_x Z_x (\sum_n \Delta\bar{\eta}\frac{\partial}{\partial\bar{\eta}}) F_{\bar{\eta}_7} + (\sum \Delta\bar{\eta}\frac{\partial}{\partial\bar{\eta}})^2 F_{\bar{\eta}_7}\}\frac{\partial f}{\partial\gamma_7} +$$
$$\{\sum_{x=8}^{13} \gamma_x Z_x (\sum_\delta \Delta\bar{\eta}\frac{\partial}{\partial\bar{\eta}}) F_{\bar{\eta}_8} + (\sum \Delta\bar{\eta}\frac{\partial}{\partial\bar{\eta}})^2 F_{\bar{\eta}_8}\}\frac{\partial}{\partial\gamma_8} + \ldots$$
$$\{\sum_{x=8}^{13} \gamma_x Z_x (\sum_\delta \Delta\bar{\eta}\frac{\partial}{\partial\bar{\eta}}) F_{\bar{\eta}_{13}} + (\sum \Delta\bar{\eta}\frac{\partial}{\partial\bar{\eta}})^2 F_{\bar{\eta}_{13}}\}\frac{\partial f}{\partial\gamma_{13}} = 0 \qquad (5.28)$$

A son tour l'intégration de cette dernière est équivalente à l'intégration des systèmes ci-dessous:

$$\frac{d\Delta\bar{\eta}_i}{dt} = (\sum_n \Delta\bar{\eta}\frac{\partial}{\partial\bar{\eta}}) F_{\bar{\eta}_i}, \qquad (i = \overline{1,7}),$$
$$\frac{d\Delta\bar{\eta}_j}{dt} = (\sum_\delta \Delta\bar{\eta}\frac{\partial}{\partial\bar{\eta}}) F_{\bar{\eta}_j}, \qquad (j = \overline{8,13}), \qquad (5.29)$$

$$\frac{d\gamma_1}{dt} = \sum_{x=1}^{7} \gamma_x Z_x (\sum_n \Delta\bar{\eta}\frac{\partial}{\partial\bar{\eta}}) F_{\bar{\eta}_1} + (\sum \Delta\bar{\eta}\frac{\partial}{\partial\bar{\eta}})^2 F_{\bar{\eta}_1},$$

$$\cdots\cdots\cdots\cdots \qquad (5.30)$$

$$\frac{d\gamma_7}{dt} = \sum_{x=1}^{7} \gamma_x Z_x (\sum_n \Delta\bar{\eta}\frac{\partial}{\partial\bar{\eta}}) F_{\bar{\eta}_7} + (\sum \Delta\bar{\eta}\frac{\partial}{\partial\bar{\eta}})^2 F_{\bar{\eta}_7},$$

$$\frac{d\gamma_8}{dt} = \sum_{x=8}^{13} \gamma_x Z_x (\sum_\delta \Delta\bar{\eta}\frac{\partial}{\partial\bar{\eta}}) F_{\bar{\eta}_8} + (\sum \Delta\bar{\eta}\frac{\partial}{\partial\bar{\eta}})^2 F_{\bar{\eta}_8},$$

$$. \quad . \qquad (5.31)$$

$$\frac{d\gamma_{13}}{dt} = \sum_{x=8}^{13} \gamma_x Z_x (\sum_\delta \Delta\bar{\eta}\frac{\partial}{\partial\bar{\eta}}) F_{\bar{\eta}_{13}} + (\sum \Delta\bar{\eta}\frac{\partial}{\partial\bar{\eta}})^2 F_{\bar{\eta}_{13}}.$$

Ainsi, pour déterminer $\gamma_1,\ldots,\gamma_{13}$ nous avons obtenu le système consistant en trois systèmes intégrables indépendamment les uns des autres.

Le système (5.29) est le système en approximation zéro dont nous parlons au-dessus. En le décomposant en deux systèmes indépendant l'un de l'autre.

Si nous trouvons la résolution du système (5.29) et faisons la substitution de cette dernière dans les systèmes (5.30) et (5.31) nous avons deux systèmes d'équations différentielles ordinaires (inhomogènes) aux coefficients variables, qui peuvent être intégrés indépendamment. L'expression pour l'opérateur S_1 est connue après que l'on avait intégré ces systèmes et par conséquent, la transformation (5.18) est sous la forme

$$\eta_i = (1 - \varepsilon \sum_{x=1}^{13} \gamma_x \frac{\partial}{\partial\bar{\eta}x})\bar{\eta}_i \qquad (i = \overline{1,13}) \qquad (5.32)$$

à l'ordre ε près. Après cela le système de départ (5.13) est aussi décomposé et il y a deux systèmes intégrables indépendants

$$\frac{d\Delta\bar{\eta}_i}{dt} = (\sum_n \Delta\bar{\eta}\frac{\partial}{\partial\bar{\eta}}) F_{\bar{\eta}_i} + \varepsilon(\sum_n \Delta\bar{\eta}\frac{\partial}{\partial\bar{\eta}})^2 F_{\bar{\eta}_i}, \quad (i = \overline{1,7})$$

$$(5.33)$$

$$\frac{d\Delta\bar{\eta}_j}{dt} = (\sum_\delta \Delta\bar{\eta}\frac{\partial}{\partial\bar{\eta}}) F_{\bar{\eta}_j} + \varepsilon(\sum_\delta \Delta\bar{\eta}\frac{\partial}{\partial\bar{\eta}})^2 F_{\bar{\eta}_j}, \quad (j = \overline{8,13}).$$

à l'ordre ε près.

6.

Etudions aussi l'exemple suivant: c'est un système hyperbolique d'équations non linéaires aux dérivées partielles. Les systèmes les plus étudiés de ce type sont les systèmes d'équations quasi linéaires à deux variables indépendantes. Comme il est connu [10], les systèmes de ce type décrivent des écoulements en dimension une (non permanents) et en dimension deus (permanents, supersoniques) des gaz et des fluides qui sont compressibles.

Nous nous fixons sur l'étude des équations décrivant un écoulement isoentropique d'un gaz polytropique [10]:

$$\begin{aligned} \frac{\partial s}{\partial t} + (\alpha s + \beta r)\frac{\partial s}{\partial x} &= \frac{\nu(\gamma - 1)(r^2 - s^2)}{4x}, \\ \frac{\partial r}{\partial t} + (\alpha r + \beta s)\frac{\partial r}{\partial x} &= -\frac{\nu(\gamma - 1)(r^2 - s^2)}{4x}, \end{aligned} \tag{6.1}$$

où on a les notations suivantes

$$\alpha = \frac{1}{2} + \frac{\gamma - 1}{4} > \frac{1}{2} > 0, \qquad \beta = \frac{1}{2} - \frac{\gamma - 1}{4}, \tag{6.2}$$

les variables r, s sont des invariants de Riemann

$$\begin{aligned} s &= u - \varphi(\rho) = u - \frac{2}{\gamma - 1}c, \\ r &= u + \varphi(\rho) = u + \frac{2}{\gamma - 1}c, \end{aligned} \tag{6.3}$$

qui sont exprimés par les paramètres du mouvement: u est une vitesse d'écoulement, ρ est une densité et la valeur $c = c(\rho, S)$ est une vitesse du son, S est une entropie,

$$\varphi(\rho) = \int_{\rho_0}^{\rho} \frac{c_0}{\rho} d\rho = c_0 \ln\frac{\rho}{\rho_0}, \tag{6.4}$$

$\gamma = \frac{c_p}{c_\nu} > 1$ est un exposant dans la formule caractérisant la pression du gaz polytropique

$$p = \frac{A^2(S)}{\gamma}\rho^\gamma, \tag{6.5}$$

et par conséquent,

$$c^2 = \frac{\partial p}{\partial \rho} = A^2(S)\rho^{\gamma-1}, \qquad c(\rho,S) = A(S)\rho^{\frac{\gamma-1}{2}}, \tag{6.6}$$

c_p et c_ν sont des coefficients caractérisant l'équation du gaz parfait.

Le paramètre ν caractérise la symétrie de plan de l'écoulement du gaz. Quand $\nu = 0$ nous avons des équations plansymétriques du mouvement

$$\begin{aligned} &\frac{\partial s}{\partial t} + (\alpha s + \beta r)\frac{\partial s}{\partial x} = 0, \\ &\frac{\partial r}{\partial t} + (\alpha r + \beta s)\frac{\partial r}{\partial x} = 0. \end{aligned} \tag{6.7}$$

Si l'exposant de la courbe polytropique γ est égal à 3 les équations (6.1) (ainsi que les équations (6.7)) se décomposent. Après cela nous avons deux équations quasi linéaires indépendantes

$$\frac{\partial s}{\partial t} + s\frac{\partial s}{\partial x} = 0, \qquad \frac{\partial r}{\partial t} + r\frac{\partial r}{\partial x} = 0. \tag{6.8}$$

Etudions des équations proches d'équations décomposées et proches des équations plan-symétriques[1]. Pour commencer posons

$$\nu = \varepsilon, \qquad \gamma = 3 + \varepsilon\Delta\gamma, \qquad 0 < \varepsilon < 1. \tag{6.9}$$

En faisant la substitution des valeurs de ν et γ par les formules (6.9) dans l'équation (6.1) nous obtenons le système

$$\begin{aligned} &\frac{\partial s}{\partial t} + s\frac{\partial s}{\partial x} = \varepsilon F_1(x,r,s), \\ &\frac{\partial r}{\partial t} + r\frac{\partial r}{\partial x} = \varepsilon F_2(x,r,s), \end{aligned} \tag{6.10}$$

---

[1] L'étude détaillée est accomplie par A.K. Lopatine.

où on désigne

$$
\begin{aligned}
F_1(x,r,s) &= \frac{r^2 - s^2}{2x} - \frac{\Delta\gamma}{4}(s - r)\frac{\partial s}{\partial x} + \varepsilon\Delta\gamma\frac{r^2 - s^2}{4x}, \\
F_2(x,r,s) &= -\frac{r^2 - s^2}{2x} - \frac{\Delta\gamma}{4}(r - s)\frac{\partial r}{\partial x} - \varepsilon\Delta\gamma\frac{r^2 - s^2}{4x},
\end{aligned}
\tag{6.11}
$$

Ainsi, le système (6.10) n'est rien d'autre que le système perturbé dont la décomposition en approximation nulle (quand $\varepsilon = 0$) consiste en deux systèmes indépendants (6.8).

Pour plus de commodité nous introduisons ci-dessous les notations suivantes

$$
\begin{aligned}
&t = x_1, \quad x = x_2, \quad s = z_1, \quad r = z_2, \\
&\frac{\partial s}{\partial x} = \frac{\partial z_1}{\partial x_2} = u_1, \quad \frac{\partial r}{\partial x} = \frac{\partial z_2}{\partial x_2} = u_2.
\end{aligned}
\tag{6.12}
$$

Alors le système (6.10) en nouvelles variables est sous la forme

$$
\begin{aligned}
&\frac{\partial z_1}{\partial x_1} = -u_1 z_1 + \varepsilon F, \qquad \frac{\partial z_2}{\partial x_1} = -u_2 z_2 + \varepsilon F_2, \\
&\frac{\partial z_1}{\partial x_2} = u_1, \qquad \frac{\partial z_2}{\partial x_2} = u_2.
\end{aligned}
\tag{6.13}
$$

Le système d'équations (6.13) peut être changé en un système aux différentielles totales

$$
\begin{aligned}
dz_1 &= (-u_1 z_1 + \varepsilon F_1)dx_1 + u_1 dx_2, \\
dz_2 &= (-u_2 z_2 + \varepsilon F_2)dx_1 + u_2 dx_2.
\end{aligned}
\tag{6.14}
$$

Comme il est bien connu, les équations (6.14), à leur tour, sont déterminées complètement par les opérateurs linéaires

$$
\begin{aligned}
Y_1 &= \frac{\partial}{\partial x_1} + (-u_1 z_1 + \varepsilon F_1)\frac{\partial}{\partial z_1} + (-u_2 z_2 + \varepsilon F_2)\frac{\partial}{\partial z_2}, \\
Y_2 &= \frac{\partial}{\partial x_2} + u_1\frac{\partial}{\partial z_1} + u_2\frac{\partial}{\partial z_2},
\end{aligned}
\tag{6.15}
$$

qui peuvent être écrits sous la forme

$$Y_1 = Y_{10} + \varepsilon\tilde{Y},$$
$$Y_2 = Y_{20}, \qquad (6.16)$$

Ici on a

$$Y_{10} = \frac{\partial}{\partial x_1} - u_1 z_1 \frac{\partial}{\partial z_1} - u_2 z_2 \frac{\partial}{\partial z_2},$$
$$Y_{20} = \frac{\partial}{\partial x_2} + u_1 \frac{\partial}{\partial z_1} + u_2 \frac{\partial}{\partial z_2}, \qquad (6.17)$$
$$\tilde{Y}_1 = F_1 \frac{\partial}{\partial z_1} + F_2 \frac{\partial}{\partial z_2}.$$

Etudions le problème de la décomposition du système (6.13). Pour abréger l'écriture nous introduisons les notations

$$\eta_1 = x_1, \quad \eta_2 = x_2, \quad \eta_3 = z_1, \quad \eta_4 = z_2, \quad \eta_5 = u_1, \quad \eta_6 = u_2. \qquad (6.18)$$

Conformément à la théorie générale exposée dans la partie 3 nous effectuons dans les opérateurs (6.16) le changement de variables d'après les formules

$$\eta = \exp(-\varepsilon S)\bar{\eta} \qquad (6.19)$$

où

$$S = S_1 + \varepsilon S_2 + \varepsilon^2 S_3 + \dots \qquad (6.20)$$

$$S_i = \sum_{j=1}^{6} S_{ij}(\eta)\frac{\partial}{\partial \eta_j}, \qquad (6.21)$$

$S_{ij}(\eta)$ sont des fonctions indéfinies pour le moment.

Après la transformation (6.19) nous avons les opérateurs (6.16) sous la forme

$$\begin{aligned} \overline{Y}_1 &= \overline{Y}_{10} + \varepsilon(\overline{Y}_{10}]S_1 + \tilde{\tilde{Y}}_1) + \varepsilon^2\ldots, \\ \overline{Y}_2 &= \overline{Y}_{20} + \varepsilon(\overline{Y}_{20}\ S_1) + \varepsilon^2\ldots, \end{aligned} \tag{6.22}$$

où

$$\begin{aligned} \overline{Y}_{10} &= \frac{\partial}{\partial\overline{x}_1} - \overline{u}_1\overline{x}_1\frac{\partial}{\partial\overline{z}_1} - \overline{u}_2\overline{z}_2\frac{\partial}{\partial\overline{z}_2}, \\ \overline{Y}_{20} &= \frac{\partial}{\partial\overline{x}_2} + \overline{u}_1\frac{\partial}{\partial\overline{z}_1} + \overline{u}_2\frac{\partial}{\partial\overline{z}_2}, \\ \tilde{\tilde{Y}}_1 &= \overline{F}_1\frac{\partial}{\partial\overline{z}_1} + \overline{F}_2\frac{\partial}{\partial\overline{z}_2} \\ \overline{F}_1 &= \frac{\overline{z}_2^2 - \overline{z}_1^2}{2\overline{x}_2} - \frac{\Delta\gamma}{4}(\overline{z}_1 - \overline{z}_2)\overline{u}_1, \\ \overline{F}_2 &= -\frac{\overline{z}_2^2 - \overline{z}_1^2}{2\overline{x}_2} - \frac{\Delta\gamma}{4}(\overline{z}_2 - \overline{z}_1)\overline{u}_2. \end{aligned} \tag{6.23}$$

En désignant comme ci-dessus le projecteur sur l'algèbre β en approximation nulle par P nous avons

$$P\tilde{\tilde{Y}}_1 = -\left(\frac{\overline{z}_1^2}{2\overline{x}_2} + \frac{\Delta\gamma}{4}\overline{z}_1\overline{u}_1\right)\frac{\partial}{\partial\overline{z}_1} - \left(\frac{\overline{z}_2^2}{2\overline{x}_2} + \frac{\Delta\gamma}{4}\overline{z}_2\overline{u}_2\right)\frac{\partial}{\partial\overline{z}_2}. \tag{6.24}$$

Pour la décomposition asymptotique des opérateurs (6.22) en approximation nulle nous avons à résoudre le système d'équations

$$\begin{aligned} \overline{Y}_{10}]S_1 &= -\tilde{\tilde{Y}}_1 + P\tilde{\tilde{Y}}_1, \\ \overline{Y}_{20}]S_1 &= 0, \end{aligned} \tag{6.25}$$

où

$$-\tilde{\tilde{Y}}_1 + P\tilde{\tilde{Y}}_1 = -\left(\frac{\overline{z}_2^2}{2\overline{x}_2^2} + \frac{\Delta\gamma}{4}\overline{z}_2\overline{u}_1\right)\frac{\partial}{\partial\overline{z}_1} - \left(\frac{\overline{z}_1^2}{2\overline{x}_2} + \frac{\Delta\gamma}{4}\overline{z}_1\overline{u}_2\right)\frac{\partial}{\partial\overline{x}_2}. \tag{6.26}$$

Ensuite nous choisissons le système des opérateurs de base ci-dessous

$$X_{11} = \frac{\partial}{\partial \bar{x}_1}, \quad X_{21} = \frac{\partial}{\partial \bar{x}_2}, \quad X_{31} = \frac{\partial}{\partial \bar{z}_1}, \quad X_{41} = \frac{\partial}{\partial \bar{z}_2},$$

$$X_{51} = \frac{\partial}{\partial \bar{u}_1}, \quad X_{61} = \frac{\partial}{\partial \bar{u}_2}, \quad X_{12} = \varphi\frac{\partial}{\partial \bar{x}_1}, \quad X_{22} = \varphi\frac{\partial}{\partial \bar{x}_2}, \tag{6.27}$$

$$X_{32} = \varphi\frac{\partial}{\partial \bar{z}_1}, \quad X_{42} = \varphi\frac{\partial}{\partial \bar{z}_2}, \quad X_{52} = \varphi\frac{\partial}{\partial \bar{u}_1}, \quad X_{2} = \varphi\frac{\partial}{\partial \bar{u}_2},$$

où

$$\varphi = \frac{\ell n \bar{z}_2}{\bar{u}_2}$$

Nous cherchons à trouver l'opérateur S_1 sous la forme

$$S_1 = \sum_{\sigma=1}^{6} \gamma_{\sigma 1} X_{\sigma 1} + \sum_{\sigma=1}^{6} \gamma_{\sigma 2} X_{\sigma 2} \tag{6.28}$$

En faisant la substitution de (6.26), (6.27) et (6.28) dans le système d'équations (6.25) nous obtenons

$$\overline{Y}_{io}]\left(\sum_{\sigma=1}^{6} \gamma_{\sigma 1} X_{\sigma 1} + \sum_{\sigma=1}^{6} \gamma_{\sigma 2} X_{\sigma 2}\right) = \sum_{\sigma=1}^{6} a_{\sigma i} X_{\sigma i}, \tag{6.29}$$

où

$$a_{11} = 0, \quad a_{21} = 0$$

$$a_{31} = -\left(\frac{\bar{z}_2^2}{2\bar{x}_2} + \frac{\Delta\gamma}{4}\bar{z}_2\bar{u}_1\right), \quad a_{41} = -\left(\frac{\bar{z}_1^2}{2\bar{x}_2} + \frac{\Delta\gamma}{4}\bar{z}_1\bar{u}_2\right),$$

$$a_{51} = 0, \quad a_{61} = 0, \quad a_{\sigma 2} = 0 \quad (\sigma = \overline{1,6})$$

Par des calculs usuels le système (6.29) se change en le système généralisé de Jacobi. Après cela on peut montres sa résolubilité si l'on profite de la théorie de E. Cartan. En trouvant sa résolution

$$\gamma_{\sigma i} = \gamma_{\sigma i}(\bar{x}_1, \bar{x}_2, \bar{z}_1, \bar{z}_2, \bar{u}_1, \bar{u}_2) \quad (\sigma, i = \overline{1,2}) \tag{6.30}$$

nous obtenons aussi une expression explicite pour le changement (6.19) à l'ordre ε près,

$$\eta = [1 - \varepsilon \sum_{\sigma=1}^{6} (\gamma_{\sigma 1}(\bar{\eta}) X_{\sigma 1} + \gamma_{\sigma 2}(\bar{\eta}) X_{\sigma 2})]\bar{\eta}, \tag{6.31}$$

où $\eta = (\eta_1, \eta_2, \ldots, \eta_6)$ sont des variables de (6.18).

Ensuite les opérateurs (6.22) sont sous la forme

$$\begin{aligned} \bar{Y}_1 &= \frac{\partial}{\partial \bar{x}_1} - [\bar{u}_1 \bar{z}_1 + \varepsilon(\frac{\bar{z}_1^2}{2\bar{x}_2} + \frac{\Delta\gamma}{4}\bar{z}_1\bar{u}_1)]\frac{\partial}{\partial \bar{z}_1} - \\ &[\bar{u}_2\bar{z}_2 + \varepsilon(\frac{\bar{z}_2^2}{2\bar{x}_2} + \frac{\Delta\gamma}{4}\bar{z}_2\bar{u}_2)]\frac{\partial}{\partial \bar{z}_2}, \\ \bar{Y}_2 &= \frac{\partial}{\partial \bar{x}_2} + \bar{u}_1\frac{\partial}{\partial \bar{z}_1} + \bar{u}_2\frac{\partial}{\partial \bar{z}_2}. \end{aligned} \tag{6.32}$$

Le système transformé aux différentielles totales, correspondant à (6.14), est exprimé de la manière suivante

$$\begin{aligned} d\bar{z}_1 &= [- \bar{u}_1\bar{z}_1 - \varepsilon(\frac{\bar{z}_1^2}{2\bar{x}_2} + \frac{\Delta\gamma}{4}\bar{z}_1\bar{u}_1)]d\bar{x}_1 + \bar{u}_1 d\bar{x}_2, \\ d\bar{z}_2 &= [- \bar{u}_2\bar{z}_2 - \varepsilon(\frac{\bar{z}_2^2}{2\bar{x}_2} + \frac{\Delta\gamma}{4}\bar{z}_2\bar{u}_2)]d\bar{x}_1 + \bar{u}_2 d\bar{x}_2 \end{aligned} \tag{6.33}$$

à l'ordre ε près.

Il est clair d'après les équations (6.33) que le système de départ aux différentielles totales est décomposé en approximation nulle. Il en résulte deux équations aux différentielles totales, indépendantes l'une de l'autre.

Nous citons en conclusion le système décomposé en approximation nulle qui correspond à celui (6.10):

$$\begin{aligned} \frac{\partial \bar{z}_1}{\partial \bar{x}_1} + \bar{z}_1\frac{\partial \bar{z}_1}{\partial \bar{x}_2} &= - \varepsilon(\frac{\bar{z}_1^2}{2\bar{x}_2} + \frac{\Delta\gamma}{4}\bar{z}_1\frac{\partial \bar{z}_1}{\partial \bar{x}_2}), \\ \frac{\partial \bar{z}_2}{\partial \bar{x}_1} + \bar{z}_2\frac{\partial \bar{z}_2}{\partial \bar{x}_2} &= - \varepsilon(\frac{\bar{z}_2^2}{2\bar{x}_2} + \frac{\Delta\gamma}{4}\bar{z}_2\frac{\partial \bar{z}_2}{\partial \bar{x}_2}). \end{aligned} \tag{6.34}$$

En achevant ma communication je voudrais souligner que j'ai cherché à attirer l'attentions sur les possibilités non encore dévoilées complètement, de la classification des groupes, appliquée aux équations différentielles à petit paramètre (ou plus précisement, de la classification en vertu de la théorie générale des groupes locaux de Lie, de la théorie des invariantes et des multiplicités invariantes). J'ai cherché à le faire voir dans les exemples concrets. Je ne me suis pas fixé pour tâche d'exposer les endroits importants et subtils de la théorie générale des groupes de Lie, des transformations ou des algèbres de Lie d'opérateurs qui sont les points d'intersection de l'algèbre et de l'analyse mathématique. L'analyse en détail de cette théorie et son application à la classification des groupes des équations différentielles est comprise dans les travaux spéciaux, par exemple, dans la monographie fondamentale de L.V. Ovsiannikov [7].

BIBLIOGRAPHIE

1. N.N. Bogoliubov: Teorija vozmučenii v nelineinoi mehanike, Sb. In-ta stroit. mehaniki AN. USSR, 1950, I 4.
2. A.A. Lebedev, L.S. Černobrovkii: Dinamika poleta, GNT, Oborongiz, Moskva, 1962.
3. A.K. Lopatin: Asimptotičeskoe rasščeplenie sistem nelineinyh obyknovennyh differencial'nyh uravnenii vysokoi razmernosti, Cb. Kibernetika i vyčislitel'naja tehnika, vyp. 39, Naukova Dumka, Kiev, 1977.
4. Ju. A. Mitropol'skii: Metod usrednenija v nelineinoi mehanike, Naukova Dumka, Kiev, 1971.
5. Ju.A. Mitropol'skii, A.K. Lopatin: O preobrazovanii, sistem nelineinyh differencial'nyh uravnenii k normal'noi forme, Respubl. mežvedomstvennyi sb. Matematičeskaja Fizika, vyp. 4, Naukova Dumka, Kiev, 1973.
6. Ju.A. Mitropol'skii, A.K. Lopatin: Asimptotičeskoe rasščeplenie sistem differencial'nyh uravnenii, In-t matematiki, preprint, 79.11, Kiev, 1979.
7. L.V. Ovsjannikov: Gruppovoi analiz differencial'nyh uravnenii, Moskva, Nauka, 1978.
8. A. Povzner: Linear Methods in Problems of Non-linear Differential Equations with Small Parameters, International Journal of Nonlin. Mehan., 9 (4), 1974.

9. P.K. Raševskiĭ : Geometričeskaja teorija uravneniĭ s častnymi proizvodnymi, OGIZ, Gostehizdat, Moskva, 1947.

10. B.L. Roždestvenskiĭ , N.N. Janenko: Sistemy kvazilineĭnyh uravneniĭ, Nauka, Moskva, 1978.

11. G. Hayashi: Nonlinear Oscillations in Physical Systems, New York, Mc-Graw Hill, 1964.

12. A.M. Fedorčenko: Metod kanoničeskogo usrednenija b teorii nelineĭnyh kolebaniĭ, Ukr. Mat. žurn., 1957, 9, b. 2.

13. A.A. Kamel: Perturbation Method in the Theory of Non-linear Oscillations, Celestial Mech., 1970, n. 3.

14. J.E. Campbell: Introductory Treatise on Lie's Theory of Finite Continuous Transformation Groups, Oxford, Clarendon Press, 1903.

BIFURCATION OF PERIODIC SOLUTIONS FOR SOME SYSTEMS WITH PERIODIC COEFFICIENTS

P. de Mottoni

Istituto per le Applicazioni del Calcolo "M. Picone"
Consiglio Nazionale delle Ricerche, Roma.

A. Schiaffino

Istituto Matematico "G. Castelnuovo"
Università di Roma

1. INTRODUCTION

The purpose of this note is to present some results on equations and systems with periodic coefficients, concerning particularly the existence and stability of periodic solutions.

One motivation for this investigation is the mathematical modelling of the time course of biological phenomena in a (time) periodic environment; thus we shall focus on the time-periodic version of some equations of population dynamics. We shall first deal with a generalization of the familiar logistic equation (Sect. 2), then briefly outline a class of systems to which essentially the same kind of analysis can be carried over (Sect. 3). In these instances, the behavior of the time-periodic systems closely parallels that of the corresponding autonomous ones. But we shall as well present a system (Sect. 4) where

ISBN 0-12-508780-2

this is not the case, since a time periodic solution arises while the corresponding autonomous time-averaged system forbids the existence of an analogous equilibrium solution. This occurs for the two - species Volterra competition system, and the onset of "unexpected" periodic solutions proves that the so - called principle of competitive exclusion, valid in the autonomous case, in violated if the environment is periodic.

2. THE GENERALIZED LOGISTIC

Consider the equation

$$\dot{u} = f(t,u)u \tag{2.1}$$

where f is a smooth function on $\mathbb{R} \times \mathbb{R}_+ \cup \{0\}$, 1-periodic in t, satisfying

f - 1) there is $M > 0$ so that $u \geq M$ implies $f(t,u) \geq 0$ for all t;

f - 2) $f(t,u)$ is decreasing in u, $u > 0$, for all t.

Notice that the right-hand side of (2.1) does not need to be concave, as sometimes assumed in the literature. Our basic result, which extends those of e.g. Cushing [2,3] is

Theorem 2.1. Let $\varepsilon = \int_0^1 f(t,0)dt$. Then if $\varepsilon \leq 0$, no positive periodic solution exists; every nonnegative solution of (1.1) converges to zero as t goes to infinity. On the other hand, if $\varepsilon > 0$, a unique positive 1-periodic solution of (2.1) exists, $u^*(t)$. This solution is globally asymptotically stable with respect to positive solutions, namely any positive solution of (2.1) approaches, as $t \to \infty$, the solution $u^*(t)$.

The basic tool for proving this theorem, as well as the results in the following, is the "period map" T , i.e., the operator which maps a solution at time t into the same solution at time t + 1 [5,8-10]. This map is well-defined on nonnegative solutions of (2.1) since, as easily verified, the (local) flow defined by (2.1) leaves the set of nonnegative solutions invariant, and, in view of f - 1), it is on then globally defined (for details, see e.g. [8]).

Proof of Theorem 2.1, case $\varepsilon > 0$: Given $u_o > 0$, denote by $u(t,u_o)$ the solution of (2.1) which is equal to u_o at $t = 0$: $u(0,u_o) = u_o$. We then have:

$$Tu_o = u(1,u_o) = \exp\left(\int_0^1 f(t,u(t,u_o)dt)u_o = \exp\left(\int_0^1 f(t,0)dt + \int_0^1 f(t,u(t,u_o)) - f(t,0))dt\right)\right.$$

Since $u_o > 0$ implies $u(t;u_o) > 0$ for all $t \geq 0$, and f is decreasing in u for $u > 0$, the second integral above is negative, so that

$$Tu_o < \exp(\varepsilon)u_o \leq u_o$$

Thus $T^n u_o$ is a decreasing sequence, and it is easily seen that its limit is exactly zero. The claim then follows by standard arguments.

To deal with the case $\varepsilon > 0$, we need the notion of 1-upper and 1-lower solution: a (smooth) function $w_u(t)$, $t \in [0,1]$ is an 1-upper solution for (2.1) if

$$\dot{w}_u \geq f(t,w_u)w_u \qquad t \in [0,1]$$

$$w_u(1) \leq w_u(0);$$

1-lower solutions are defined similarly, by just reversing the inequality signs.

The following result is very easy to prove [5,9]:

<u>Lemma 2.2</u>. Suppose w_u, w_ℓ are 1-upper, respectively 1-lower solutions of (1.1), satisfying $w_u(t) \geq w_\ell(t)$, $t \in [0,1]$. Then there is a minimal and a maximal 1-periodic solutions $\underline{u}(t)$, $\bar{u}(t)$: $w_\ell(t) \leq \underline{u}(t) \leq \bar{u}(t) \leq w_u(t)$. Any solution of (2.1) with initial datum $u_o \in [w_\ell(0),\underline{u}(0)]$ approaches, as $t \to \infty$, the solution $\underline{u}(t)$, and any solution with initial datum $u_o \in [\bar{u}(0),w_u(0)]$ approaches, as $t \to \infty$, $\bar{u}(t)$.

Proof of Theorem 1.1, case $\varepsilon > 0$: We first establish existence using the above Lemma. To this end, notice that any constant $N \geq M$ (see Assumption f - 1)) is an 1-upper solution. To find a 1-lower solution, pick $\sigma > 0$, and define $u_\sigma(t) = u(t,\sigma)$: then

$$u_\sigma(1) = T(\sigma) = T'(0) + o(\sigma) = (\exp\int_0^1 f(s,0)ds)\sigma + o(\sigma) > \sigma + o(\sigma),$$

so that, for σ small, $u_\sigma(1) > u_\sigma(0)$, which proves that $u_\sigma(t)$ is a lower solution, provided σ is small enough. On the strength of the Lemma 2.2, the existence is proved. To prove the uniqueness, which entails, by the last claim of the same Lemma, the global asymptotic stability, we simply need to show that if u_1, u_2 are two 1-periodic solutions with $u_1 \geq u_2 > 0$, then $u_1 \equiv u_2$. To this end, write $v(t) = u_1(t) - u_2(t) \geq 0$. Suppose $v > 0$: then v satisfies

$$\dot{v}(t) = [f(t,u_1) - f(t,u_2)]u_1 + f(t,u_2)v$$

Since $f(t,u_1) < f(t,u_2)$, we conclude

$$v(t) < w(t),$$

where

$$\dot{w} = f(t,u_2)w$$

$$w(0) = v(0) > 0.$$

Since $t \to f(t,u_2(t))$ is 1-periodic, and $f(t,u_2(t)) = \dot{u}_2/u_2$ has mean zero, w is 1-periodic ad well. Thus $v(1) < v(0)$, which contradicts the fact that v, as difference of two 1-periodic solutions, is 1-periodic, thus completing the proof.

3. EXTENSION TO A CLASS OF SYSTEMS

The ideas underlying the previous section carry over to systems which share with the generalized logistic the following properties

i) The flows generated by the system and by its linearization on any positive solution leave invariant the positive cone K.

ii) The flow generated by the system leaves invariant a bounded set within K.

iii) The uniqueness of the 1-periodic solution in the interior of the cone K is guaranteed.

More specifically, we shall consider systems of the form

$$\dot{y} = F(t,y)y \qquad y \in K = [0,\infty)^n, \tag{3.1}$$

$F(t,y)$ being a $n \times n$ matrix, depending smoothly on $t \in \mathbb{R}$, $y \in K$, 1-periodic in T. Denoting by $F_y(t,y)\cdot y$ the matrix with entries

$$(F_y(t,y)\cdot y)_{ij} = \sum_{k=1}^{n} \cdot[\frac{\partial}{\partial y_j} F_{ik}(t,y)]y_k,$$

we shall require

F - 0) The matrices $F(t,y)$, $F_y(t,y)\cdot y$ have positive off-diagonal entries.

F - 1) There is an $y_M \in \overset{\circ}{K}$ such that $F(t,y) \leq 0$ elementwise whenever $y - y_M \in K$, for all $t \in [0,1]$.

F - 2) $F(t,y_1) - F(t,y_2) > 0$ (elementwise) whenever $y_2 - y_1 \in \overset{\circ}{K}$, $t \in [0,1]$.

<u>Remark</u>: With respect to the logistic, the quasi-monotonicity assumption F - 0) [6] is of course new, while F - 1), F - 2) parallel f - 1), f - 2); the items F - 0), F - 1), F - 2) ensure that i), ii), iii) are fulfilled. Note that concavity of the r.h.s. of (3.1) is not required. Observe further

that F - 2) implies

F - 2') $F_y(t,y)\cdot y > 0$ elementwise, whenever $y \in \overset{o}{K}$, $t \in [0,1]$.

The results concerning (3.1) are similar to those regarding (2.1), since the existence and uniqueness of a (globally asymptotically stable) nontrivial 1-periodic solution can be established by looking at the stability character of 0 relative to the linear system $\dot{w} = F(t)w$. On the other hand, the latter property may be difficult to ascertain, since for linear systems with periodic coefficients no necessary and sufficient conditions, in general, are known for establishing wheter 0 is stable or not.

Our main result is

Theorem 3.1. a) Suppose all solutions of the linear differential system

$$\begin{aligned} &\dot{w} = F(t,0)w \\ &w(0) \in K\setminus\{0\} \end{aligned} \tag{3.2}$$

satisfy

$$|w(1)| \leq |w(0)| \tag{3.3}$$

Then all solutions of (3.1) with initial data in K approach 0 as $t \to \infty$. In particular, no periodic solution exists in K except 0.

b) Suppose there is a solution w of (3.2) satisfying

$$w(1) - w(0) \in K\setminus\{0\}. \tag{3.4}$$

Then there is a unique 1-periodic solution $u(t) \in \overset{o}{K}$.

Moreover, any solution $y(t)$ of (3.1) approaches $u(t)$, as $t \to \infty$, provided the initial datum $y_o = y(0)$ satisfies

$$y_o - \sigma w(0) \in K, \text{ for } s \in (0,\sigma_o),\ \sigma_o \text{ some positive number,}$$

$$\text{if } w(1) - w(0) \in \overset{o}{K},$$

or

$$y_o > 0, \text{ if } w(1) - w(0) \in \partial K \setminus \{0\}.$$

The proof involves 1-upper and 1-lower solutions, and is similar to that of Theorem 2.1. Let us sketch, for the sake of brevity, only the proof of uniqueness. Suppose u_1, u_2 are 1-perioric solutions of (3.1) lying in $\overset{o}{K}$, and satisfying $u_1(0) - u_2(0) \in K$. Then the difference $x(t) = u_1(t) - u_2(t)$ obeys the equations

$$\dot{x} = [F(t,u_1) - F(t,u_2)]u_1 + F(t,u_2)x \tag{3.5}$$

Since $x(0) \in K$, F - 0) implies $x(t) \in \overset{o}{K}$ for all $t > 0$, whence $(F(t,u_1) - F(t,u_2)\ z \in \overset{o}{K}$ for $z \in \overset{o}{K}$, $t > 0$, too. By comparison, we infer

$$z(t) - x(t) \in \overset{o}{K}, \quad t > 0 \tag{3.6}$$

where

$$\dot{z}(t) = F(t,y_2)z. \tag{3.7}$$

Consider now the period map T_2 relative to the linear equa tion (3.7). Of course, $T_2y_2 = y_2$, since $y_2 \in \overset{o}{K}$ ia a 1-periodic solution of (3.1). By the Perron-Frobenius theorem it therefore follows $spr(T_2) = 1$. Denote next by $\tilde{T}$ the period map relative to the (linear inhomogeneous) equation (3.5). Then, (3.6) implies

$$spr(\tilde{T}) < 1,$$

so that, for n large enough, $|x(n)| < |x(1)|$, which contradicts the fact that x, as difference of two 1-periodic solutions, should be 1-periodic. This completes the proof.

Remark: Examples of systems of this form are the so-called positive feedback systems, studied, in the absence of periodicity, in [7] and references therein. In case of periodic perturbations of an autonomous system of this type, an existence theorem for periodic solutions in proved in [1]. Using our results, this solution can be proved to be unique and globally asymptotically stable.

4. PERIODIC COMPETITION SYSTEMS

In this section we shall present some results on a simple system, which does not fall within the class explored above:

$$\begin{aligned} \dot{u}_1 &= a_1 u_1 (1 - u_1 - b_1 u_2) \\ \dot{u}_2 &= a_2 u_2 (1 - u_2 - b_2 u_1), \end{aligned} \tag{4.1}$$

where the coefficients are smooth positive 1-periodic functions. System (4.1) is a kind of standard form, to which more general competition systems can be reduced by an appropriate change of variables. It satisfies F - 3) and a quasi-monotonicity property holds, since $F_y \cdot y$ has both off-diagonal elements negative: so the linearized equation on any solution in the positive cone $\overset{\circ}{K}$ leaves invariant the cones $\tilde{K}$, $-\tilde{K}$, where $\tilde{K} = \{u_1, u_2 \in \mathbb{R}^2, u_1 \leq 0, u_2 \geq 0\}$.
However, F - 2) is violated, and the analysis is therefore much more involved. In the following, we state some results on (4.1) [10]. We may note that some of them parallel those valid for the corresponding autonomous system, while others markedly deviate from the autonomous case.

4.1. Any nonnegative solution of (4.1) approaches, as $t \to \infty$, some 1-periodic solution, the values of which lie between zero and one. Thus the 1-periodic solutions fully describe the asymptotics of our system: no subharmonic solutions exist.

4.2. The local asymptotic stability character of the "trivial" solutions (1,0) and (0,1) is determined by the sign of $\overline{a_i(1 - b_i)}$ (i = 1,2), where overbar denotes time average. Moreover if $\overline{a_i(1 - b_i)} > 0$ for i = 1,2 (which entails instability of both (1,0) and (0,1)) and if the 1-periodic solutions are finite in number, there is at least one of them which is nontrivial and locally asymptotically stable.

In contrast with the autonomous case, the local stability character of the trivial solutions is no longer sufficient for describing the asymptotics of all solutions; the main reason being that "new" nontrivial 1-periodic solutions may arise:

4.3. Nontrivial 1-periodic solutions may occur even when the autonomous system obtained from (4.1) by time averaging has no other stationary points than the trivial ones.

The role played by the 1-periodic solutions motivates investigating the general properties of these solutions. To this end, we use again the period map T, the fixed points of which correspond to 1-periodic solutions, and for the discrete dynamical system defined by T^n we develop a geometric theory, which sheds light on the structure of the set of all fixed points of T and of their attractivity domains. In particular,

4.4. All fixed points of T (except (0,0)) lie on a separatrix, joining (1,0) with (0,1) in a monotonically decreasing way.

4.5. No fixed point of T (except (0,0)) is a full repeller. The domain of attraction of any fixed point (except (0,0)) lies between the graphs of two monotonic (possibly coinciding) functions, vanishing at zero.

4.6. If the coefficient functions are analytic, and the trivial solutions are not degenerate, then only a finite number of 1-periodic solutions occurs.

Note, in particular that the result 4.3 above tells us that the standard tenet of competitive exclusion is violated in the case of time-periodic environment, since the two species can coexist, whilst periodically oscillating, even when, considering time averages, either of the two's should be forced to extinction. Related results were obtained by Cushing, who investigated system (4.1) from a different point of view and with different techniques [4].

Detailed proofs are contained in [10]: for the sake of illustration, let us outline the proof of result 4.1.

For any $\hat{P} \in K = \mathbb{R}^{+2}$, $\hat{P} \equiv (\hat{u}_1, \hat{u}_2)$, let us define the relative (open) quadrants

$$[\hat{P}]_1 = \{u_1, u_2 \in K, u_1 - \hat{u}_1 > 0, u_2 - \hat{u}_2 > 0\}$$
$$[\hat{P}]_2 = \{u_1, u_2 \in K, u_1 - \hat{u}_1 < 0, u_2 - \hat{u}_2 > 0\}$$
$$[\hat{P}]_3 = \{u_1, u_2 \in K, u_1 - \hat{u}_1 < 0, u_2 - \hat{u}_2 < 0\}$$
$$[\hat{P}]_4 = \{u_1, u_2 \in K, u_1 - \hat{u}_1 > 0, u_2 - \hat{u}_2 < 0\}$$

Defining $G(t,s)$ as the solution operator for (4.1), namely the operator mapping the solution at time s into the solution at time t, $t \geq s$, we prove the following

Lemma 4.7. For any $P \in \mathbb{R}^{+2}$, $i = 2,4$, $t > 0$, the following inclusion holds

$$G(t,0)(\overline{[P]}_i \setminus \{P\}) \subset [G(t,0)P]_i.$$

This is a stronger form of the already observed invariance of the cones $\tilde{K}$, $-\tilde{K}$ under the flow generated by the linearized equation.

The proof involves comparison arguments applied to the equations for $h_i(t)$ $(i = 1,2)$, where $(u_1(t), u_2(t))$ denotes the coordinate of $G(t,0)P$ and $(u_1(t) + h_1(t), u_2(t) + h_2(t))$ those of $G(t,0)Q$, Q being a point within $\overline{[P]}_i$, $Q \neq P$, for $i = 2$ or 4.

The next step is represented by

Lemma 4.8. Let $P \in \mathbb{R}^{+2}$, and suppose $Q \in [P]_k$, with $k = 1$ (respectively, $k = 3$). Then exactly one of the following alternatives is met:

i) there is some $t' > 0$ such that $G(t',0) \in \overline{[G(t',0)P]}_j \setminus \{G(t',0)P\}$ for $j = 2$ or 4, in which case $G(t,0)Q \in [G(t,0)P]_j$ for all $t > t'$;

ii) $G(t,0)Q \in [G(t,0)P]_k$ for all $t \geq 0$, with $k = 1$ (respectively, $k = 3$).

Proof: The implication in item i) is an obvious consequence of Lemma 4.7. Let us show that other possibilities are ruled out: suppose, for definiteness, $k = 1$, and assume that, for $t_1 > 0$

$$G(t_1,0)Q \in [G(t_1,0)P]_3. \tag{4.3}$$

This implies that there is a $t'' \in (0,t_1)$ such that either $G(t'',0)Q \in [G(t'',0)P]_j$ with $j = 2$ or 4, or $G(t'',0)Q = G(t'',0)P_i$ but both these possibilities contradict (4.3), the first one because of Lemma 4.7, the second one because of the uniqueness theorem for the Cauchy problem for (4.1). This completes the proof.

To establish the claim 4.1 we proceed as follows: pick $P \in \mathbb{R}^{+2}$, and consider TP. If $TP = P$, the proof is finished. If not, $TP \in \overline{[P]_j} \setminus \{P\}$ for some $j = 1,\ldots,4$. Applying Lemmas 4.7 and 4.8, we see that there is a $\hat{j}$ and a $\hat{n}$ such that

$$T^nP \in [T^{n-1}P]_{\hat{j}} \quad \text{for all } n > \hat{n}.$$

This entails that both components of $\{T^nP\}$ form a monotone sequence. Since every solution in $\mathbb{R}^{+2}$ obeys an a-priori estimate, T^nP converges to a fixed point of T, which completes the proof.

Acknowledgements: It is a pleasure to thank Dr. V. Capasso, who provided the initial stimulus for the research described in Sect. 3.

REFERENCES

1. V. Capasso: Periodic Solutions for a System of Nonlinear Differential Equations Modelling the Evolution of Oro-faecal Disease, these Proceedings.
2. J.M. Cushing: Stable Positive Periodic Solutions of the Time-Dependent Logistic Equation under Possible Hereditary Influences, Journ. Math. Anal. Appl. 60, 747-754, (1977).

3. J.M. Cushing: Periodic Time-Dependent Predator - Prey Systems, SIAM J. Appl. Math. 32, 82-95, (1977).

4. J.M. Cushing: Two Species Competition in a Periodic Environment, Journ. Math. Biol., to appear.

5. Ju.S. Kolesov: Periodic Solutions of Quasi-linear Parabolic Equations of the Second Order, Trudy Mosk. Math. Obšč. 21, 103-133 (1970); transl. Trans. Moscow Math. Soc. 21, 114-146 (1970).

6. V. Lakshmikantham, S. Leela: Differential and Integral Inequalities, Vols. I and II, New York, Academic Press, 1969.

7. R.H. Martin: Asymptotic Stability Analysis and Critical Points for Nonlinear Quasimonotone Systems, Journ. Diff. Equations 30, 391-423 (1978).

8. P. de Mottoni, A. Schiaffino: On Logistic Equation with Time Periodic Coefficients, Pubbl. IAC (Roma), n. 192, (1979).

9. P. de Mottoni, A. Schiaffino: Bifurcation Results for a Class of Periodic Quali-linear Parabolic Equations, Math. Meth. in the Appl. Sci., to appear.

10. P. de Mottoni, A. Schiaffino: Competition Systems with Periodic Coefficients: a Geometric Approach, Journ. Math. Biol., to appear.

GLOBAL ATTRACTIVITY FOR DIFFUSION DELAY LOGISTIC EQUATIONS

A. Schiaffino

Istituto Matematico "G. Castelnuovo"
Università di Roma
Roma, Italy

A. Tesei

Istituto per le Applicazioni del Calcolo "M. Picone"
Consiglio Nazionale delle Ricerche
Roma, Italy

1. INTRODUCTION

We want to study the following integro-partial differential equation of parabolic type:

$$\partial_t u(t,x) = \Delta u(t,x) + \lambda u(t,x) - bu^2(t,x) - u(t,x)\int_{-\infty}^{t} ds\, k(t-s)u(s,x) \quad (t > 0,\ x \in \Omega), \tag{0}$$

supplemented with initial and Dirichlet homogeneous boundary conditions:

$$u(t,x) = u^\circ(t,x) \quad (t \leq 0,\ x \in \Omega), \tag{0_I}$$

$$u(t,x) = 0 \quad (t > 0,\ x \in \partial\Omega); \tag{0_B}$$

ISBN 0-12-508780-2

here $\Omega \subset R^d$ is an open bounded domain with smooth boundary $\partial\Omega$; $\lambda \in R$, b is a real positive constant; k, u^o are given nonnegative functions. The equation (0) has to be regarded as representative of a wider class of integro-differential equations to which the following considerations apply; in particular, space dependence in the coefficients of (0) and more general (homogeneous) boundary conditions can be dealt with similarity. The motivation comes from population theory: in fact, (0) generalizes the well known logistic equation so as to include past hystory and space diffusion effects. Let us consider in a heuristic way the space-clamp problem associated with (0), namely the initial value problem for the classical Volterra population equation:

$$\begin{cases} u'(t) = \mu u(t) - bu^2(t) - u(t)\int_{\infty}^{t} ds\ k(t-s)u(s) & (t > 0) \\ u(t) = u^o(t) & (t \leq 0) \end{cases} \qquad \text{(SC)}$$

where $\mu \in R$. Let $\mu > 0$ and $k \in L^1(0,+\infty)$; then the following results hold:

- if $b > \hat{k} := \int_0^{+\infty} dt\ k(t)$, the nontrivial equilibrium $u^* := \mu/(b + \hat{k})$ is globally attractive with respect to nonnegative solutions of (SC) [1];
- if the equation $z + bu^* + u^*\tilde{k}(z) = 0$ ($\tilde{k}$ denoting the Laplace transform of k) doesn't have any root in the closed complex half-plane Re $z \geq 0$, u^* is asymptotically stable; if there exists a root in the open half-plane Re $z > 0$, u^* is unstable [2].

The proof of the first result is fairly involved and requires showing that u^* is the only solution of the associated limiting equation which is bounded on R (observe that the limiting equation is nothing but the equation itself, due to the translation invariance). As for the second result, the proof relies on a classical Paley-Wiener result for the linear renewal equation, via a standard linearization argument.

2. STATEMENT OF THE RESULTS

Let X denote the Banach space $C_o(\bar{\Omega})$ of continuous functions on $\bar{\Omega}$ which vanish on $\partial\Omega$, endowed with the supremum norm $(|\cdot|_X)$; let us define

$$\begin{cases} D(A) = \{u \in X: \Delta u \in X\} \\ Au = -\Delta u \quad (u \in D(A)). \end{cases}$$

Then the problem $(0)-(0_B)$ can be rewritten as the following abstract initial value problem for maps taking values in X:

$$\begin{cases} u'(t) = -Au(t) + \lambda u(t) - bu^2(t) - u(t)\int_{-\infty}^{t} ds\, k(t-s)u(s) \\ \qquad\qquad (t > 0) \\ u = u^o \qquad (t \leq 0). \end{cases} \tag{P}$$

For any nonnegative u^o in the space $BC(-\infty,0;X)$ of continuous bounded maps from $(-\infty,0]$ to X, there exists a unique nonnegative global solution of (P). If $\lambda > \lambda_o$ (λ_o denoting the principal eigenvalue of $-A$), there exists a unique nontrivial equilibrium solution u^* of problem (P), which is strictly positive in Ω [3], namely:

$$-Au^* + \lambda u^* - (b + \hat{k})u^{*2} = 0. \tag{E}$$

We can now state the main result we whish to prove.

Theorem 1 [6]. Let $k \in L^1(0,+\infty)$, $b > \hat{k}$ and $\lambda > \lambda_o/(1-\hat{k}/b)$. Then u^* is globally attractive in the X-norm with respect to nonnegative solutions of (P).

A similar result in the case of Neumann homogeneous boundary conditions has been proven by Schiaffino [4]; clearly, Theorem 1 extends the results of [1] to the present situation. For the sake of completeness, let us state in an informal way the following result, which parallels those of [2].

Theorem 2 [5]. Let $d \leq 3$ and $\lambda > \lambda_o$; moreover, let $k \in L^1(0,+\infty)$ and satisfy suitable decrease properties at $t = +\infty$. If $[zI + A_1 + u^*\tilde{k}(z)]^{-1}$, $A_1 := A + (2b + \hat{k})u^*I$, exists as a bounded operator on X for any z, Re $z \geq 0$, then u^* is X-asymptotically stable; if the above condition is violated for some complex z, Re $z > 0$, u^* is X-unstable.

The analogous result for the Neumann case [6] can be more easily proven due to the commutativity of the terms in the square brackets (a special instance of the widely accepted tenet, according to which "Dirichlet is more difficult than Neumann as for stability problems"). Due to noncommutativity and the lack of Plancherel's theorem in the present infinite-dimensional frame, use has to be made of a projection procedure onto invariant subspaces and Sobolev's embedding properties.

3. PROOF OF THEOREM 1

Monotone methods are well suited to prove attractivity results [3]; unfortunately, however, they can't be applied to problem (P) due to the presence of the delay term. We shall see that this difficulty can be removed so as to make use of monotone methods in an indirect way. The method of the proof seems to be interesting in itself and applicable to a class of delay problems, including e.g. competition models; the underlying philosophy is that global attractivity properties are preserved, provided "the delay is not too large". We shall make use, in the following, of the following hypothesis:

$$k \in L^1(0,+\infty),\ b > \hat{k} \text{ and } \lambda > \lambda_o/(1 - \hat{k}/b). \tag{H}$$

Let us introduce the family of coupled problems:

$$\begin{cases} -A\bar{u}_{n+1} + [\lambda - \hat{k}\bar{v}_n]\bar{u}_{n+1} - b\bar{u}_{n+1}^2 = 0 \quad (n=0,1,2,\ldots;\bar{v}_o := 0) & (EP_n) \\ -A\bar{v}_n + [\lambda - \hat{k}\bar{u}_n]\bar{v}_n - b\bar{v}_n^2 = 0 \quad (n=1,2,\ldots). & (EQ_n) \end{cases}$$

The following propositions, concerning nontrivial solutions of (EP_n), (EQ_n), can be proven.

Proposition 1. Let (H) hold. Then $\{\bar{u}_n\}$ is nonincreasing, $\{\bar{v}_n\}$ is nondecreasing and $\bar{u}_n \geq u^* \geq \bar{v}_n$ for any $n > 0$.

Proposition 2. Let (H) hold. Then $\bar{u}_n$ in globally attractive in the X-norm with respect to nonnegative solutions of the parabolic equation

$$u_n'(t) = -Au_n(t) + [\lambda - \hat{k}\bar{v}_{n-1}]u_n(t) - bu_n^2(t) \quad (t > 0,\ n > 0);$$

a similar result holds for $\bar{v}_n$.

As a consequence of Propositions 1 and 2 we have the following result.

Proposition 3. Let (H) hold. Then for any $u^0 \geq 0$ $(u^0 \not\equiv 0)$, for any $\varepsilon > 0$ and $n > 0$, there exists $T_{\varepsilon,n}$ such that, for any $t > T_{\varepsilon,n}$ and $x \in \bar{\Omega}$

$$\bar{v}_n(x) - \varepsilon < u(t,x) < \bar{u}_n(x) + \varepsilon,$$

$u(t,\cdot)$ denoting the solution of (P) with initial data $u^0(\cdot)$.

Let $\bar{u} := \lim \bar{u}_n$, $\bar{v} := \lim \bar{v}_n$ (uniform by Dini's theorem). As one should hope, we have the following result.

Proposition 4. Let (H) hold. Then $\bar{u} = \bar{v} = u^*$.

Proof of Theorem 1: If follows from Propositions 3 and 4 by an $\varepsilon/3$ argument.

Proof of Proposition 1: Observe that the equation $-A\bar{u}_1 + \lambda\bar{u}_1 - b\bar{u}_1^2 = 0$ admits a unique solution $\bar{u}_1 > 0$ such that $|\bar{u}_1|_X \leq \lambda/b$. Then the equation (EQ_1) admits a unique solution $\bar{v}_1 > 0$ because of (H). As

$$-A\bar{u}_{n+1} + [\lambda - \hat{k}\bar{v}_{n+1}]\bar{u}_{n+1} - b\bar{u}_{n+1}^2 = \hat{k}[\bar{v}_n - \bar{v}_{n+1}]\bar{u}_{n+1},$$

$$-A\bar{v}_n + [\lambda - \hat{k}\bar{u}_{n+1}]\bar{v}_n - b\bar{v}_n^2 = \hat{k}[\bar{u}_n - \bar{u}_{n+1}]\bar{v}_n,$$

by the uniqueness of the (nontrivial) solution to (EP_n), (EQ_n) it follows that $\bar{v}_1 > 0$; then $(\bar{u}_1$ is an upper solution for

(EP_1), hence) $\bar{u}_1 \geq \bar{u}_2$, which implies ($\bar{v}_1$ to be a lower solution for (EQ_1), hence) $\bar{v}_1 \leq \bar{v}_2$ and so on. The other claims follow from the equalities

$$-Au^* + [\lambda - \hat{k}\bar{v}_n]u^* - bu^{*2} = \hat{k}[u^* - \bar{v}_n]u^* \quad (n \geq 0),$$

$$-Au^* + [\lambda - \hat{k}\bar{u}_n]u^* - bu^{*2} = \hat{k}[u^* - \bar{u}_n]u^* \quad (n \geq 1);$$

again $u^* > 0$ implies $u^* \leq \bar{u}_1$, thus $u^* \geq \bar{v}_1$ and so on.

Proof of Proposition 2: It follows from (H) and the uniqueness of the solution to (EP_n), (EQ_n) by standard comparison arguments [3].

Proof of Proposition 3: As $u(t) \geq 0$ for any $t \geq 0$, we have

$$u'(t) \leq -Au(t) + \lambda u(t) - bu^2(t),$$

thus $u(t) \leq u_1(t)$ $(t \geq 0)$, where

$$u_1(0) = u^o(0), \quad u_1'(t) = -Au_1(t) + \lambda u_1(t) - bu_1^2(t) \quad (t \geq 0).$$

By Proposition 2 $|u_1(t) - \bar{u}_1|_X \to 0$ as t goes to infinity, thus for any $\varepsilon > 0$ there exists $\tilde{T}_{\varepsilon,1}$ such that for any $t > \tilde{T}_{\varepsilon,1}$ and $x \in \bar{\Omega}$, $u(t,x) < \bar{u}_1(x) + \varepsilon/2$. Then it is easily checked that

$$\int_{-\infty}^{t} ds\, k(t - s)u(s) \leq f(t) + \hat{k}[u_1 + \varepsilon/2],$$

where $|f(t)|_X$ goes to zero as t diverges; as a consequence

$$u'(t) \geq -Au(t) + [\lambda - \hat{k}(\bar{u}_1 + \varepsilon)]u(t) - bu^2(t)$$

for any $t > \tilde{T}_{\varepsilon,1}$. Hence $u(t) \geq v_1(t)$, where

$$v_1(T_{\varepsilon,1}) = u(T_{\varepsilon,1}), \quad v_1'(t) = -Av_1(t) + [\lambda - \hat{k}(\bar{u}_1 + \varepsilon)]v_1(t) - bv_1^2(t) \quad (t > T_{\varepsilon,1});$$

here $T_{\varepsilon,1}$ is a suitable constant greater then $\tilde{T}_{\varepsilon,1}$. Using again

Proposition 2 we prove the claim for n-1 and, by iteration, for any $n \geq 1$.

Proof of Proposition 4. Follows from the equality $\int_\Omega \bar{u}\bar{v}(\bar{u} - \bar{v}) = 0$ as $\bar{u} \geq \bar{v}$.

REFERENCES

1. R.K. Miller: On Volterra's Population Equation, S.I.A.M. J. Appl. Math., 14, 446-452 (1966).
2. R.K. Miller: Asymptotic Stability and Perturbation for Linear Volterra Integrodifferential Systems. In: Delay and Functional Differential Equations and Their Applications, Ed. by K. Schmitt, New York - London, Academic Press, (1972).
3. D.H. Sattinger: Monotone Methods in Nonlinear Elliptic and Parabolic Boundary Value Problems, Indiana Univ. Math. J., 21, 979-1000 (1972).
4. A. Schiaffino: On a Diffusion Volterra Equation, Nonl. Anal.: T.M.A., 3, 595-600 (1979).
5. A. Schiaffino, A. Tesei: Asymptotic Stability Properties for Nonlinear Diffusion Volterra Equations, Rend. Acc. Naz. Lincei, Serie VIII, 67, 227-232 (1979).
6. A. Schiaffino, A. Tesei: Monotone Methods and Attractivity Results for Volterra Integro-Partial Differential Equations, preprint (1980).
7. A. Tesei: Stability Properties for Partial Volterra Integrodifferential Equations, Ann. Mat. Pura Appl. (to appear).

ON SUITABLE SPACES FOR STUDYING FUNCTIONAL EQUATIONS USING SEMIGROUP THEORY

Rosanna Villella-Bressan

Seminario Matematico
Università di Padova, Italy

The purpose of this note is to study the functional equation $x(t) = F(x_t)$ by associating with it a semigroup of translations in the space of initial data. The space is weighted in order to make the generator m-accretive and to deduce information on existence, regularity and asymptotic behaviour of solutions. First the case that the initial data φ is continuous is considered, then the case that φ belongs to L^1. The results are applied to a nonlinear age dependent population problem.

1.

Consider the functional equation

$$x(t) = F(x_t), \quad t \geq 0, \quad x_0 = \varphi \tag{FE}$$

where $\varphi \in C = C(-r,0;X)$, X a Banach space, $0 < r < +\infty$; here $x_t \in C$ it the history of $x(t)$ at time t, $x_t(\theta) = x(t + \theta)$ [4]. Let $F\colon C \to X$ be continuous and suppose that $x(t;\varphi)$, $t \geq -r$, is the unique continuous solution of (FE). Then set

$$T(t)\varphi = x_t(\cdot;\varphi), \quad t \geq 0$$

ISBN 0-12-508780-2

$T(t)$ is a semigroup in C whose generator is the operator

$$-A\varphi = \varphi', \quad D_A = \{\varphi \in C^1;\ \lim_{t\to 0} \frac{F(x_t) - \varphi(0)}{t} \text{ exists and}$$
$$\varphi'(0) = \lim_{t\to 0} \frac{F(x_t) - \varphi(0)}{t}\} \tag{1}$$

We consider instead of the operator (1), whose definition involves the solution $x(t;\varphi)$, the following

$$A\varphi = -\varphi', \quad D_A = \{\varphi \in C^1;\quad \varphi(0) = F(\varphi)\} \tag{2}$$

which extends (1). In order to use nonlinear semigroup theory to get information on the solutions of (FE) we look for conditions on F which make A m-accretive. It $F(\varphi) = H(\varphi(-r))$, $H: X \to X$, then it can be proved that A is m-accretive if and only if H is Lipschitz continuous with constant 1 [6]. So it would seem natural to require: F is Lipschitz continuous with constant $\gamma \leq 1$. This condition is too restrictive; it is not satisfied even for the equation $x(t) = \alpha x(t - r)$, $\alpha > 1$. So in order to apply the results to a sufficiently wide class of equa tions we consider the weighted space C_σ, $\sigma \in \mathbb{R}$, that is the space $C(-r,0;X)$ of X-valued continuous functions on the interval $[-r,0]$ with norm $\|\varphi\|_\sigma = \sup_{-r\leq\theta\leq 0} e^{-\sigma\theta}|\varphi(\theta)|$, where $|\cdot|$ denote the norm in X. Note that $\|\varphi\|_\sigma$ increases with σ; hence if $G: C \to X$ is Lipschitz continuous and $\beta_\sigma = \inf\{\gamma_\sigma,\ \gamma_\sigma$ a Lipschitz constant for $G: C_\sigma \to X\}$, then β_σ decreases with σ. In [2] it has been proved that it $F(\varphi) = H(\varphi(-r))$, then $A + \sigma I$ is m-accretive in C_σ if and only if H is Lipschitz continuous with Lipschitz constant $e^{\sigma r}$. It follows that

<u>Proposition 1</u>. Let $F(\varphi) = H(\varphi(-r)$, where $H: X \to X$. If $A + \sigma I$ is m-accretive in C_σ for some $\sigma \in \mathbb{R}$, then, for all $\omega > \sigma$, $F: C_\omega \to X$ is Lipschitz continuous with constant $\gamma_\omega < 1$.

<u>Proof</u>: Let $\omega > \sigma$ and $\varphi,\ \psi \in C$. Then

$$|F(\varphi) - F(\psi)| = |H(\varphi(-r)) - H(\psi(-r))| \leq e^{\sigma r}|\varphi(-r) - \psi(-r)|$$
$$= e^{(\sigma-\omega)r}e^{\omega r}|\varphi(-r) - \psi(-r)| \leq e^{(\sigma-\omega)r}\|\varphi - \psi\|_\omega$$

and $\gamma_\omega = e^{(\sigma-\omega)r}$ is a Lipschitz constant of $F: C_\omega \to X$ which satisfies $\gamma_\omega < 1$.

So we are led to require:

(H.1) there exists a $\omega \in \mathbb{R}$ such that $F: C_\omega \to X$ is Lipschitz continuous with constant $\gamma_\omega < 1$.

The following proposition characterizes the functions F which satisfy condition (H.1).

Proposition 2. The function $F: C \to X$ satisfies (H.1) if and only if

(H.2) there exist $\alpha \geq 0$, p, $0 < p \leq r$, β, $0 \leq \beta < 1$ such that

$$|F(\varphi) - F(\psi)| \leq \alpha \sup_{-r\leq\theta\leq-p} |\varphi(\theta) - \psi(\theta)| + \beta\|\varphi - \psi\|_0$$

for all $\varphi, \psi \in C$

Moreover $\gamma_\omega < 1$ for all $\omega > \max\{0, \frac{1}{p}\log \frac{\alpha}{1-\beta}\}$.

Proof: Let (H.1) hold. If $\omega \leq 0$, then (H.2) is satisfied with $\alpha = 0$ and $\beta = \gamma_\omega$. If $\omega > 0$, then let p, $0 < p \leq r$, be such that $\gamma_\omega e^{\omega p} < 1$. Then

$$|F(\varphi) - F(\psi)| \leq \gamma_\omega\|\varphi - \psi\|_\omega \leq \gamma_\omega e^{\omega r} \sup_{-r\leq\theta\leq-p} |\varphi(\theta) - \psi(\theta)| + \gamma_\omega e^{\omega p}\|\varphi - \psi\|_0,$$

hence (H.2) is satisfied with $\alpha = \gamma_\omega e^{\omega r}$ and $\beta = \gamma_\omega e^{\omega p}$.

Suppose now that (H.2) holds. Let $\omega > 0$, then

$$|F(\varphi) - F(\psi)| \leq \alpha e^{-\omega p} \sup_{-r\leq\theta\leq-p} e^{-\omega\theta}|\varphi(\theta) - \psi(\theta)| + \beta\|\varphi - \psi\|_\omega \leq (\alpha e^{-\omega p} + \beta)\|\varphi - \psi\|_\omega$$

and $\gamma_\omega \leq \alpha e^{-\omega p} + \beta < 1$ provided $\omega > \frac{1}{p}\log \frac{\alpha}{1-\beta}$.

The following theorem summarizes the existence and asymptotic results obtained when (H.1) is satisfied.

Theorem 1. Suppose (H.1) holds. Then $A + \omega I$ is m-accretive in C_ω and therefore generates a semigroup of type ω $T(t)$, $t \geq 0$, in C_ω; $T(t)$ is a translation semigroup in the sense that

$$(T(t)\varphi(\theta) = \varphi(t+\theta), \qquad t+\theta \le 0$$

$$= (T(t+\theta)\varphi)(0), \quad t+\theta > 0.$$

Hence set

$$x(t;\varphi) = \varphi(t), \qquad -r \le t \le 0$$

$$= (T(t)\varphi)(0), \quad t > 0,$$

$x_t(\cdot;\varphi) = T(t)\varphi$. Il follows that $x(t;\varphi)$ is a solution of (FE) for all $\varphi \in \bar{D}_A$, where

$$\bar{D}_A = \{\varphi \in C,\ \varphi(0) = F(\varphi)\},$$

and that

$$|x(t;\varphi) - x(t;\psi)| \le e^{\omega t}\|\varphi - \psi\|_\omega,\ \varphi,\ \psi \in \bar{D}_A;$$

$x(t;\varphi)$ is the unique solution of (FE).
If $\omega < 0$, then there exist a unique "equilibrium" solution $\varphi_o = \varphi_o(0)$ and φ_o is asymptotic exponentially stable, in the sense that

$$|x(t;\varphi) - \varphi_o| \le e^{\omega t}\|\varphi - \varphi_o\|_\omega$$

for all $\varphi \in \bar{D}_A$.

For the proof of existence and asymptotic results see [2]. The uniqueness follows easily from condition (H.1).

Note that if $\gamma_o < 0$, then ω can be choosen negative, for any $\omega > \frac{1}{r}\log\gamma_o$ makes $\gamma_\omega < 1$.

As $T(t)$ is a translation semigroup, the subsets of D_A

$$D_{A,\sigma} = \{\varphi \in C \text{ is Hölder continuous with exponent } \sigma;\ \varphi(0) = F(\varphi)\}$$

are flow-invariant [8]. This yields regularity results for (FE).

2.

It is obvious how to adapt the definitions and results of section 1 to the case of infinite delay. The initial data φ now belongs to the space

$$C_\omega = \{\varphi: (-\infty,0] \to X;\ \varphi(\theta)e^{-\omega\theta} \text{ is continuous and bounded}\}$$

endowed with the norm $\|\varphi\|_\omega = \sup_{\theta\in(-\infty,0]} e^{-\omega\theta}|\varphi(\theta)|$. The generator of the semigroup associated with (FE) is

$$A_\omega\varphi = -\varphi',\ D_{A_\omega} = \{\varphi \in C_\omega,\ \varphi' \in C_\omega \text{ and } \varphi(0) = F(\varphi)\}, \qquad (2)'$$

again if F satisfies (H.1) then $A_\omega + \omega I$ generates a semigroup $T_\omega(t)$ of type ω in

$$\bar{D}_{A_\omega} = \{\varphi \in C_\omega,\ e^{-\omega\theta}\varphi(\theta) \text{ is uniformly continuous and } \varphi(0) = F(\varphi)\},$$

$T_\omega(t)\varphi$ is a translation semigroup and gives the segments of solutions of (FE) for all $\varphi \in \overline{D}_{A_\omega}$. Proposition 2 now becomes:

<u>Proposition 2'</u>. Let $F: C_{\tilde{\sigma}} \to X$ be Lipschitz continuous for some $\tilde{\sigma} \in \mathbb{R}$. Then F satisfies (H.1) if and only if

(H.2)' there exist $\sigma \geq \tilde{\sigma}$, $\alpha_\sigma \geq 0$, $p > 0$ and β_σ, $0 \leq \beta_\sigma < 1$, such that

$$|F(\varphi) - F(\psi)| \leq \alpha_\sigma \sup_{\theta\leq -p} e^{-\sigma\theta}|\varphi(\theta) - \psi(\theta)| + \beta_\sigma\|\varphi - \psi\|_\sigma$$

for all $\varphi, \psi \in C_\sigma$.

Moreover $\gamma_\omega < 1$ for all $\omega > \sigma + \max\{0, \frac{1}{p}\log \frac{\alpha_\sigma}{1-\beta_\sigma}\}$.

As an example consider the Volterra equation

$$x(t) = \int_{-\infty}^{t} a(t - s)Hx(s)ds \qquad (3)$$

where $a: [0,\infty) \to \mathbb{R}$ is such that $e^{\sigma\xi}a(-\xi)\in L^1(-\infty,0;X)$ for some σ, $H: X \to X$ is Lipschitz continuous with constant $\tilde{\beta}$ and $H(0)=0$.

Here $F(\varphi) = \int_{-\infty}^{0} a(-\xi)H(\varphi(\xi))d\xi$ and (H.2)' is satisfied with $\alpha_\sigma = \tilde{\beta}\int_{-\infty}^{0} |a(-\xi)|e^{\sigma\xi}d\xi$ and $p > 0$ small enough so that $\beta_\sigma = \tilde{\beta}\int_{-p}^{0} |a(-\xi)|e^{\sigma\xi}d\xi < 1$. If $a(-\xi) = e^{\delta\xi}$ and $\delta > \tilde{\beta}$, then (H.1) is satisfied for all $0 > \omega > \tilde{\beta}-\delta$ and hence the zero solution of (3) is asymptotic exponentially stable.

It is interesting to note that Theorem 1 yields existence, uniqueness, regularity and asymptotic results for the problem

$$\lim_{h\to 0} \frac{u(t+h,a+h) - u(t,a)}{h} = 0 \qquad t > 0,\ a > 0$$

$$u(0,a) = u_o(a) \qquad a \geq 0 \tag{4}$$

$$u(t,0) = G(u(t,\cdot)) \qquad t \geq 0$$

where $u_o \in C(0,\infty;X)$, $G: C(0,\infty;X) \to X$. Define $F: C(-\infty,0;X) \to X$, $F(\varphi(\cdot)) = G(\varphi(-\cdot))$, and set $\varphi_o(\cdot) = u_o(-\cdot)$; if F satisfies (H.1), then $u(t,a) = T_\omega(t)\varphi_o(-a)$, $t \geq 0$, $a \geq 0$, is a continuous solution of (4) for all u_o such that $u_o(-\cdot) \in \bar{D}_{A_\omega}$. If $G(u) = \int_0^\infty \beta(a,u(a))da$, (4) is an age-dependent population problem ([3],[11]).

3.

Suppose now that F is continuously Frechet differentiable; then $x(t)$ is a continuously differentiable solution of (FE) if and only if it is a solution of the neutral equation

$$x'(t) = F'(x_t)x_t', \quad x_o = \varphi \tag{NFE}$$

and satisfies $x(0) = F(\varphi)$. Suppose that for all $\varphi \in C^1$ (NFE) has a unique solution $x(t;\varphi)$. Set $S(t)\varphi = x_t(\cdot;\varphi)$, $t \geq 0$, the generator of the semigroup $S(t)$ in C^1 is the operator $-B$: $D_B \subset C^1 \to C^1$

$$B\varphi = -\varphi', \; D_B = \{\varphi \in C^2, \; \varphi'(0) = F'(\varphi)\varphi'\}.$$

Following [6], we require F to satisfy

(H.3) there exist $\beta \geq 0$, $\sigma \in \mathbb{R}$, $\gamma_\sigma < 1$ such that

$$|F'(\varphi)\varphi' - F'(\psi)\psi'| < \beta \, \|\varphi - \psi\|_o + \gamma_\sigma \|\varphi' - \psi'\|_\sigma.$$

Then the following result can be proved [6]:

Theorem 2. Let F satisfy (H.3). Set $\omega = \max\{\frac{\beta}{1-\gamma_\sigma}, \sigma\}$, $B + \omega I$ is m-accretive in C^1 endowed with the norm $\|\varphi\|_{1,\omega} = \max\{\omega\|\varphi\|_o, \|\varphi'\|_o\}$ and therefore generates a semigroup, $S(t)$, in $\bar{D}_B = \{\varphi \in C^1, \; \varphi'(0) = F'(\varphi)\varphi'\}$; $S(t)$ is a translation semigroup and set

$$x(t;\varphi) = \varphi(t) \qquad t \leq 0$$
$$= (S(t)\varphi)(0) \qquad t > 0$$

$x(t;\varphi)$ is a continuously differentiable solution of (NFE) for all $\varphi \in \bar{D}_B$. Such a solution is unique.

Hence if $\varphi \in \bar{D}_A \cap \bar{D}_B$, $x(t;\varphi)$ is a continuously differentiable solution of (FE) and set $u(t,a) = S(t)\varphi(-a)$, $u(t,a)$ is a classical solution of the problem

$$\begin{aligned} & u_t(t,a) + u_a(t,a) = 0 \qquad t \geq 0, \; 0 \leq a \leq r, \\ & u(0,a) = u_o(a) \\ & u(t,0) = G(u(t,\cdot)) \end{aligned} \tag{4'}$$

where $u_o(a) = \varphi(-a)$ and $G(\varphi(-\cdot)) = F(\varphi(\cdot))$.

In some interesting cases (NFE) is in fact a functional differential equation; consider as an example the Volterra equation

$$x(t) = \int_{-\infty}^{t} g(t-\tau, x(\tau))d\tau, \quad x_0 = \varphi \in C \tag{5}$$

that is the functional equation $x(t) = F(x_t)$, where $F(\varphi) = \int_{-\infty}^{0} g(-s,\varphi(s))ds$. Suppose the nonlinear kernel $g(\xi,x)$: $[0,\infty) \times X \to X$ is such that we can differentiate (5) and obtain

$$x'(t) = g(0,x(t)) + \int_{-\infty}^{t} g_1(t-\tau,x(\tau))d\tau \quad x_0 = \varphi \tag{6}$$

where g_1 is the derivative of g with respect to the first variable; (6) is a functional differential equation of type

$$x'(t) = f(x(t)) + h(x_t)$$

where $f(x) = g(0,x)$ and $h(\varphi) = \int_{-\infty}^{0} g_1(-s,\varphi(s))ds$. As in [4] we associate with it the operator in C

$$A_1\varphi = -\varphi', \quad D_{A_1} = \{\varphi \in C^1, \ \varphi'(0) = f(\varphi(0)) + h(\varphi)\}$$

and suppose that $A_1 + \omega I$ is m-accretive in C, for some $\omega \in \mathbb{R}$, so that it generates a semigroup of translations $T_1(t)$ in $\bar{D}_{A_1}$ [5]. Let now X be a reflexive Banach space. As the set D_{1,A_1} = $\{\varphi \in C, \ \varphi$ is Lipschitz continuous$\}$ is flow-invariant [8] and $T_1(t)$ is a translation semigroup, it follows that set $u(t,a) = T_1(t)\varphi(-a)$, $u(t,a)$ satisfies the problem

$$\begin{aligned} &u_t(t,a) + u_a(t,a) = 0 && \text{a.e. } t > 0, \ a > 0 \\ &u_t(t,0) = g(0,u(t,0)) + \int_0^\infty g_1(a,u(t,a))da && t \geq 0 \\ &u(0,a) = \varphi(-a) && a \geq 0 \end{aligned}$$

for all $\varphi \in D_{1,A_1}$. If moreover φ is such that $\varphi(0) = F(\varphi)$, that is $\varphi(0) = \int_0^\infty g(a,\varphi(-a))da$, then $u(t,a)$ is a solution of the age dependent population problem

$$\begin{aligned} & u_t(t,a) + u_a(t,a) = 0 && \text{a.e. } t > 0,\ a > 0 \\ & u(t,0) = \int_0^\infty g(a,u(t,a))da && t \geq 0 \\ & u(0,a) = \varphi(-a) && a \geq 0. \end{aligned} \tag{7}$$

4.

Consider again (FE), where $F: C \to X$ is continuous and $\varphi \in C(-r,0;X)$. If we want the solution $x(t;\varphi)$, $t \geq -r$, be continuous at $t = 0$ we must require $F(\varphi) = \varphi(0)$, that is the initial data φ must belong to $\bar{D}_A$. Moreover if $\varphi, \psi \in \bar{D}_A$, then

$$|F(\varphi) - F(\psi)| = |\varphi(0) - \psi(0)| = e^{-\sigma 0}|\varphi(0) - \psi(0)| \leq \|\varphi - \psi\|_\sigma,$$

and $F_{|\bar{D}_A}: \bar{D}_A \subset C_\sigma \to X$ must be Lipschitz continuous with constant $\gamma_\sigma \leq 1$ for all $\sigma \in \mathbb{R}$.

Hence, in order to relax condition (H.1) we require continuity of solutions at $t = 0$ only from the right and set (FE) in the space $L^1 = L^1(-\infty,0;\mu;X)$, where $d\mu(\theta) = e^{-\omega\theta}d\theta$ and X a Hilbert space. Again we consider the particular case that (FE) is the Volterra equation

$$x(t) = \int_{-\infty}^t g(t-s,x(s))ds \qquad x_0 = \varphi \in L^1, \tag{5'}$$

suppose we can differentiate and obtain the functional differential equation

$$x'(t) = f(x(t)) + h(x_t), \quad x_0 = \varphi \tag{8}$$

where $f(x) = g(0,x)$, $h(\varphi) = \int_{-\infty}^0 g_1(-s,\varphi(s))ds$. As in [9], see also [1], we associate with (8) the operator in $L^1 \times X$,

$$A_2\{\varphi,k\} = \{-\varphi',-f(k) - h(\varphi)\} = \{-\varphi',-g(0,k) - \int_{-\infty}^{0} g_1(-s,\varphi(s))ds\}$$

$$D_{A_2} = \{\{\varphi,k\} \in L^1 \times X,\ \varphi \in W^{1,1}(-\infty,0;\mu;X),\ \varphi(0) = k\}.$$

We require the kernel g to satisfy suitable conditions (see [10] and [11]) so that $A_2 + \omega I$, $\omega \in \mathbb{R}$, generates a semigroup $T_2(t)$ in $L^1 \times X$ which satisfies the following translation property: set

$$x(t) = \varphi(t) \qquad t < 0$$

$$= \pi_2 T_2(t)\{\varphi,k\} \qquad t \geq 0$$

then $x_t = \pi_1 T_2(t)\{\varphi,k\}$, π_1 and π_2 being the projections of $L^1 \times X$ onto L^1 and X respectively; and such that x(t) is a solution of (8) for all $\{\varphi,k\} \in W^{1,1} \times X$. If we choose $k = F(\varphi) = \int_{-\infty}^{0} g(-s,\varphi(s))ds$, x(t) is in fact a solution of (5)'. From the translation property it follows that set $u(t,a) = \pi_1 T_2(t) \cdot \{\varphi,k\}(-a)$, $t \geq 0$, $a \geq 0$, u(t,a) is a solution of problem (7). See [11] for the semigroup approach to age-dependent population problems in a more general situation. Here we want to stress the fact that when studying (FE) by associating with it a semigroup of translations in the space of initial data we are in fact studying problem (4).

REFERENCES

1. J. Dyson, R. Villella-Bressan: Nonlinear Functional Differential Equations in L^1 Spaces, J. Nonlinear Analysis TMA, 1, 383-395, (1977).
2. J. Dyson, R. Villella-Bressan: Semigroup of Translations Associated with Functional and Functional Differential Equations, Proc. Roy. Soc. Edinburgh, 82A, 171-188, (1979).

3. M.E. Gurtin, R.C. MacCamy: Non-linear Age-dependent Population Dynamics, Arch. Rat. Mech. Analysis, 54 281-300, (1974).
4. J. Hale: Theory of Functional Differential Equations, Berlin, Springer, (1977).
5. A.T. Plant: Nonlinear Semigroups of Translations in Banach Spaces Generated by Functional Differential Equations, J. Math. Anal. Appl., to appear.
6. A.T. Plant: Nonlinear Semigroups Generated by Neutral Functional Differential Equations, Essex Fluid Mech. Res. Inst. Rep. 72 (1976).
7. R. Villella-Bressan: Equazioni di evolutione con condizioni ai limiti non lineari, Rend. Mat., 4 (1972).
8. R. Villella-Bressan: Flow-invariant Sets for Functional Differential Equations, to appear.
9. G.F. Webb: Functional Differential Equations and Nonlinear Semigroups in L^p Spaces, J. Differential Equa. 20, 71-89, (1976).
10. G.F. Webb: Volterra Integral Equations as Functional Differential Equations on Infinite Intervals, Hiroshima Math. J., 1, 61-70, (1977).
11. G.F. Webb: Age-dependent population problems, Seminari tenuti all'Università di Padova, to appear.